01-CTS-2a

UPSET EARLY WARNING SYSTEMS FOR BIOLOGICAL TREATMENT PROCESSES: SOURCE AND EFFECT RELATIONSHIPS

by:

Nancy G. Love

Inês D.S. Henriques

Richard T. Kelly II

Virginia Polytechnic Institute and State University,
Department of Civil and Environmental Engineering

2005

WERF

The Water Environment Research Foundation, a not-for-profit organization, funds and manages water quality research for its subscribers through a diverse public-private partnership between municipal utilities, corporations, academia, industry, and the federal government. WERF subscribers include municipal and regional water and wastewater utilities, industrial corporations, environmental engineering firms, and others that share a commitment to cost-effective water quality solutions. WERF is dedicated to advancing science and technology addressing water quality issues as they impact water resources, the atmosphere, the lands, and quality of life.

For more information, contact:
Water Environment Research Foundation
635 Slaters Lane, Suite 300
Alexandria, VA 22314-1177
Tel: (703) 684-2470
Fax: (703) 299-0742
www.werf.org
werf@werf.org

This report was co-published by the following organization. For non-subscriber sales information, contact:

IWA Publishing
Alliance House, 12 Caxton Street
London SW1H 0QS, United Kingdom
Tel: +44 (0) 20 7654 5500
Fax: +44 (0) 20 7654 5555
www.iwapublishing.com
publications@iwap.co.uk

Library of Congress Catalog Card Number: 2004114170

IWAP ISBN: 978-1-84339-742-7

Virginia Polytechnic Institute and State University

ACKNOWLEDGMENTS

We would like to acknowledge the management and staff at Blacksburg-VPI Wastewater Treatment Plant for access to their facility during this study. In addition to the project team members who prepared the report, listed below, we acknowledge the contributions and hard work of the following individuals who contributed to this study: Jennifer Dauphinais, Mike Hair, Katherine Linares, and Kelli Mattson. Finally, the Project Subcommittee members and Project Managers (listed below) gave very helpful and constructive assistance throughout this project and we are grateful for their support.

Report Preparation

Principal Investigator:
Nancy G. Love, Ph.D., *Virginia Polytechnic Institute and State University*

Project Team:
Inês D. S. Henriques, M.S., *Virginia Polytechnic Institute and State University*
Richard T. Kelly II, Ph.D., *Virginia Polytechnic Institute and State University*

Project Subcommittee

John Barber, Ph.D., P.E., *Eastman Chemical*
Paul Bowen, Ph.D., *Coca-Cola USA*
Glen T. Daigger, Ph.D., PE, DEE (Chair), *CH2M Hill, Inc.*
Ben Horenstein, *East Bay Municipal Utilities District*
Robert A. Reich, P.E., *DuPont Company*
Bill Winters, P.E., *Metropolitan Sewer District of Greater Cincinnati*

Water Environment Research Foundation Staff

Director of Research: Daniel M. Woltering, Ph.D.
Project Managers: Jennifer Simmons
Mary Strawn
Amit Pramanik, Ph.D.

ABSTRACT AND BENEFITS

Abstract:

A laboratory study was conducted to elucidate the source-effect relationships for seven chemicals (sources) that can cause activated sludge process upset (effect). These chemicals were studied over a range of concentrations using both nitrifying and non-nitrifying laboratory-scale activated sludge sequencing batch reactors. Effects were characterized according to traditional methods of evaluating process effluent and mixed liquor quality. A range of process effects were observed for both biomass sources. Overall impact was assessed. In determining the relative order of chemical impact, a chemical that caused the substantial deterioration of one aspect of process performance was considered to be more detrimental than a chemical that caused multiple processes to deteriorate moderately or slightly. The order in which chemicals caused adverse effects for the nitrifying biomass was: pH 5 ammonium < octanol < 2,4-dinitrophenol (DNP) < cyanide < 1-chloro-2,4-dinitrobenzene (CDNB) < cadmium and pH 11. For the non-nitrifying biomass, the order in which the chemicals caused adverse effects was: pH 5 octanol < ammonia < DNP < cyanide < CDNB < cadmium < pH 11. Almost all chemicals caused multiple process effects, but the intensity and type of process effect was not always predictable based on the chemical applied. The findings show that there are multiple ways that chemicals can impact activated sludge plants and suggest that corrective action practices need to be tailored based on the nature of the chemical causing the upset.

Benefits:

- Determines the source-effect relationships for seven chemicals from distinct chemical classes for nitrifying and non-nitrifying activated sludge biomasses.
- Shows that overall patterns of process effect are hard to predict for any chemical.
- Provides a baseline for other studies to determine appropriate corrective action practices to apply to upset systems that can lead to rapid process recovery.
- Serves as a basis for prioritizing development and commercialization of upset early warning devices.

Keywords: inhibition, deflocculation, upset, shock, pH, ammonia, octanol, 2,4-dinitrophenol cyanide, 1-chloro-2,4-dinitrobenzene cadmium, stress, activated sludge, nitrifying, non-nitrifying.

TABLE OF CONTENTS

LIST OF TABLES

LIST OF FIGURES

LIST OF ACRONYMS

AMO	ammonia monooxygenase
AOB	ammonia oxidizing bacteria
BOD	biochemical oxygen demand
CDNB	1-chloro-2,4-dinitrobenzene
COD	chemical oxygen demand
CST	capillary suction time
CSTR	Continuous-flow stirred tank reactor
DNA	deoxyribonucleic acid
DNP	2,4-dinitrophenol
EPS	extrocellular polymeric substances
GGKE	glutathione-gated potassium (K^+) efflux
GSSG	oxidized glutathione
GSX	conjugated glutathione
HPLC	High performance liquid chromatography
M:D	monovalent:divalent cation ratio
MLSS	mixed liquor suspended solids
MLVSS	mixed liquor volatile suspended solids
NEM	N-ethylmaleimide
NOB	nitrite oxidizing bacteria
PSC	Project Subcommittee
POTW	publicly owned treatment works
SBR	sequencing batch reactor
SNGR	specific nitrate generation rate
SOUR	specific oxygen uptake rate
SRT	solids residence time
SVI	sludge volume index
TSS	total suspended solids
VSS	volatile suspended solids

EXECUTIVE SUMMARY

ES.1 Introduction

A literature search revealed that few studies have been conducted under appropriately controlled conditions to allow source-cause-effect relationships to be established for activated sludge treatment systems that experience process upset due to the influx of toxins (Love and Bott, 2000). Although source-cause-effect relationships that are experimentally established ultimately need to be validated at full-scale treatment systems, the limited state of current knowledge justifies conducting basic laboratory experiments at this time. The goal of this project was to elucidate the inherent source-cause-effect relationship for several key upset events that are experienced in activated sludge treatment processes. Seven source conditions were studied over a range of concentrations to evoke process related effects in laboratory-scale activated sludge sequencing batch reactors (SBRs). These effects were characterized according to traditional methods of evaluating process effluent or mixed liquor quality; this information was used to establish source-effect relationships. The results from the source-effect experiments are presented in this report, and summarized in the source-effect matrices presented in Tables ES-1 and ES-2. A second report will be released that addresses additional experiments that evaluate a series of hypotheses that were established based on potential mechanisms of cause (i.e., the root cause) for selected source-effect responses reported here.

ES.2 Project Objectives

The objectives established for the project were to:

- Experimentally determine the process effects associated with a broad range of realistic upset events in order to establish source-effect relationships.
- Develop hypotheses that establish biological, chemical, and/or physical causes for the effects observed that explain source-cause-effect relationships.
- Experimentally test a short list of the hypotheses developed under the previous objective.

This report addresses the first project objective. A second final report will address objectives 2 and 3. This second report will focus on phenomena that associate with two process effects that were found in this study to be particularly prevalent: nitrification inhibition and increased effluent turbidity. It is hypothesized that understanding these phenomena will assist with the development of useful upset early warning strategies that can be linked with useful corrective actions that operators can implement at plants impacted by toxic upsets.

ES.3 Project Approach

Experiments were conducted using biomass collected from two parent sequencing batch reactor (SBR) systems that contained either a nitrifying biomass maintained with a 10-day solids residence time (SRT), or a non-nitrifying biomass maintained at a 2-day SRT. The parent SBRs were 180 liters each and were fed sewage collected from a manhole located on the Virginia Polytechnic Institute and State University (Virginia Tech) campus. The biomass in the pilot plant systems was not intentionally exposed to toxic inputs during the study. The source-effect experiments were conducted by collecting biomass from these parent reactors and returning the biomass to the laboratory. Four 4-liter SBR reactors were established with either 10-day or 2-day SRT biomass, and exposed to three concentrations of each chemical toxin under study (one reactor served as an unstressed control). The toxins studied include: ammonium, cadmium, 1-chloro-2,4-dinotrobenzene (CDNB), cyanide, 2,4-dinitrophenol (DNP), 1-octanol, and acid and base pH. Laboratory reactors were operated for up to 3X SRT (30 days for the 10-day SRT biomass, 6 days for the 2-day SRT biomass) and monitored intensely to determine the impact of the toxic shock event on treatment process performance.

ES.4 Major Conclusions

A summary of results from the 14 source-effect experiments completed for this study are provided in Tables ES-1 and ES-2, and discussed thoroughly in Chapter 10.0. As is apparent from these tables, a range of process effects were observed. Based on overall impact, the order in which chemicals caused adverse effects on process performance parameters for the nitrifying, 10-day SRT biomass was (least detrimental to most detrimental): pH 5< ammonium < octanol < DNP < cyanide < CDNB < cadmium and pH 11. Hypotheses of presumed causal mechanisms for inhibition of nitrification in activated sludge systems, which was observed in response to shocks by all chemicals tested in the 10-day SRT study, are provided in Chapter 10.0. For the non-nitrifying 2-day SRT biomass, the order in which the chemicals caused adverse effects was: pH 5 < octanol < ammonia < DNP < cyanide < CDNB < cadmium < pH 11. The pH 5 shock event had a minimal impact on process performance indicators in both the 2- and 10-day biomasses and, therefore, is not represented in Tables ES-1 and ES-2. In determining the relative order of chemical impact, a chemical that caused the substantial deterioration of one aspect of process performance was considered to be more detrimental than a chemical that caused multiple processes to deteriorate moderately or slightly. This approach was used because it only takes one element of a process to fail in order to wreak havoc on an activated sludge plant, and long recovery periods may be required.

These findings suggest that there are multiple ways in which chemicals can impact an activated sludge plant, and suggest that corrective action practices need to be tailored based on the nature of the chemical causing the upset. Identifying the appropriate corrective action is often difficult for operators because they do not know the composition of the chemical or chemicals that caused problems until days later, if ever. Furthermore, upset events can also occur with mixtures of chemicals that may produce different results from this study, which only looked at single chemical impacts. The study results strongly suggest that further work needs to be done to define appropriate corrective actions for operators to implement after treatment plants have been exposed to toxic shock events. To facilitate information flow as part of a corrective action/decision support system, robust, inexpensive and field-compatible upset early warning devices need to be developed and implemented.

Table ES-1. Summary of Process Effects for Seven Source Chemicals for the 10-Day SRT Biomass Using a Pseudo-Quantitative Scale[1].

Measurement	Observed Problematic Process Effect Relative to Control	Hypothesized Causal Mechanism	Ammonium ~ 400 mg/L	Cadmium	CDNB	Cyanide	DNP	Octanol	pH 11
Effluent TSS/VSS	Increase in effluent TSS/VSS	Flocs deteriorate and cause smaller, less dense particles (deflocculation)	0	↓↓↓↓	↓↓↓	0	↓↓	+	↓↓↓↓
Effluent COD	Decrease in COD removal	Inhibition of metabolic pathways	0	↓↓↓↓	↓↓↓	↓	↓↓	+	↓↓↓↓
SOUR	Decrease in SOUR	Inhibition of catabolic pathways	0	↓↓↓↓	↓↓↓	↓↓	↓↓/++	↓↓	↓↓↓
Soluble Potassium	Increase in soluble K^+ concentration	Glutathione-gated K^+ efflux	0	↓↓	↓↓↓	0	0	0	X
Inorganic N effluent concentrations and NGR	Nitrification inhibition	See Table 10-4	↓	↓↓	↓↓↓	↓↓↓	↓↓	↓↓	↓↓↓↓
SVI	Increase in SVI	Poor biosolids compression, or settleability	0	++	+	↓	0	0	+
CST	Increase in CST	Retention of bound water, leading to poor dewaterability	0	0	0	0	0	0	0

[1] The qualitative scale reflects the intensity of the effect for the IC_{50}-shocked reactors and the indicated NH_3 and pH shock level, in comparison to a negative control. The intensity scale ranges from ↓↓↓↓ (most intense process deterioration effect), 0 (no effect), and ++++ (most intense process improvement effect).
X means inconclusive results

Table ES-2. Summary of Process Effects for Seven Source Chemicals for the 2-Day SRT Biomass Using a Pseudo-Quantitative Scale[1].

Measurement	Observed Problematic Process Effect Relative to Control	Hypothesized Causal Mechanism	Ammonia 300 mg/L-N	Cadmium	CDNB	Cyanide	DNP	Octanol	pH 11
Effluent TSS/VSS	Increase in Effluent TSS/VSS	Flocs deteriorate and cause smaller, less dense particles (deflocculation)	0	↓↓↓↓	↓↓↓	0	0	↓	↓↓↓↓
Effluent COD	Decrease in COD removal	Inhibition of metabolic pathways	0	↓↓↓↓	↓↓↓↓	↓↓	0	↓	↓↓↓↓
SOUR	Decrease in SOUR	Inhibition of catabolic pathways	↓↓	↓↓↓↓	↓↓↓↓	↓↓↓↓	↓↓↓	↓	↓↓↓↓
Soluble Potassium	Increase in soluble K^+ concentration	Glutathione-gated K^+ efflux	↓↓	↓↓↓↓	↓↓	0	0	0	↓↓↓
SVI	Increase in SVI	Poor biosolids compression, or settleability	0	++	++	0	0	0	++
CST	Increase in CST	Retention of bound water, leading to poor dewaterability	0	↓	X	0	0	↓	↓↓↓

[1] The qualitative scale reflects the intensity of the effect for the IC_{50}-shocked reactors and the indicated NH_3 and pH shock level, in comparison to a negative control. The intensity scale ranges from ↓↓↓↓ (most intense process deterioration effect), 0 (no effect), and ++++ (most intense process improvement effect). X means inconclusive results.

CHAPTER 1.0

INTRODUCTION

1.1 Background and Research Need

Worldwide, biological wastewater treatment systems are critical for pollution control. These systems must be designed and operated properly in order to effectively prevent environmental contamination or to avoid problems in downstream water reuse operations. Even when biological treatment systems are well designed, sudden inputs of inhibitory chemicals upstream can occur and lead to moderate-to-severe disruption of the biological treatment process. In publicly owned treatment works (POTWs) and especially in industrial wastewater treatment systems, sudden discharges of toxic or inhibitory chemicals can pass through treatment systems with little effective removal (Bhattacharya et al., 1996; Galil et al., 1988; Janssens et al., 1997; Monteith et al., 1995; O'Brien and Teather, 1995; Paxeus, 1996; Rebhun and Galil, 1988). Additionally, changes in wastewater composition can impose a range of process responses. Those responses most often reported by operators in response to toxic upset events were determined through a survey conducted by the Water Environment Research Foundation, and included poor 5-day biochemical oxygen demand (BOD_5) removal, disrupted nitrification, and deflocculation of activated sludges (Love and Bott, 2000). Reductions in bioprocess treatment efficiency can produce effluents that cause unacceptable levels of acute or chronic damage to natural environments and public health. In the case of deflocculation, loss of biomass out the secondary clarifier effluent can lead to a reduced solids residence time (SRT) and poorer process performance. Furthermore, sorption of toxins to deflocculated particles can cause a significant flux of toxicity into the receiving environment. The operational problems that occur in biological treatment processes as a result of upset events may require a significant amount of time to correct (e.g., recovering acceptable sludge-settling characteristics) (Berthouex and Fan, 1986; Bott and Love, 2001; Bott and Love, 2002), thereby extending the time over which the damage occurs.

Efforts have been made over the past decade to provide operators with diagnostic tools that allow them to respond proactively rather then reactively during process operation. Several devices have been designed that are capable of detecting changes in wastewater composition that result in process upset. For example, on-line respirometers can detect many upset events in a relatively short time frame (several minutes) (Love and Bott, 2000). Other devices, such as the on-line Microtox bioassay, can detect the presence of many toxins, but the relevance to biological wastewater treatment cultures is questionable. Furthermore, neither of these devices is designed to identify the *source* leading to a change in wastewater composition nor, more importantly, the *effect* that a given upset event may have on treatment processes. Without a correlation between toxin source and effect, operators often do not know what control actions to implement in response to toxic influxes in order to most effectively mitigate their environmental impacts.

The purpose of this study was to establish a matrix that explains which process effects are experienced when activated sludge is exposed to different classes of chemical toxins. The approach to the study involved performing laboratory experiments with sequencing batch reactors (SBRs) that were either nitrifying (10-day SRT) or non-nitrifying (2-day SRT). Chemical toxins were selected and used for performing shock experiments. Single pulse shock events were applied for three concentrations of a given toxic chemical and compared to an unshocked control for a time period that lasted up to three times the SRT of the system (30 days for 10-day SRT biomass and six days for 2-day SRT biomass). Both conventional and novel process effect indicators were monitored over time. This report presents the results from this matrix of source-effect experiments.

To develop a better understanding of how source and effect are related, it is necessary to identify the biochemical, chemical or physical *causes* that link source and effect. By understanding the mechanisms that cause process disruptions, it will be possible to (i) conceive of and design intelligent sensor systems that relay relevant information to operators, (ii) establish appropriate corrective actions to minimize the environmental impact of upset events, (iii) strengthen pretreatment programs by allowing utilities to screen industrial contributors during or before the permitting process, and (iv) integrate source-cause-effect information with process control and decision support systems that will allow treatment plant operations to be optimized. It is necessary to take a more careful look at physiological, microscopic- and molecular-level mechanisms that occur within individual cells or microbiological communities in order to elucidate causal mechanisms that link source chemicals with specific process effects. Additional experiments are being conducted to elucidate causal mechanisms more clearly in response to the results presented in this report, and those results will be presented in a separate report.

1.2 Selection of Source Chemicals

Seven source chemicals from seven distinct chemical classes were selected for this project, and are listed in Table 1-1 along with their physical/chemical characteristics. It was presumed that selecting chemicals that represent a range of chemical properties would, in turn, yield a range of inhibitory responses that would manifest into distinct source-effect patterns. In concert with this assumption is the notion that by selecting these chemicals, the project would cover a broad range of potential process effects that operators may experience due to toxic shock events. The scope of the experimental plan for the project mandated that only single chemical shocks could be tested; however, it is understood that mixtures of chemicals could cause distinctly different process effect patterns and should be evaluated further.

Table 1-1. Source Chemicals that Cover a Broad Range of Chemical Classes, Physical/Chemical Properties, and Known Toxic Features.

Chemical	Chemical Class	Solubility[a]	phobicity[a]	Volatility	Toxic Features
Cadmium	Heavy Metal and Electrophile	Highly insoluble as phosphate (K_{sp} = 2.53 x 10^{-33}), carbonate (K_{sp} = 1 x 10^{-12}) and hydroxide (K_{sp} = 7.20 x 10^{-15}) salts.	Precipitates sorb to biomass.	Not volatile, vapor pressure of solid $CdCl_2$ at 412°C = 1 Pa.	Inactivates enzymes, complexes with phosphates in biomolecules to inactivate their function, and can cause DNA mutagenesis.
Octanol	Hydrophobic Organic	Slightly soluble to 540 mg/L in water at 25°C.	Strongly hydrophobic, Log K_{OW} = 3.07.	Slightly volatile, vapor pressure at 30°C = 10 Pa.	Diffuses readily into phospholipids membranes and causes a narcotic (lethargic) effect.
Ammonia/um	Non-Heavy Metal Inorganic	Ammonium sulfate is soluble to 764 g/L in water at 25°C. Ammonium carbonate is soluble to 1000 g/L in water at 25°C.	This is a polar compound and not hydrophobic.	Ammonium not volatile. Ammonia is highly volatile, vapor pressure at 28°C = 1062 kPa.	Free ammonia (dissolved NH_3) form is inhibitory to nitrifying bacteria.
Cyanide	Inorganic Metal Chelator	Sodium cyanide is soluble to 582 g/L in water at 20°C. Zinc cyanide is soluble to 4.7 mg/L in water at 20°C.	This is a very polar compound and not hydrophobic.	HCN is volatile, vapor pressure at 28°C = 105.9 kPa.	The primary means of toxicity is through respiration inhibition.
1-chloro-2,4-dinitrobenzene (CDNB)	Chlorinated Organic Solvent and Electrophile	Sparingly soluble to 9.2 mg/L in water at 25°C. We found ~400 mg/L in filtered mixed liquor at 100°C.	Hydrophobic, Log K_{OW} unknown.	Not volatile, vapor pressure unknown.	Preferentially attacks proteins and DNA. Functions as an oxidative stressor.
2,4-Dinitrophenol (DNP)	Oxidative Uncoupler	Slightly soluble to 460 mg/L at 25°C.	Hydrophobic, Log K_{OW} = 1.53.	Not volatile, vapor pressure at 20°C = 0.052 Pa.	An uncoupler of oxidative phosphorylation.
pH	Acid/Base	Highly soluble	NA	NA	Base dissolves proteins. Acid hydrolyzes biomolecules.

[a] K_{sp} = solubility product K_{OW} = octanol:water partition coefficient

1.3 Selection of Process Effects to Monitor

Several water quality parameters and assays were measured and served as the basis for determining if a treatment process effect had occurred. Those parameters and assays are identified in Table 1-2 along with the treatment-process effect that they represent.

Chemical oxygen demand (COD) was used as a reliable indicator of organics removal efficiency. It was used in lieu of biochemical oxygen demand (BOD) because of the unknown impact that toxins can have on BOD measurements. If a toxin that is inhibitory to bacterial respiration is present in a BOD sample, the toxin will reduce the oxygen consumed during the time period of the assay and give the impression that the sample BOD concentration is very low. In the case of monitoring treatment process effluents, this will give the false sense that the treatment efficiency has improved. For this reason, BOD could not be used as an indicator and COD served as a surrogate treatment efficiency in removing organic compounds.

Suspended solids analysis of effluents and mixed liquors are routinely performed in full-scale treatment plants and were, likewise, performed here to indicate effluent quality and to assist with interpreting observed performance shifts that are a function of the mixed liquor concentration. Additionally, the effluent total suspended solids (TSS) and mixed liquor suspended solids (MLSS) are both used to calculate the solids residence time of a reactor. Furthermore, the mixed liquor volatile suspended solids (MLVSS) was used to normalize kinetic assays, as discussed below. An increase in effluent TSS is interpreted to mean that biological flocs have disaggregated for whatever reason, and small floc particles that do not settle during the settling period are removed with the effluent; we refer to this phenomenon as deflocculation. Deflocculation is particularly bad if it occurs in response to toxins that tend to sorb to particles, because the effluent toxicity increases in parallel with the increased TSS.

The glutathione-gated potassium efflux (GGKE) stress response is believed to be a causal stress mechanism that results in deflocculation when biomass is exposed to oxidative (or electrophilic) chemicals. It was previously shown experimentally that K^+ levels increase significantly and rapidly (in minutes) in the bulk water phase and concurrently decrease in the floc-associated phase of mixed liquors exposed to sublethal levels of electrophilic organic and inorganic toxins (Bott and Love, 2002; Bott and Love, 2004). This increase in K^+ is believed to be due to an efflux of potassium from the cytoplasm of cells via the electrophile-activated GGKE system (Bott and Love, 2004). These authors proposed that high localized levels of K^+ result in weakened binding between extracellular polymeric substance (EPS) ligands due to high, localized monovalent:divalent (M:D) cation ratios. It has been shown that increases in the M:D cation ratio in wastewater can lead to poor settling and dewatering properties (Higgins and Novak, 1997b; Novak et al., 1998) because monovalent cations build weaker bridges between negatively charged EPS ligands and, in some cases, repulsive forces can be established. In the presence of mixing-induced shear forces, floc structure deteriorates and deflocculation results. Because of this known phenomenon, K^+ was routinely monitored in reactor effluents after shocks were applied.

Table 1-2. Water Quality Parameters and Assays that were Measured and the Associated Process Effect Indicated by Each.

Water Quality Parameter or Assay	Location of Sample	Deleterious Process Effect Indicated if Parameter Value:	
		Increases	Decreases
Chemical Oxygen Demand (COD)	Effluent	Inhibition of catabolic pathways and, therefore, deterioration in BOD removal efficiency.[a]	
Total Suspended Solids (TSS)/Volatile Suspended Solids (VSS)	Effluent	Loss of floc strength and formation of smaller floc (deflocculation), yielding increased effluent turbidity.	
	Mixed Liquor		Inhibition of anabolic pathways, resulting in a decreased mixed liquor suspended solids concentration.
Ammonia-Nitrogen	Effluent	Decrease in activity of ammonia oxidizing bacteria, resulting in poor ammonia or nitrogen removal.[b]	
Nitrite-Nitrogen	Effluent	Decrease in activity of nitrite oxidizing bacteria, resulting in poor nitrogen removal and probable high chlorine demand.[b]	
Nitrate-Nitrogen	Effluent		Decrease in activity of nitrite (and probably ammonia) oxidizing bacteria, resulting in poor ammonia or nitrogen removal.[b]
Alkalinity	Effluent	Indicates a reduction in the degree of nitrification and implies inhibition.[b]	
Potassium	Effluent	Indicates oxidative toxin and correlates with increase in effluent TSS/VSS (deflocculation).	
Specific Oxygen Uptake Rate (SOUR)	Mixed Liquor	Implicates presence of an oxidative phosphorylation uncoupler and causes low dissolved oxygen levels in basin without corresponding improvement in BOD removal or increase in mixed liquor concentration.	Inhibition of catabolic pathways and, therefore, deterioration in BOD removal efficiency.[a]
Nitrate Generation Rate (NGR)	Mixed Liquor		Nitrification inhibition, resulting in poor ammonia or nitrogen removal.[b]
Sludge Volume Index (SVI)	Mixed Liquor	Poor sludge compaction, resulting in rising sludge blankets in secondary clarifiers.	
Capillary Suction Time (CST)	Mixed Liquor	Poor biosolids dewatering properties, resulting in demand for conditioning agent.	
Microscopic Evaluations	Mixed Liquor		Loss of sentinel species indicates presence of unfavorable conditions for bacterial biomass.

[a] After COD is corrected for by subtracting contribution from added organic toxins.
[b] Only pertains to nitrifying biomass.

The biological nitrification process is a two-step process that includes ammonia oxidation to nitrite via ammonia oxidizing bacteria (AOB) and nitrite oxidation to nitrate via nitrite oxidizing bacteria (NOB). The first step produces acid and, therefore, corresponds stoichiometrically with alkalinity consumption. To evaluate the performance of this biological process, soluble inorganic nitrogen species (ammonia, nitrite, and nitrate) and alkalinity were monitored in reactor effluents to determine if nitrification inhibition occurred in response to chemical shocks. Nitrification is known to be highly vulnerable to chemical toxins (Blum and Speece, 1991; Daigger and Saddick, 1998; Grunditz and Dalhammar, 2001; Nowak and Svardal, 1993; Wood et al., 1981), but the relationship between exposure to the toxin and the point in the metabolic process where inhibition occurs is not clear. Therefore, monitoring for ammonia, nitrite, and nitrate allowed for some inferences to be drawn regarding the point in nitrification where inhibition was occurring. Alkalinity was monitored as a confirmation check on the results from analyzing the nitrogen species. Furthermore, recovery of nitrification after an inhibitory event can take time (Stasinakis et al., 2003), leaving the treatment system vulnerable to permit violations and the downstream environment vulnerable to ecological damage. Therefore, monitoring of soluble inorganic nitrogen species and alkalinity occurred for extended periods of time throughout this study to determine the time required to recover nitrification for each chemical shock event.

Specific oxygen uptake rate (SOUR) and nitrate generation rate (NGR) are both kinetic assays performed with mixed liquor that directly evaluate the respirometric and nitrification rates at which biomass performs. Rapid decreases in these rates relative to control assays imply inhibition. SOUR results can be compared to relative changes in the effluent soluble COD, and NGR can be compared to relative changes in the effluent concentrations of oxidized soluble inorganic nitrogen species and alkalinity. The effluent measurements are a less direct measure of inhibition but serve as an independent measure of inhibition relative to the kinetic assays. Therefore, whenever effluent measurements detected inhibition, it was expected that the kinetic assays would validate this observation. However, the kinetic assays are a more sensitive indicator of inhibition. Therefore, it is possible for moderate inhibition to be detected by the kinetic assays but not by effluent measurements if the react phase of the SBR is long enough for complete degradation of COD or soluble inorganic nitrogen compounds to occur, even if at a slower rate.

Sludge volume index (SVI) and capillary suction time (CST) are indicators of sludge compaction and dewatering qualities, respectively. Although these indicators have limitations, they provide a relative sense of the impact of toxic shock on settling quality, which can be compared to effluent TSS measurements and dewatering potential.

CHAPTER 2.0

METHODS

2.1 Pilot Plant

2.1.1 Design and Setup

The pilot plant facility received wastewater from the Virginia Tech campus and was designed to serve as a consistent source of mixed liquor for laboratory stress experiments. The facility consisted of two SBRs (System A and System B) maintained at two different SRTs to discourage nitrification in one reactor but to allow it to occur in the other. The reactor configurations and operating parameters are summarized in Table 2-1 below. Pilot plant operation began on March 23, 2001, and monitoring started 5 days later. The reactors were reseeded on October 7, 2001, and again on October 24, 2001, due to filamentous bulking problems. The 2-day SRT reactor (System A) was reseeded in August 2002 because of the same problem. The reactors were augmented with mixed liquor from the Blacksburg/VPI Wastewater Treatment Facility, which achieves both nitrification and denitrification.

Table 2-1. Configurations and Operating Parameters for Pilot Plant Systems.

	System A	System B
Reactor Working Volume	180 L	180 L
Target Temperature	20°C	20°C
Target SRT	2 days	10 days
Target Nominal Hydraulic Retention Time (HRT)	1 day	1 day
Cycles per Day	4	4
Feed Time per Cycle (with mixing and aeration on)	~ 6 min	~ 6 min
React Time per Cycle	5 hours	5 hours
Settling Time per Cycle	1 hour	1 hour

2.1.2 Daily Maintenance and Operation

Wastewater was pumped from an adjacent manhole to two 200 L tanks. Starting September 2001, a supplementation program was initiated to boost the COD load to the pilot plant reactors and, therefore, achieve higher MLSS concentrations. The supplement added 400 mg/L COD and contained equal COD quantities of acetate and glucose as well as a 15:30 fertilizer (K-Mart GrowBest All Purpose Plant Food). The amount of plant fertilizer added was determined based on an average value of influent ammonia of 18 mg/L N (before supplement addition), an average total COD value of 768 mg/L (including supplement COD) in the influent, and an assumed growth yield of 0.5 mg COD biomass/mg COD. It was determined that 20 g of fertilizer/day should be added to each reactor, which is equivalent to 15 mg/L N. In addition to nitrogen, the fertilizer also added 13.1 mg/L phosphorus, 12.5 mg/L potassium, 0.15 mg/L iron, 0.07 mg/L copper, 0.06 mg/L zinc, 0.05 mg/L manganese, 0.02 mg/L boron, and 0.0005 mg/L molybdenum. Calcium sulfate was also added to the supplement daily at 20 g $CaSO_4 \cdot \frac{1}{2} H_2O$ per reactor in order to maintain the M:D cation ratio in the influent below four (on a milliequivalent basis); the amount added was based on average cation concentrations measured in the influent with ion chromatography prior to supplement addition.

It was difficult to block nitrification in the 2-day SRT reactor so a new supplement program was implemented on August 7, 2002 for the 2-day SRT reactor. The COD load was doubled (400 mg/L glucose as COD plus 400 mg/L sodium acetate as COD), and the amount of fertilizer was increased by a factor of three. Moreover, the influent was pumped into the 2-day SRT reactor on a step-feed mode, whereby the full influent volume was added in equal volume aliquots at three points during the reaction cycle: when the cycle began; 1-½ hours after the cycle began; and 3 hours after the cycle began. This new feed design allowed for higher COD concentrations to be maintained in the reactor, thereby favoring the growth of heterotrophic organisms and limiting the development of a significant nitrifying population. The higher COD loading allowed for higher biomass concentrations to be achieved. Due to the increase in COD loading, the amount of calcium sulfate added to control the M:D cation ratio was increased accordingly; this increase resulted in high concentrations of sulfides being pumped into the reactor, which quickly led to an uncontrolled growth of filamentous organisms. The reactor was reseeded and calcium carbonate was used instead as the source of divalent cations. A total of 36 g $CaCO_3$ was added to the 2-day SRT influent tank each day.

Although there were few problems achieving nitrification in the 10-day SRT reactor system, a change was made to this system in August 2002 as well to establish a higher level of nitrifying bacteria in the 10-day SRT biomass. The ammonia supplement to the sewage was increased to yield a total average influent ammonia-N concentration around 45 mg/L, which was 50% higher than the previous supplementation period. The source of additional divalent cations was also changed to calcium carbonate, to a total of 14 g $CaCO_3$ per day.

In addition to supplement addition, the sewage was settled for 30 minutes prior to being pumped into the reactors at the beginning of a cycle, was mechanically mixed for 0.5 hour before the settling period, and was aerated at all other times to prevent the development of septic conditions. Wastage was collected and measured daily for each of the two reactors and corrected weekly to maintain the proper SRT. Due to the low wastage volumes from the 10-day SRT reactor, daily manual wastage from this reactor was initiated in August 2002 to ensure higher accuracy in the wasted volumes. At the same time, the 2-day SRT reactor wastage volume was manually adjusted daily, if needed. Attached biomass was scraped daily from the reactor walls back into the suspension. The ambient temperature in the shed housing the pilot plant was monitored to ensure that it remained between 20°C and 25°C in the winter for proper operation of the nitrifying and non-nitrifying reactors. Similarly, the shed was air-conditioned during the summer to maintain an ambient temperature of 20°C. Controlling the ambient temperature ensured that the temperature of the mixed liquor in the reactors stayed at around 20°C. On August 5, 2002, the influent tanks, which were initially located in a little shed adjacent to the SBRs shed, were moved inside the larger shed, allowing for both the influent and SBRs tanks to be maintained at the same temperature; overall, this improved temperature stability of the system.

The timers were checked daily for proper operation of the air, influent, wastage, and effluent pumps. Additionally, the decant and wastage tube levels located inside the tanks were checked daily to make sure that effluent and wastage volumes were correct.

2.1.3 Sampling Schedule

One-liter grab samples of influent, effluent, and mixed liquor were collected for monitoring twice per week during the morning cycle. Near the end of July 2002, this frequency was increased to three times per week. The influent and effluent samples were analyzed for soluble COD, TSS/VSS, and ammonia levels, whereas the mixed liquor was tested for MLSS and MLVSS. In addition, total COD was analyzed for influent samples.

2.2 Laboratory-Scale Source-Effect Experiment Reactors

2.2.1 Inhibitory Concentrations (IC_{XX})

Inhibitory concentrations (IC_{XX}) for the stressing chemicals were determined one day prior to initiating each source-effect experiment to ensure that the composition of the mixed liquor used for the test was representative of the biomass that would be used to inoculate the source-effect reactors. The procedure was based on inhibition of respiration and reflects the percent of respiratory inhibition for a given concentration of stressing chemical. Tests were performed in duplicate using an Acumet dual-channel pH/ion meter (model AR25) and two Orion model 97-08 oxygen electrodes. Dissolved oxygen readings were recorded every 6 seconds using LabView 6i software operating on a Dell Dimension XPS P100c personal computer. Mixed liquor from the 2- and 10-day SRT pilot plant reactors was used in the IC_{XX} assays. The mixed liquor was aerated for four hours prior to testing to remove any biodegradable materials, and both MLSS and MLVSS were determined. Mixed liquor was then added to 300-mL BOD bottles and spiked with varying masses of contaminant. Soluble COD was composed of one-third protein (beef extract, phytone, bacto-casitone, yeast extract), one-third carbohydrate (fructose, galactose, glucose), and one-third organic acids/alcohols (glacial acetic acid and glycerol) on a COD basis. The soluble COD was added to the mixed liquor in the BOD bottle to achieve a starting concentration of 100 mg/L as COD to ensure that unrestricted respiration could occur throughout the 10-minute test. Oxygen consumption rate tests were conducted to determine the oxygen uptake rate (OUR) of the mixed liquor exposed to the contaminant and were performed according to section 2710 B of Standard Methods (APHA, 1998). SOURs were calculated by dividing the OUR by the MLVSS concentration. The inhibitory concentrations were determined by plotting SOUR versus contaminant concentration or percent inhibition versus contaminant concentration and fitting a second order polynomial or exponential function through the data (best fit determined by least squares analysis, Excel 2000, or Sigma Plot). By looking at the inhibition relative to the control reactor, the IC_{15}, IC_{25}, and IC_{50} concentrations were determined for the laboratory source-effect experiments.

This procedure was modified for 1-chloro-2,4-dinitrobenzene (CDNB), due to the slow kinetics of solubility. The mixed liquor was concentrated by 75%, and the supernatant was removed. A known amount of CDNB was melted and dissolved into heated supernatant using sonication (Branson Sonifier 250 with microtip) at a continuous pulse for 5 minutes at 140 watts. This mixture was cooled to room temperature and the appropriate amount of concentrated mixed liquor was added, to achieve the same starting biomass concentration. Oxygen consumption tests were then run as previously described.

In the case of the pH shock experiment, a 1 N solution of sulfuric acid was used to adjust the mixed liquor pH prior to running the low pH SOUR assays and a 1 N NaOH + 1 N $Ca(OH)_2$ mixture (balanced M:D ratio to avoid causing deflocculation due to cation imbalance) was used to adjust the biomass pH prior to running the higher pH SOUR assays.

The ammonia (NH_4^+ + NH_3-N) concentrations used to shock the source-effect reactors during the ammonia shock experiment were based on the average ammonia concentration in the influent. At the time the 10-day SRT experiment was run, the influent was being supplemented with COD and nutrients. The target ammonia in the influent was 40 mg/L as N, based on previous estimates of approximately 20 mg/L as N in the raw wastewater plus 20 mg/L added in the supplement. Actual ammonia concentrations in the influent to the pilot plant systems averaged 39 mg/L as N, between the date at which supplement was started and the start of the 10-day SRT source-effect laboratory reactors. The corresponding standard deviation was 10 mg/L, which shows significant variation in the ammonia levels but with an average close to the desired target concentration of 40 mg/L. To mimic an ammonia shock event, concentrations equal to 2X (influent ammonia + 40 mg/L as N), 5X (influent ammonia + 160 mg/L as N), and 10X (influent ammonia + 360 mg/L as N) were applied to the 10-day SRT laboratory-scale reactors. The ammonia (NH_4^+ + NH_3-N) concentrations used to shock the 2-day SRT reactors were based on a target ammonia concentration in the influent of 30 mg/L, based on previous estimates of approximately 15 mg/L as N in the raw wastewater plus 15 mg/L as N added in the supplement. However, actual ammonia concentrations in the influent of the laboratory 2-day SRT reactors were lower than the target value and averaged 15.4 ± 4 mg/L N, which is about half the target concentration. Therefore, to mimic an ammonia shock event, concentrations equal to 2.9X (influent ammonia + 30 mg/L as N), 8.8X (influent ammonia + 120 mg/L as N), and 18.5X (influent ammonia + 270 mg/L as N) were applied to the 2-day SRT laboratory-scale reactors, based on the average ammonia-N level of 15.4 mg/L. To simplify, 3X, 9X, and 18X are used to identify each source condition. Ammonium bicarbonate was used for shocking in both systems. The corresponding alkalinity added with the shock allowed for pH effects due to intense nitrification to be neutralized and, therefore, for ammonia/ammonium effects to be studied without interferences by dramatic pH swings. Although nitrification was not expected in the 2-day SRT reactors, the same ammonia source was used as for the 10-day SRT experiments for comparison purposes.

2.2.2 Reactor Configurations

The laboratory-scale reactors were designed to mimic the pilot-scale SBRs. The laboratory system consisted of four beakers maintained at the 10-day target SRT and four maintained at the 2-day target SRT. Within each SRT, one reactor was used as a control reactor, and the other three were shocked with varying concentrations of contaminants. Reactor configurations and operating parameters are summarized in Table 2-2 below. Times for SBR phases were chosen to closely coincide with pilot plant values.

Table 2-2. Configurations and Operating Conditions for Laboratory-Scale Reactors.

Operating Condition	Setting
Reactor Working Volume	3.5 L
Target Nominal Hydraulic Retention Time (HRT)	1 day
Cycles per Day	4
Feed Time per Cycle (with mixing and aeration)	~ 2-3 min
React Time per Cycle	5 hours
Settling Time per Cycle	1 hour

2.2.3 Daily Maintenance and Operation

To perform toxic shock experiments, soluble contaminants were manually added with the influent during the first cycle of operation. For CDNB the test concentrations were dissolved in the influent by heating and sonicating the contaminant into the influent 1 hour before the experiments. For all subsequent cycles, influent was pumped over a period of approximately 2 to 3 minutes. The influent pump flow rates were checked daily. Influent used to feed the laboratory reactors was obtained daily from the same manhole that supplied the pilot plant. It was settled for 30 minutes to mimic primary clarification before being decanted to a 20-L bucket. The settled influent received the same supplemental feed as the pilot plant influent. The influent for the laboratory reactor experiments was mixed and refrigerated throughout the toxic shock experiments.

During one cycle each day but before wastage, attached growth biomass was scraped off of the reactor walls and back into suspension to maintain a suspended-growth system. Mixed liquor was wasted manually once daily from the 10-day reactors due to the low-wastage volumes required. Wastage was automatically controlled for the 2-day reactors. Wastage volumes were corrected to account for sampling but were not corrected to account for solids in the effluent (rather, the true SRT was calculated and considered the measured effluent TSS). Timers used to control the mixers, air compressors, influent, effluent, and wastage pumps were checked daily to ensure proper operation. J-shaped tubes were used to waste biomass from the 2-day reactors and to decant the treated effluent from all reactors. The height of these tubes was inspected daily to ensure the proper volumes were maintained.

2.2.4 Monitoring Schedule

Samples were collected to monitor influent, effluent, and mixed liquor quality characteristics. Tables 2-3 and 2-4 list the frequency and name of tests performed during operation of the reactors for both the 2- and 10-day SRT systems, respectively. For all influent tests, samples were pulled from the influent bucket during mixing to ensure uniform sampling.

For all effluent tests, the total effluent volume for a cycle was collected and mixed before sampling to ensure that uniformly representative samples were collected and to eliminate the influence of changes in effluent quality over the decant time. Mixed liquor samples were pulled at the beginning of the cycles under mixed and aerated conditions to ensure that solids were properly mixed and to enable comparison of data over time. The reactors were operated for a minimum of one SRT and no more than three SRTs. If recovery (defined as a return of the quality parameters to control reactor levels over three sample points) was noted within this time, reactor operation was discontinued.

Table 2-3. 2-Day SRT Laboratory-Scale Reactor Monitoring Schedule.

	Test	1st Day	2nd, 3rd and 4th Days	5th to 7th Day
Effluent	Soluble Contaminant	every cycle	every other cycle	every day
	COD	every cycle	every other cycle	every day
	pH	every other cycle	every day	every other day
	TSS	every cycle	every other cycle	every day
	Ammonia	every cycle	every other cycle	every day
	Nitrite/Nitrate	every cycle	every other cycle	every day
	Alkalinity	every other cycle	every day	every other day
Influent	COD	every day	every day	every day
	Ammonia	every day	every day	every day
	TSS/VSS	every day	every day	every day
	pH	Every other day	every other day	every other day
	Alkalinity	Every other day	every other day	every other day
	TKN	Every 3 days	every 3 days	every 3 days
Mixed Liquor	SVI	1st cycle	every day	every day
	Carbonaceous SOUR	1st cycle	every day	every other day
	NGR	1st cycle	every day	every other day
	MLSS/MLVSS	1st cycle	every day	every day
	CST	1st cycle	—	last cycle
	Microscopic Evaluation	1st cycle	every day	every other day
	Total Contaminant	every cycle	every day	every day

Table 2-4. 10-Day SRT Laboratory-Scale Reactor Monitoring Schedule.

	Test	1st Day	2nd and 3rd Days	4th to 10th Day	11th to 30th Day
Effluent	Soluble Contaminant	every cycle	every other cycle	every day	every other day
	COD	every cycle	every other cycle	every day	every other day
	pH	every other cycle	every day	every other day	every 4 days
	TSS	every cycle	every other cycle	every day	every other day
	Ammonia	every cycle	every other cycle	every day	every other day
	Nitrite/Nitrate	every cycle	every other cycle	every day	every other day
	Alkalinity	every other cycle	every day	every other day	every 4 days
Influent	COD	every day	every day	every day	every day
	Ammonia	every day	every day	every day	every day
	TSS/VSS	every day	every day	every day	every day
	pH	every other day	every other day	every other day	every other day
	Alkalinity	every other day	every other day	every other day	every other day
	TKN	every 3 days	every 3 days	every 3 days	every 3 days
Mixed Liquor	SVI	1st cycle	every day	every day	every other day
	Carbonaceous SOUR	1st cycle	every day	every other day	every 4 days
	NGR	1st cycle	every day	every other day	every 4 days
	MLSS/MLVSS	1st cycle	every day	every day	every other day
	CST	1st cycle	—	—	last cycle
	Microscopic Evaluation	1st cycle	every day	every other day	every other day
	Total Contaminant	every cycle	every day	every day	every other day

2.3 Analytical Procedures

Constituents were analyzed as described below. In all cases, soluble is defined as that which passes through a 0.45-μm filter.

2.3.1 Alkalinity

Samples for alkalinity were run in duplicate due to the high degree of reproducibility. Alkalinity was performed using a potentiometric titration to a pH of 4.5 as outlined in section 2320 B of Standard Methods (APHA, 1998).

2.3.2 Ammonia

Ammonia-nitrogen was analyzed either by distillation/titration (methods 4500-NH_3 B and 4500-NH_3 C), as described in Standard Methods (APHA, 1998), or by ion chromatography. For the latter, 5 mL of soluble sample was injected into a Dionex DX120 suppressed conductivity ion chromatograph using an AS40 autosampler. The DX120 used a CS12 cation separation column with a CG12 guard column for cation analysis of NH_3. Samples were carried in an eluent of 20 mM methane sulfonic acid. All samples were filtered and stored at -20°C for no more than 28 days prior to ion chromatography analysis.

Ammonia concentrations measured over the time of the ammonia shock experiment were tested using ion chromatography. The measured concentrations are, therefore, equivalent to the total (ammonia + ammonium) concentrations, since the eluent used to carry the sample through the column is acidic (pH ~ 1-2) and converts all the free ammonia present in the sample to ammonium. At 20°C the dissociation constant for NH_4^+ is 3.84 x 10^{-10} (pKa is 9.42). It is possible to calculate the free ammonia concentration based on the measured total ($NH_4^+ + NH_3$) in solution, as long as the pH and temperature are known (equation 2-1).

$$[NH_3] = \frac{10^{-pK_a} \times ([NH_3] + [NH_4^+])}{10^{-pH} + 10^{-pK_a}} \qquad \text{[equation 2-1]}$$

2.3.3 Capillary Suction Time (CST)

CST is an indicator of mixed liquor dewaterability. CST was performed according to section 2710 G of Standard Methods (APHA, 1998).

2.3.4 Chemical Oxygen Demand (COD)

COD tests were performed on total and soluble fractions of the influent and on soluble fractions of the effluent for each reactor. The test was performed following the procedure outlined in section 5220 C of Standard Methods (APHA, 1998), with minor modifications. The ferrous ammonium sulfate titrant concentration was changed to 0.05 M to increase sensitivity, and the mercuric sulfate concentration was decreased to 1.33 g/L because the samples contained low levels of chloride. This action was implemented to reduce the volume of hazardous waste generated.

2.3.5 Soluble CDNB and DNP Determination

Soluble CDNB and 2,4-dinitrophenol (DNP) concentrations were determined using high performance liquid chromatography (HPLC). Samples were run on an HP 1090 liquid chromatograph using an ALTECH model ECONOSPHERE C18, 250 mm x 4.6 mm column. The carrier eluent consisted of 20% acetonitrile and 80% phosphoric acid solution (0.7 mL/L concentrated H_3PO_4), the flowrate was 2.0 mL/min, the oven temperature was 35°C, and the absorbance was measured at 210 nm. The CDNB peak occurred at 3.5 minutes and the DNP peak occurred at 2.5 minutes. DNP samples were acidified to pH < 2 for storage prior to analysis.

2.3.6 Microscopic Observations

Microscopic evaluations were performed with approximately 0.4 mL of mixed liquor from each reactor using wet mounts with a cover slip. Floc morphology was characterized as per Jenkins et al. (1993) and floc diameter was also measured. The following parameters were measured qualitatively on a 0-4 scale: shelled amebae, small flagellates, large flagellates, free-swimming ciliates, carnivorous ciliates, crawling ciliates, wide-mouthed stalked ciliates, and narrow-mouthed stalked ciliates. Metazoa (e.g., rotifers, nematodes) were monitored for general abundance. Filament abundance was measured using the subjective scoring scale given by Jenkins et al. (1993). The microscope used for general evaluations was a bright-field Olympus model CH-2 with 10X eyepieces. Filament evaluations were performed using a phase contrast brightfield Olympus model BH-2 with 15X eyepieces. All evaluations were performed using the 10X objective lenses located on each microscope.

2.3.7 Nitrate/Nitrite

Samples for nitrate and nitrite were analyzed in duplicate using ion chromatography (method 4110 C) as described in Standard Methods (APHA, 1998). All samples were filtered and stored at –20°C for no more than 28 days. Upon thawing the samples, 5 mL of each was loaded into a vial and placed into an AS40 autosampler. Samples were injected into a Dionex DX120 suppressed conductivity ion chromatograph containing an AS14 anion separation column with an AG14 guard column for anion analysis. Samples were carried in an eluent of 3.5 mM Na_3CO_3 and 1.0 mM $NaHCO_2$.

2.3.8 Specific Nitrate Generation Rate (SNGR)

For the NGR tests, 300 mL samples were pulled from the reactors and placed in well-aerated, 500-mL, wide-mouth Erlenmeyer flasks. These samples were spiked with 100-mg/L soluble COD (same as that described in Section 2.2.1) and 20 mg/L $(NH_4)_2CO_3$ as N to achieve the maximum nitrate generation rate. Starting at time 0, 10 mL samples were collected every 10 minutes. Samples were centrifuged for three minutes on a Fisher Scientific Centrific centrifuge at 3400×g to pellet the solids. The supernatant was removed and stored at -20°C for no more than 28 days. For analysis, these samples were thawed and filtered through a 0.45 μm filter and analyzed for nitrate by ion chromatography as described in Section 2.3.7. The NGR (mg/L-min as oxygen demand) was determined from the nitrate profile by calculation, as described for the IC_{xx} determination procedure. The NGR was normalized by MLVSS to yield a specific NGR (SNGR). The residual mixed liquor used during the NGR assay was reintroduced to the reactors as only a small quantity of readily degradable COD and ammonia were added, and no toxin was added. Therefore, it was possible to use as much biomass as was needed to achieve reliable results.

2.3.9 Sludge Volume Index (SVI)

SVI analysis was conducted on mixed liquor samples as per section 2710 D in Standard Methods (APHA, 1998). A 250-mL graduated cylinder was used for the test so that more accurate volume readings could be taken.

2.3.10 Specific Oxygen Uptake Rate (SOUR)

Mixed liquor samples were collected and analyzed in duplicate for each reactor to determine the SOUR. The samples were aerated for five minutes before being placed in a 300-mL BOD bottle and spiked with 100 mg/L COD (defined in Section 2.2.1) to ensure that respiration was not substrate limited. OUR tests followed the procedure outlined in section 2710 B of Standard Methods for oxygen consumption rate tests. These samples were re-introduced to the reactors after SOURs were completed as only a small quantity of readily degradable COD was added.

An automatic data acquisition system was used to run the oxygen consumption rate tests. This system consisted of two Orion oxygen electrodes, model 97-08, attached to an Acumet dual-channel pH/ion meter, model AR25. Dissolved oxygen readings were recorded every 6 seconds using LabView 6i software operating on a Dell Dimension XPS P100c personal computer.

2.3.11 pH

pH measurements were taken according to Section 4500-H^+ B in Standard Methods. pH measurements were made using an Acumet pH meter model 910 with an Acumet liquid-filled polymer body combination electrode and a Ag/AgCl reference element.

2.3.12 Potassium

Samples for soluble potassium were run in duplicate using ion chromatography or atomic absorption spectrophotometry. Filtered samples were stored at –20°C for no more than 28 days. Upon thawing the samples, 5 mL of each was loaded into a vial and placed into an AS40 autosampler, and analyzed as for ammonia (Section 2.3.2). After August 2002, IC analysis for potassium was found to incur larger variability, and all subsequent shock experiments utilized atomic absorption spectrophotometry. These included all of the 2-day SRT shock experiment re-runs (ammonia, pH, cadmium, CDNB, octanol, DNP, and cyanide) as well as the 10-day SRT cyanide shock experiment. The atomic adsorption spectrophotometer used was a Perkin Elmer 5100.

2.3.13 Suspended Solids

TSS and volatile suspended solids (VSS) were analyzed using the method outlined in sections 2540 D and 2540 E of Standard Methods (APHA, 1998). TSS and VSS were performed on mixed liquor samples, whereas TSS only was analyzed for effluent samples due to the low concentration of suspended solids in the samples and limited sample volumes available. Volumes to be filtered were selected based on ease of filtration (less than two minutes to filter at 20 psi vacuum).

2.3.14 Total Kjeldahl Nitrogen (TKN)

TKN samples were taken in duplicate and preserved by adding concentrated H_2SO_4 to a pH < 2 and refrigerated for no more than 28 days, as described by Table 1060:I of Standard Methods. Samples were analyzed using a volume modification of the macro-Kjeldahl method outlined by section 4500-N_{org} B of Standard Methods. Volumes were cut proportionally to minimize waste generation.

2.4 Data Presentation and Statistical Analysis

For most of the quality parameters monitored during this study, duplicate or triplicate measurements were performed so that statistical analyses could be performed (Table 2-5). Lines are drawn through the average of triplicate or duplicate measurements with error bars representing one standard deviation (for triplicates) or the distance between measured values (for duplicates).

Table 2-5. Number of Replicates for Each of the Quality Parameters Monitored During the Source-Effect Experiments.

Effluent Quality Parameter	# Replicates	Mixed Liquor Quality Parameter	# Replicates	Influent Quality Parameter	# Replicates
Ammonia	2	MLSS/MLVSS	3	COD	3
Nitrite/Nitrate	2	SOUR	2	TSS	3
Potassium	2	NGR	1	pH	2
Soluble COD	3	SVI	1	Alkalinity	2
TSS	3	CST	3	TKN	2
pH	2	Microscopic Evaluation	1	Ammonia	2
Alkalinity	2				

To assess recovery of the shocked reactors to control levels, statistics were performed on the duplicate and triplicate sample tests using Dunnett's method for multiple comparisons with a control. In all cases, the significance level tested was $\alpha = 0.05$. When an effect was significantly different, the probability was less than 0.05. Also, the criterion applied to declare recovery was three consecutive data points showing no significant difference from the control. In the text, this is designated by stating that $\alpha = 0.05$. Dunnett's critical t-values used in the calculation were determined using SAS statistical software and are 3.62 for $n = 2$ and 2.88 for $n = 3$. Microsoft Excel was used to carry out the calculations.

CHAPTER 3.0

CADMIUM SHOCK EXPERIMENTS

3.1 Literature Review

Cadmium is a prevalent industrial contaminant. The electroplating industry is responsible for approximately half the cadmium consumed in the United States every year. Cadmium is also used as a color pigment, so it can also be found in industrial wastewaters from textiles, plastics, paints, or printing inks industries. Another source that contributes largely to the emissions of cadmium in the environment is fossil fuel burning (Weber and Sherrard, 1980). As a prevalent pollutant in the environment, scientific studies to investigate its effects in different environmental systems are of great importance.

From the standpoint of wastewater treatment, cadmium has been shown to cause significant upset in activated sludge systems. Studies have found that the most detrimental effect of cadmium in activated sludge systems is deflocculation. Neufeld (1976) reported that deflocculation was the main process effect of shock by cadmium and other heavy metals, although this effect did not appear rapidly. In the case of cadmium and zinc, significant deflocculation was observed three to four days after the shock, while for mercury the effect was observed within 5 hours of the shock. Bott and Love (2001) demonstrated that cadmium induced significant deflocculation within 5 hours of shock loading in SBR reactors using activated sludge and primary effluent feed from a full-scale facility.

Other treatment process effects of cadmium shock have been reported in the literature. Weber and Sherrard (1980) saw a slight decrease in COD removal efficiency in CSTR reactors fed with 5 and 10 mg/L cadmium and a decrease in nitrification rate as the influent cadmium concentration increased. Madoni et al., (1999) studied the response of SOUR and ammonia uptake rate to heavy metal contamination and found that cadmium had a negative impact on both heterotrophic- and nitrifying-bacteria respiration rates. Zarnovsky et al., (1994) concluded that cadmium toxicity to non-acclimated sludge decreased as the SRT increased. This implies that higher MLSS concentrations yield lower levels of toxicity at a given cadmium concentration.

Previous work conducted by the project team showed that cadmium shocks invoke a rapid potassium efflux response due to the thiol-reactive nature of the heavy metal (Bott and Love, 2002). Data collected by the project team indicated that the soluble form of the metal is the toxic agent, which is consistent with work done by others (Collins and Stotzky, 1989; Hughes and Poole, 1989). Additionally, a predominant insoluble form of cadmium in natural waters tends to be cadmium carbonate (otavite), and its formation potential is heavily dependent upon the alkalinity of the water, as previously determined through speciation modeling with MINEQ+ (Bott and Love, 2001). Therefore, metal speciation is a critical issue that must be considered when assessing cadmium toxicity and process effects in wastewater treatment plants. Finally, previous studies conducted in our laboratory suggest that cadmium precipitation and adsorption to biomass can be very significant and fast, with precipitation and adsorption occurring at approximately the same rate (Sonzini, 2002).

At the molecular level, cadmium has been shown to bind nucleic acids and induce DNA mutagenesis (Hughes and Poole, 1989). Cadmium has also been associated with enzyme inactivation. Additionally, it has been suggested that cadmium can complex with phosphate groups in membrane phospholipids, thereby adversely impacting metabolic pathways that rely upon phosphorylation by these functional groups (Collins and Stotzky, 1989). Bochner et al., (1984) indicated that cadmium chloride induced a stress response with the production of heat shock proteins and adenylylated nucleotides in *Salmonella typhimurium*. This result is consistent with the observation reported by Bott and Love (2001) who measured rapid induction of GroEL, a heat shock protein, in activated sludge cultures stressed with cadmium at concentrations as low as 5 mg/L.

3.2 Results

3.2.1 Inhibitory Concentrations (IC_{XX})

The impact of varying cadmium concentrations on the SOUR of the 2- and 10-day SRT biomass was determined using respirometric inhibition tests. IC_{XX} values were calculated from equations derived from best fit curves to the data (shown in Appendix B), which reflected a percentage inhibition relative to the control tests (zero contaminant added). The results of these calculations are listed on Table 3-1, and were the final concentrations added to each of the reactors at the beginning of the first cycle of the source-effect experiment.

Table 3-1. IC_{15}, IC_{25} and IC_{50} Values for Cadmium in 2- and 10-Day SRT Mixed Liquor.

	2-Day SRT		10-Day SRT	
	mg/L	mg Cd/g MLVSS	mg/L	mg Cd/g MLVSS
IC_{15}	21	44	17	14
IC_{25}	36	75	27	22
IC_{50}	84	178	52	42

In contrast to the results of Zarnovsky et al. (1994), the IC_{XX} values on a mg/L basis for the 10-day biomass were found to be lower (i.e., more sensitive) than for the 2-day SRT biomass. The same trend was observed when IC_{XX} values were reported on a mg/g biomass basis. Since it is expected that a larger nitrifier population existed in the mixed liquor from the 10-day SRT reactor, the IC_{XX} was measured based on respirometry, and nitrifiers consume 4.6 g O_2 per g ammonia-N oxidized, the difference could be due to the disproportionately higher impact of cadmium on nitrifying bacteria. Nitrifiers are generally more sensitive to shock loads of metal toxins than heterotrophs (Bagby and Sherrard, 1981) and are known to be sensitive to cadmium (Hu et al., 2002). Collectively, this may explain why the IC_{XX} for the nitrifying 10-day SRT culture was lower overall than for the 2-day SRT biomass.

3.2.2 Influent Characteristics

The average influent wastewater characteristics over the course of the cadmium source-effect experiments are given in Table 3-2. The influent quality was relatively constant over the time of each experiment with the exception of the suspended solids for the 2-day SRT experiment, which varied significantly as seen by the high standard deviation. Both experiments occurred after influent supplementation was initiated. Accordingly, the influent composition reveals that the wastewater used as the feed for the source-effect experiments with cadmium was medium to strong in strength (Metcalf and Eddy, 2003).

Table 3-2. Characteristics of Wastewater Used During Cadmium Source-Effect Experiments.

Characteristic	2-Day SRT Biomass Average $\pm$ std dev	10-Day SRT Biomass Average $\pm$ std dev
Alkalinity (mg/L as $CaCO_3$)	445 $\pm$ 73.1	179 $\pm$ 12
Ammonia (mg/L as N)	16.1 $\pm$ 1.3	17.6 $\pm$ 4.0
COD, soluble (mg/L)	746 $\pm$ 67	123 $\pm$ 44
COD, total (mg/L)	836 $\pm$ 151	368 $\pm$ 97
pH	7.6 $\pm$ 0.2	7.8 $\pm$ 0.3
TSS (mg/L)	194 $\pm$ 78	173 $\pm$ 30
VSS (mg/L)	64 $\pm$ 50	148 $\pm$ 27

3.2.3 Effluent Characteristics

3.2.3.1 Soluble COD

Effluent soluble COD data showed that cadmium impacted treatment efficiency for both the 2- and 10-day SRT reactors, as shown in Figures 3-1 and 3-2. For the 2-day SRT biomass, the effluent soluble COD remained significantly elevated in all the cadmium-stressed reactors relative to the control (α=0.05) until the end of the experiment. Although partial recovery was seen, complete recovery back to the concentrations measured in the control was not achieved during a 3 SRT period (24 cycles). Similarly, all cadmium-exposed reactors containing nitrifying, 10-day SRT biomass showed poor COD removal until the 15th, 23rd and 51st cycle (α=0.05), respectively, for the IC_{15}, IC_{25} and IC_{50} reactors, after which the performance of each reactor was similar to the unshocked control. In both cases, the degree to which COD removal deteriorated corresponded with increased cadmium concentrations. However, the peak effluent COD concentrations in the 2-day SRT experiment were much larger than in the 10-day SRT experiment, suggesting that factors associated with shorter SRTs made the biomass more susceptible. Finally, it can also be seen that loss of COD removal efficiency reached a peak in all cadmium-shocked reactors 18 (three cycles) to 24 hours (four cycles) after the shock occurred, which indicates that the toxic effects were not instantaneous and delayed toxicity was observed.

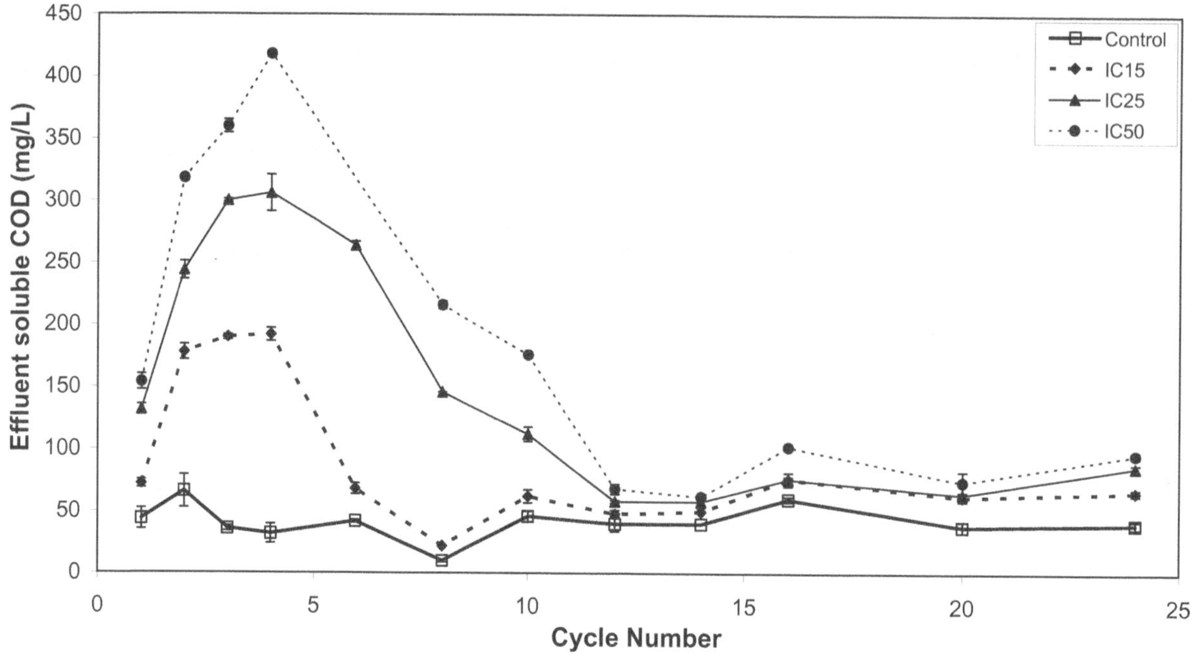

Figure 3-1. Effluent Soluble COD for the 2-Day SRT Reactors Exposed to Cadmium Shock.

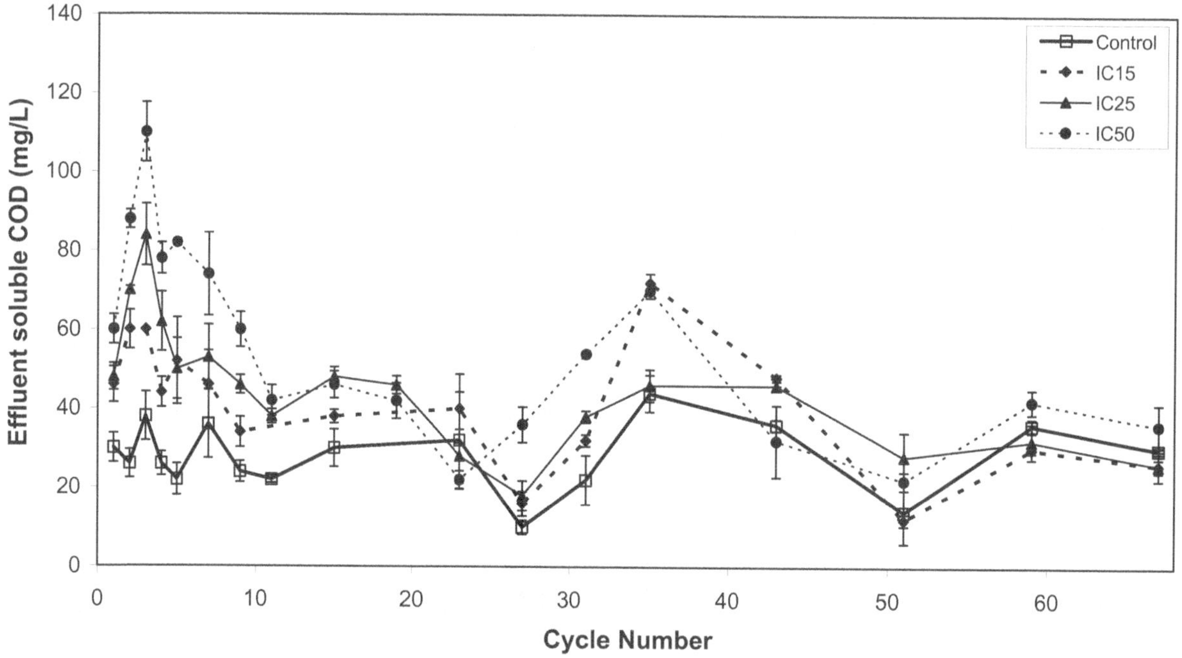

Figure 3-2. Effluent Soluble COD for the 10-Day SRT Reactors Exposed to Cadmium Shock. Data through Cycle 67 Are Shown.

Cadmium speciation played an important role in process deterioration. It is known that cadmium is highly insoluble in wastewater, especially in the presence of high alkalinity concentrations since cadmium precipitates as otavite ($CdCO_3$, which has a solubility constant around 10^{-12}). Despite significant differences in wastewater alkalinity concentrations between the two experiments and the higher total cadmium dose applied to the 2-day SRT biomass, the soluble cadmium concentrations measured for both the 2- and 10-day SRT systems were similar (data in Appendix B). The soluble concentrations in the reactor effluents after the first cycle were approximately 3-11% of the total cadmium added in both cases. Soluble cadmium concentrations were detected for the 2-day SRT biomass throughout the experiment whereas it was below detection after cycle 51 for the 10-day SRT biomass. Interestingly, the presence of soluble cadmium corresponded with the time period when poor effluent COD removal was observed in all cadmium shocked reactors. This observation is consistent with the notion that it is the soluble form of cadmium that is believed to cause inhibition in cells (Collins and Stotzky, 1989; Hughes and Poole, 1989). Measured total cadmium concentrations are presented in Appendix B and show that most of the cadmium ended up in the insoluble phase, enmeshed within the mixed liquor. Although this insoluble cadmium slowly washed out of the system, it was detectable in all shocked reactors for the duration of the experiments.

3.2.3.2 TSS and Potassium Concentrations

Effluent TSS was significantly impacted by cadmium shock, as was expected based on previous studies conducted by the project team (Bott and Love, 2001). In all cases, the increases in effluent TSS were due to high turbidity that was presumably caused by deflocculation. Deflocculation was observed beginning with the first cycle after shock for reactors at both sludge ages and for all cadmium concentrations, resulting in effluent TSS concentrations that were two to three times higher than the values observed for the control reactors (Figures 3-3 and 3-4). The effluent TSS in the stressed 2-day SRT reactors increased immediately after shock, but there was no significant difference between the effluent TSS concentrations measured in these reactors, despite being exposed to different cadmium concentrations. Additionally, while the difference in effluent TSS between the stressed and control reactors decreased with time, only the IC_{15} reactor recovered to control levels ($\alpha = 0.05$), which occurred between cycles eight and 12 (days two and three); however, this recovery was temporary, since toward the end of the experiment the effluent TSS in this same reactor deteriorated again. The IC_{25} and IC_{50} reactors did not show statistical recovery over the course of the experiment ($\alpha = 0.05$); however, the IC_{25} reactor appeared to be reaching recovery when the experiment ended. For the 10-day SRT biomass, increasing concentrations of cadmium corresponded with increasing effluent turbidity during the first few cycles; however, after eight cycles (two days), there was no distinction between the effluent TSS concentrations for the stressed reactors, which remained at a higher concentration than the control reactor until the 80th cycle (20 days).

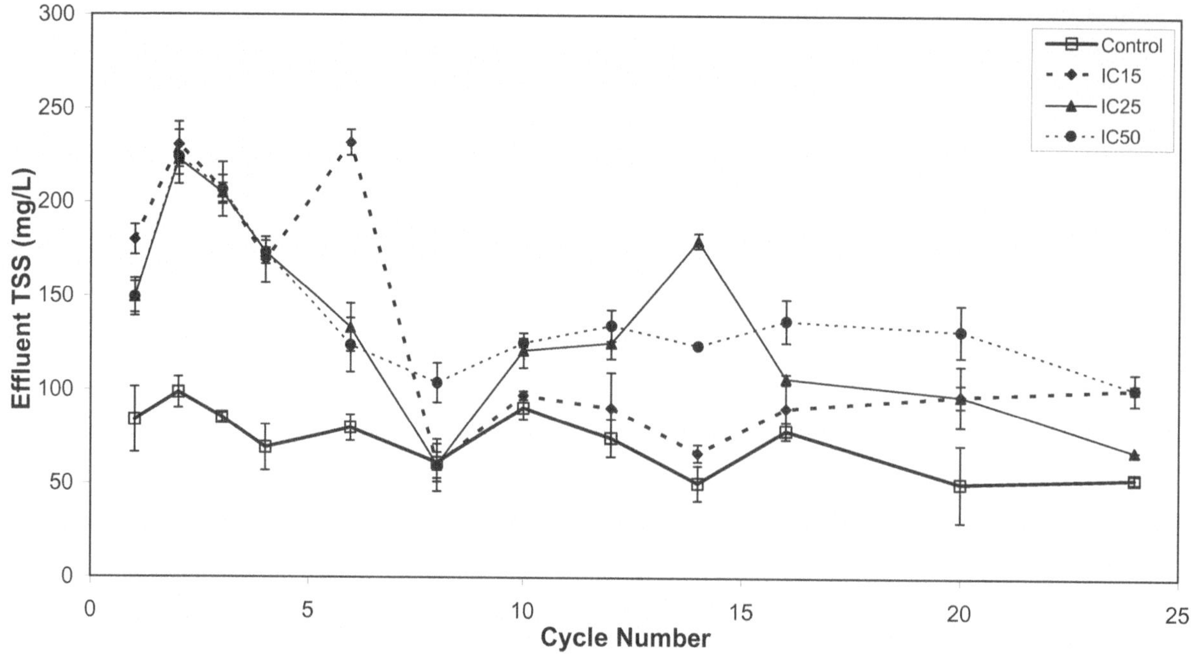

Figure 3-3. Effluent TSS for the 2-Day SRT Reactors Exposed to Cadmium Shock.

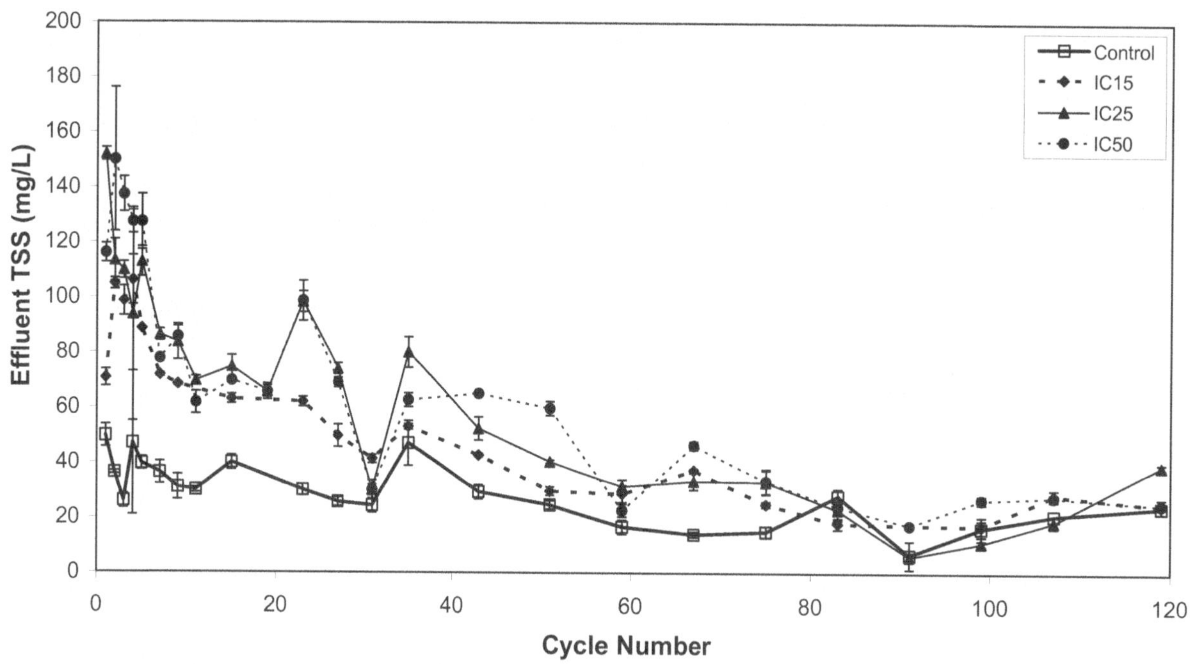

Figure 3-4. Effluent TSS for the 10-Day SRT Reactors Exposed to Cadmium Shock.

Effluent potassium (K^+) concentrations are shown for the first 10% of experimental run time (three cycles for 2-day SRT and 12 cycles for 10-day SRT) because all effluent K^+ levels recovered to effluent levels after this time. The effluent K^+ concentration increased significantly (α=0.05) during the first two cycles for the 2-day SRT reactors, as shown in Figure 3-5. The change in liquid phase K^+ concentration was around 9 mg/L for the IC_{25} and IC_{50}-stressed biomass, which is higher than what was seen previously by the project team with cadmium (Bott, 2001). For the 10-day SRT shocked reactors, effluent K^+ concentrations increased above those in the control to a more modest degree beginning with and after the first cycle (Figure 3-6). The change in liquid phase K^+ concentration peaked around 2 mg/L for the IC_{50}-stressed biomass in the 10-day SRT system, which is consistent with previous observations by the project team with cadmium (Bott, 2001). It is believed that electrophilic (thiol-reactive) chemicals such as cadmium activate the GGKE protective response to stress (Bott and Love, 2002). The increases in potassium concentration observed here are consistent with that theory. Additionally, these authors showed that the increase is significant and rapid after stress, and they hypothesized that it is high intrafloc concentrations that lead to such severe deflocculation in response to electrophilic contaminants (Bott and Love, 2004). Effluent turbidity remained elevated for several days after cadmium shock, even though the soluble K^+ concentrations returned to background levels (data not shown). This is consistent with the notion that deflocculated sludges need to "regrow" into flocculant sludges after cation-induced deflocculation events (Novak et al., 1999).

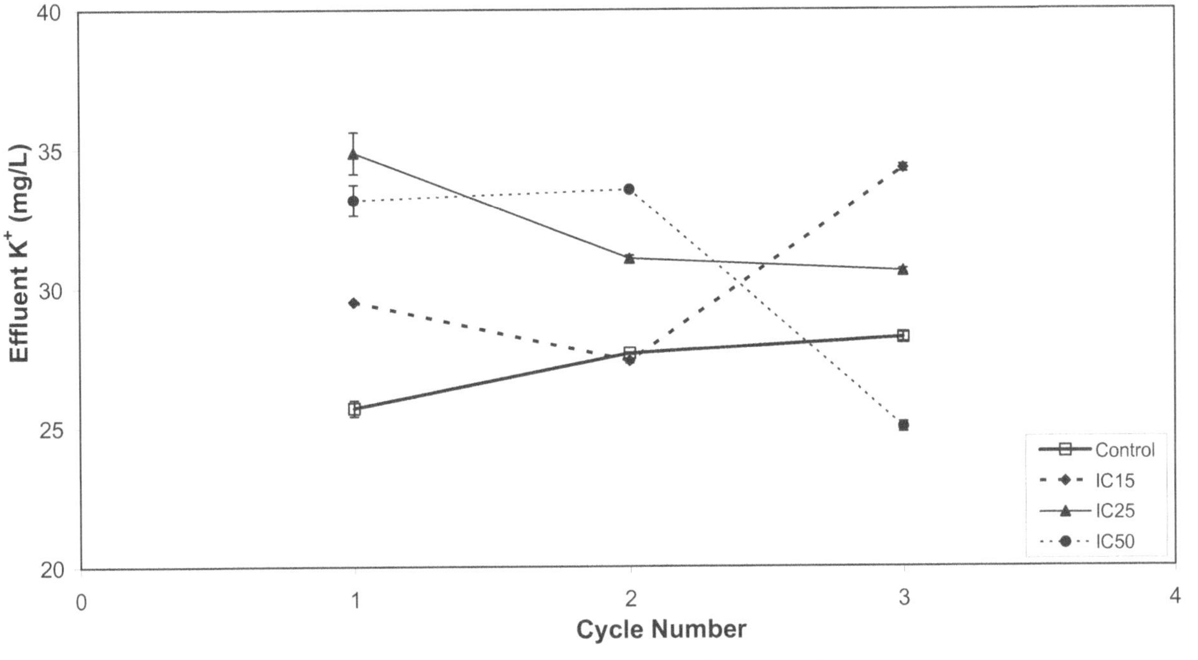

Figure 3-5. Effluent Potassium for the 2-Day SRT Reactors Exposed to Cadmium Shock. Data through Cycle 3 Are Shown.

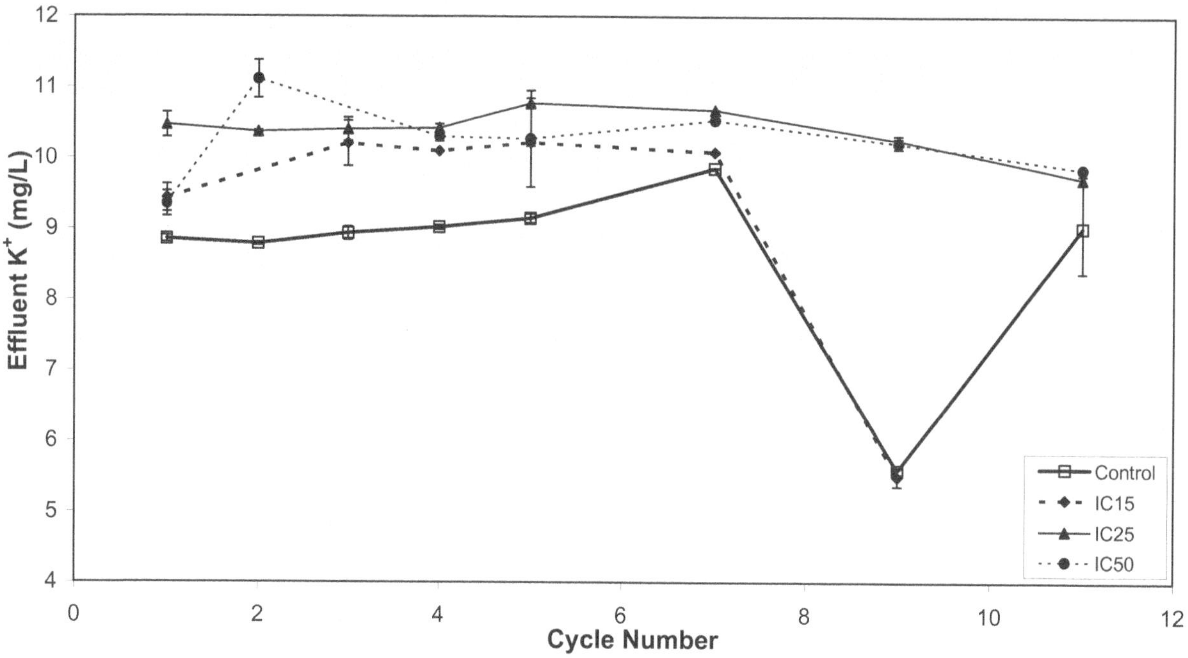

Figure 3-6. Effluent Potassium for the 10-Day SRT Reactors Exposed to Cadmium Shock. Data through Cycle 11 Are Shown.

3.2.3.3 Nitrification Inhibition

The 2-day SRT biomass used in the cadmium shock experiments was not nitrifying during most of the experiment, and only modestly during the middle phase. The ammonia levels remained elevated throughout the experiment except during the middle phase (Figure 3-7), but nitrite and nitrate levels remained below 1.0 and 0.5 mg/L as N, respectively, during this phase (data not shown). Therefore, the data strongly indicate that an insignificant nitrifying population was present in the 2-day SRT biomass, as planned. It is interesting to note that the effluent ammonia concentrations of the shocked reactors were significantly higher than in the control effluent during the first five cycles after the cadmium shock. Based on mass balance calculations, the increase in supernatant ammonia-N in the shocked reactors was greater than that expected based on the loss of biomass through the turbid effluent. Therefore, the increase in ammonia-N in the shocked reactors is probably due to a decrease in assimilative nitrogen requirements because of both biomass loss via the turbid effluent and reduced biomass growth overall due to toxicity effects. This reduction in biomass growth is consistent with the reductions in specific respiration capacity by the biomass, as discussed in Section 3.2.4.3. The decrease in ammonia concentrations between cycles eight and 12 (days two and three) corresponds with the time period when effluent sCOD concentrations decreased most significantly (Figure 3-1). Dissimilation of the accumulated sCOD would normally correspond with an increase in MLSS concentration due to growth. This effect did occur (discussed below, Figure 3-13), but the extent of MLSS increase was probably moderated by the continued loss of biomass in the effluent through deflocculation. Therefore, the decrease in effluent ammonia-N concentration during the middle of the experiment seems to correspond with recovery of biomass growth.

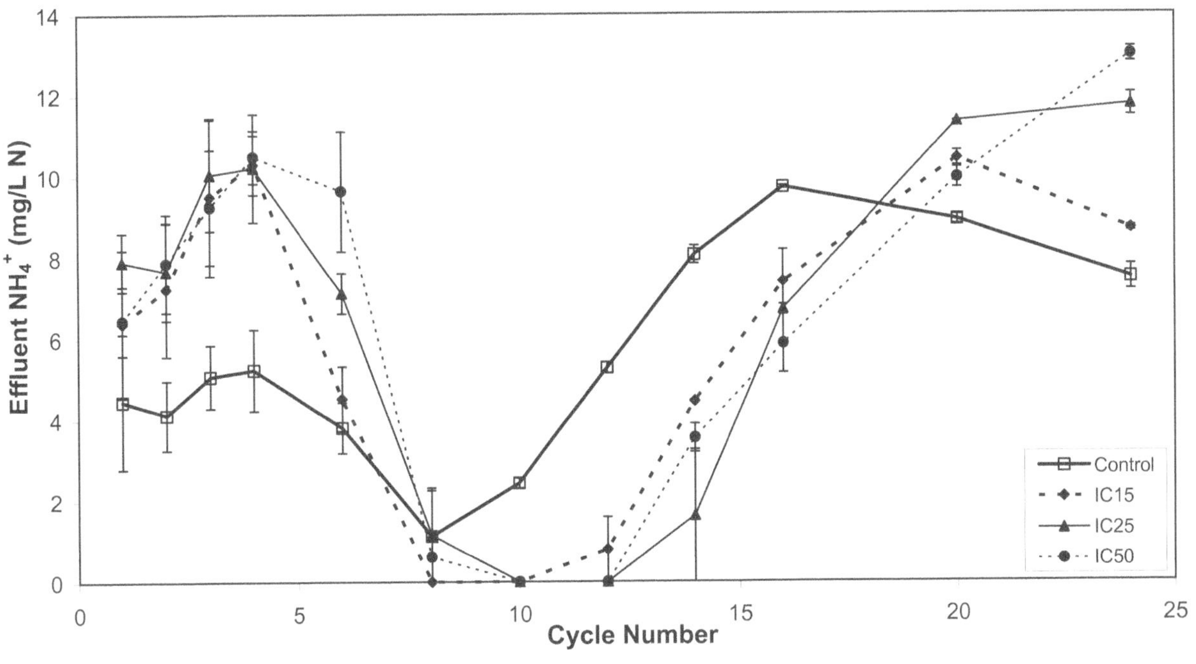

Figure 3-7. Effluent Ammonia-Nitrogen for the 2-Day SRT Reactors Exposed to Cadmium Shock.

Nitrification in the 10-day SRT reactors was significantly affected by cadmium shock. As shown in Figure 3-8, effluent ammonia concentrations increased with increasing cadmium concentrations beginning with the first cycle. Nitrification recovered in the 10-day SRT reactors after six, 10, and 15 cycles for the IC_{15}-, IC_{25}- and IC_{50}-stressed systems, respectively. Therefore, the rate of recovery from nitrification inhibition was more rapid than for deflocculation or effluent COD removal inhibition. Variations in effluent nitrate concentrations corresponded with observed ammonia-removal trends (Figure 3-9). Effluent nitrate concentrations took longer to return to levels measured in the control reactor because of the time required for accumulation of nitrate to steady-state levels after inhibition. Nitrite concentrations remained low to nondetectable in the stressed reactors (data not shown), so nitrification inhibition appeared to be primarily targeted against the AOBs.

Effluent alkalinity concentrations are affected by both the degree of nitrification occurring and the amount of cadmium carbonate (otavite) formed. Since the 2-day SRT system was not nitrifying at the time of the experiment, the second factor should determine the alkalinity variations in these reactors. The decrease in alkalinity in the IC_{25}- and IC_{50}-stressed reactors (as shown in Figure 3-10) and low soluble cadmium concentrations measured from the bioreactors (Appendix B) are consistent with the formation of otavite. This result also indicates that precipitation, rather than sorption to negatively charged sites within the biomass flocs, was the main process affecting cadmium speciation in this system. Effluent pH remained quite stable around 8.3 throughout the experiment (data shown in Appendix B), which is consistent with the alkalinity trends observed.

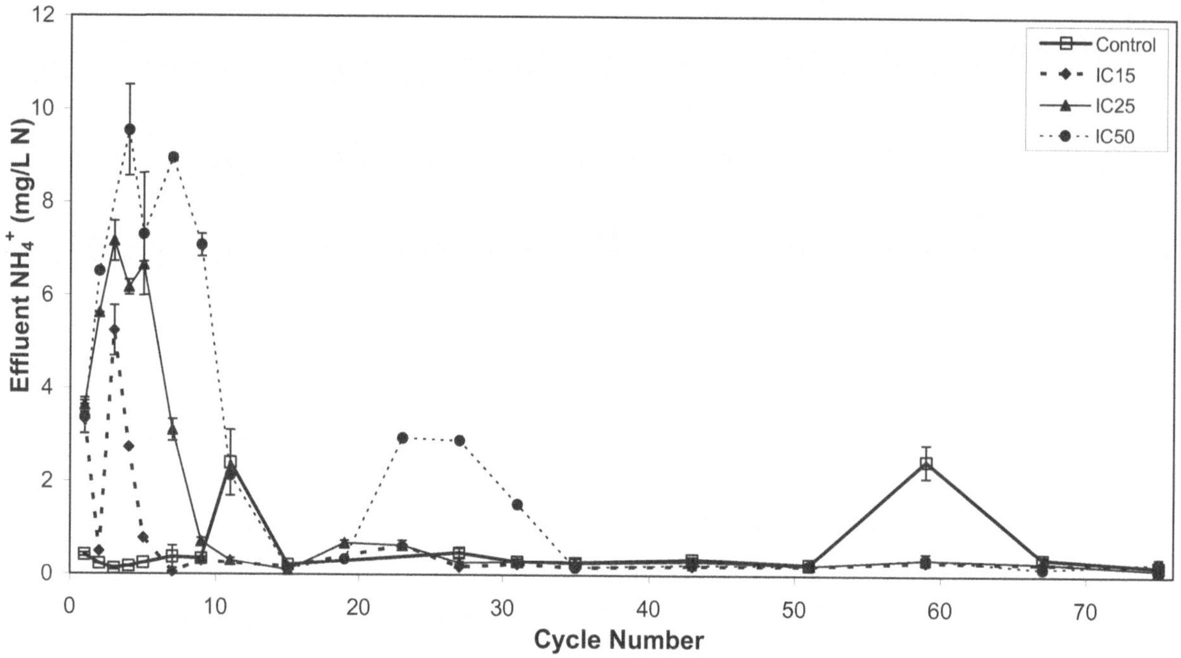

Figure 3-8. Effluent Ammonia-Nitrogen for the 10-Day SRT Reactors Exposed to Cadmium Shock. Data Through Cycle 75 Are Shown.

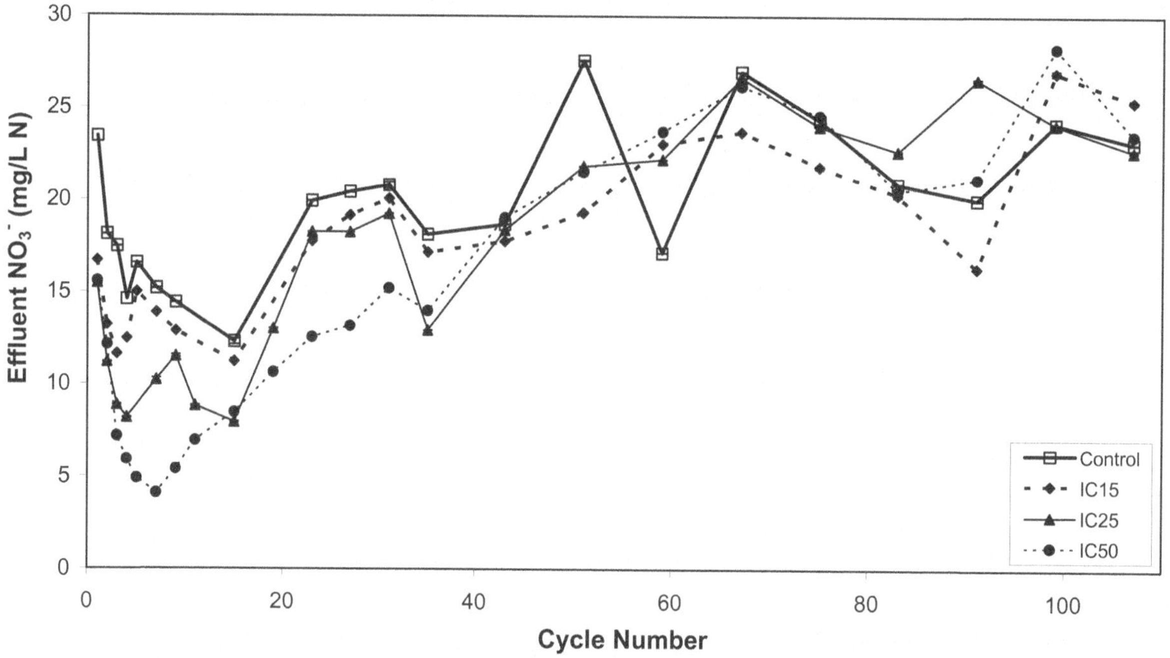

Figure 3-9. Effluent Nitrate-Nitrogen for the 10-Day SRT Reactors Exposed to Cadmium Shock.

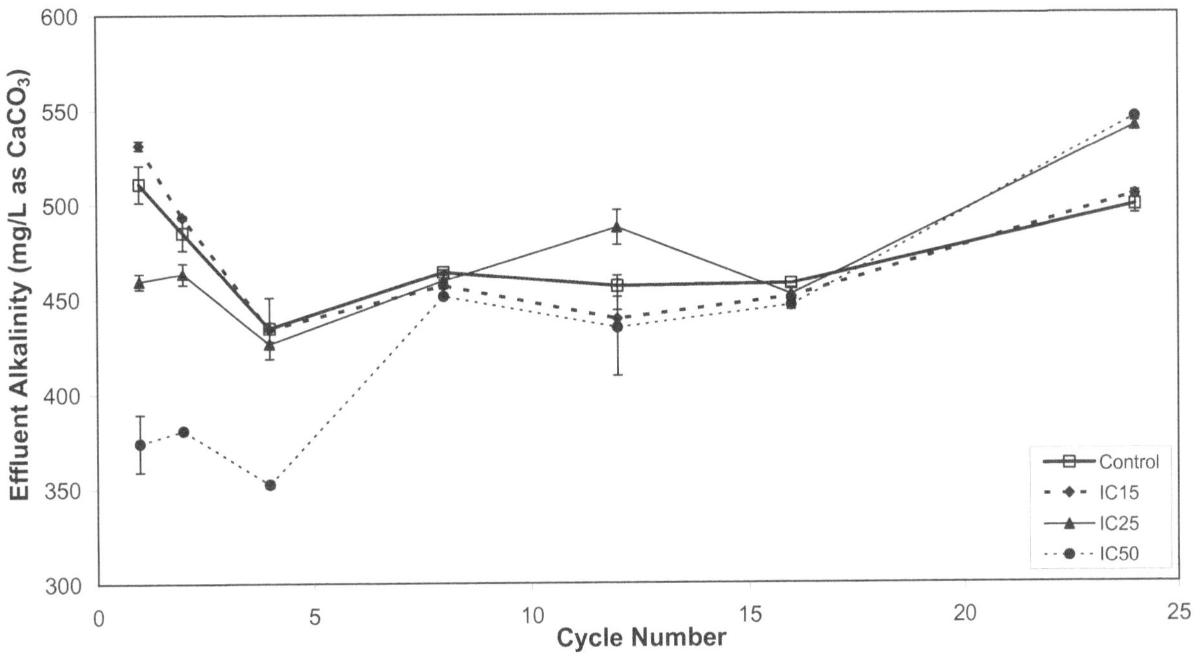

Figure 3-10. Effluent Alkalinity for the 2-Day SRT Reactors Exposed to Cadmium Shock.

In contrast to the non-nitrifying 2-day SRT biomass, effluent alkalinity concentrations in the 10-day SRT bioreactors were more heavily influenced by the degree of nitrification than otavite formation. As can be seen in Figure 3-11, alkalinity increased in the stressed reactors, which correlates with the patterns observed with the soluble nitrogen species. Assuming 7.23 mg alkalinity as $CaCO_3$ is consumed per mg NO_3^--N formed, the theoretical increase in alkalinity for the IC_{50} shocked experiment in the 10-day SRT reactor should have been 55 mg/L as $CaCO_3$ during the first cycle after the shock, which is exactly what was observed in Figure 3-11. This suggests that loss of alkalinity to insoluble otavite precipitation (which would have reduced the increase in alkalinity) was not large. Furthermore, the opposing patterns of change to alkalinity in the non-nitrifying and nitrifying systems need to be considered for facilities that choose to use alkalinity-based titration systems as monitors for upset early warning (Love and Bott, 2000). pH trends for the 10-day SRT system corresponded with alkalinity levels (data shown in Appendix B). As expected, severe decreases in pH were observed when alkalinity decreased to below 25 mg/L as $CaCO_3$ between cycles 40 and 80 in the 10-day SRT system.

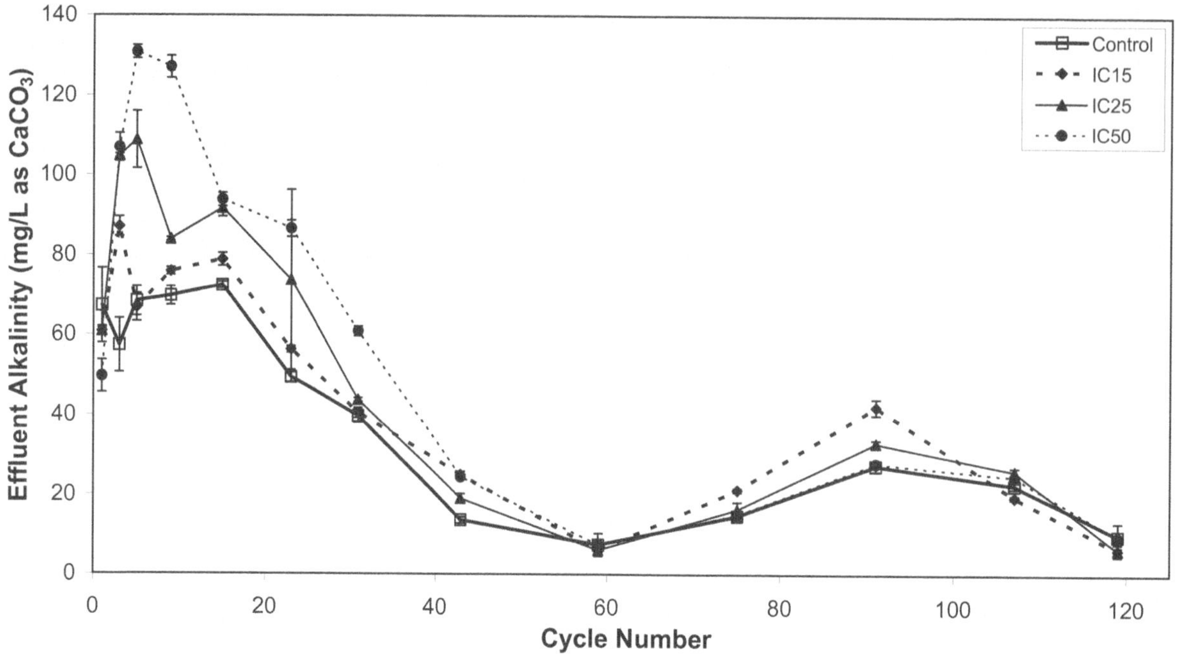

Figure 3-11. Effluent Alkalinity for the 10-Day SRT Reactors Exposed to Cadmium Shock.

3.2.4 Mixed Liquor Characteristics

3.2.4.1 Settling and Dewatering Characteristics

CST and SVI measurements indicated that there was a modest to insignificant impact on biosolids dewaterability or sludge blanket settleability due to cadmium shock. CST values for all biomass samples from both the 2- and 10-day SRT experiments were always less than 60 seconds and frequently less than 25 seconds for the 10-day SRT biomass, which indicates that cadmium had moderate to little impact on dewatering properties (data not shown). The cadmium-exposed 2-day SRT biomasses showed much lower SVI values as compared to the control during the entire experimental period (Figure 3-12). It is possible that the insoluble cadmium precipitates integrated into the flocs and increased the density of the 2-day biomass, thereby enhancing compaction during the SVI test. Furthermore, the supernatant above the settled biomass was turbid for the cadmium-stressed reactors in a manner consistent with the effluent TSS data. The cadmium-exposed biomasses in the 10-day SRT reactors had SVI measurements that were in the excellent range (< 100 mL/g, data not shown), indicating that sludge compaction was unaffected by cadmium. As with the 2-day SRT system, however, the supernatant above the biomass blanket during the test was turbid. The difference in SVI response between the 2-day and 10-day SRT biomasses may be due to the fact that the 2-day SRT biomass did not compact well to start with (control SVI was higher than 150 mL/g) and the shocked reactors received much higher doses of cadmium per unit mass of MLVSS than the 10-day SRT biomass (Table 3-1). Microscopic observation of the mixed liquor on the shocked and control reactors revealed no trends in terms of reduced filament composition, indicating that cadmium did not improve settling by selectively killing filaments.

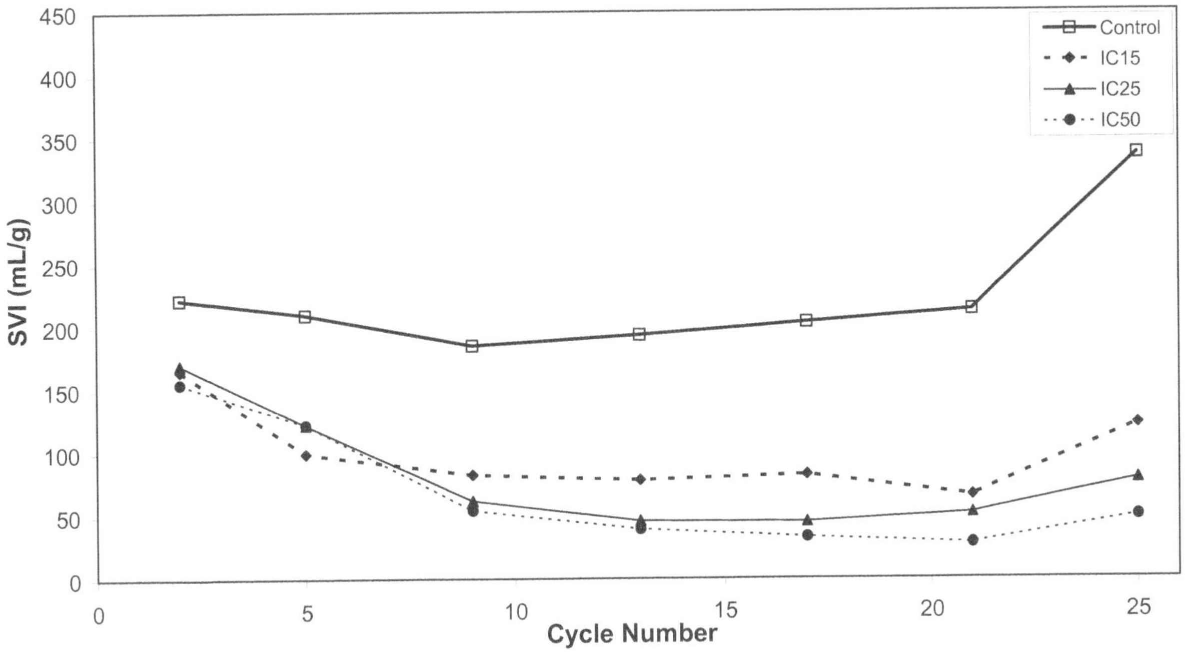

Figure 3-12. SVI for the 2-Day SRT Reactors Exposed to Cadmium Shock.

3.2.4.2 Mixed Liquor Solids Trends

MLSS and MLVSS data are shown in Table 3-3 for the 2- and 10-day SRT reactors. The concentrations for the 10-day SRT systems reflect an unsupplemented COD load to the reactors, whereas the 2-day SRT data reflect a supplemented COD load. The 2- and 10-day SRT-shocked reactors saw a 20%–35% decrease in mixed liquor concentration for the first few cycles of each experiment (Figures 3-13 and 3-14, respectively). For both the 10-day SRT (IC_{25} and IC_{50} reactors) and the 2-day SRT system, a solids mass balance indicated that loss of biomass through the turbid effluent only accounted for part of the decrease in MLSS, and slowed growth due to cadmium inhibition undoubtedly also contributed significantly to the observed, sustained decrease. This is consistent with the decreased sCOD removal efficiency observed. Finally, the MLSS in the 10-day SRT reactors increased toward the end due to higher influent total COD loads, but the 2-day SRT reactors did not recover to control levels by the end of the experiment.

Table 3-3. Mixed Liquor Concentrations in 2- and 10-Day SRT Reactors Exposed to Cadmium Shock.

	MLSS (mg/L)	MLVSS (mg/L)	MLVSS:MLSS Ratio
2-day SRT reactors			
Control	1,182 ± 49	892 ± 63	0.75 ± 0.045
IC_{15}	980 ± 78	722 ± 41	0.74 ± 0.050
IC_{25}	820 ± 125	586 ± 67	0.72 ± 0.061
IC_{50}	793 ± 152	587 ± 99	0.75 ± 0.050
10-day SRT reactors			
Control	1,248 ± 161	1,043 ± 156	0.83 ± 0.03
IC_{15}	1,251 ± 157	1,035 ± 155	0.83 ± 0.04
IC_{25}	1,239 ± 176	1,030 ± 170	0.83 ± 0.04
IC_{50}	1,147 ± 182	940 ± 170	0.82 ± 0.04

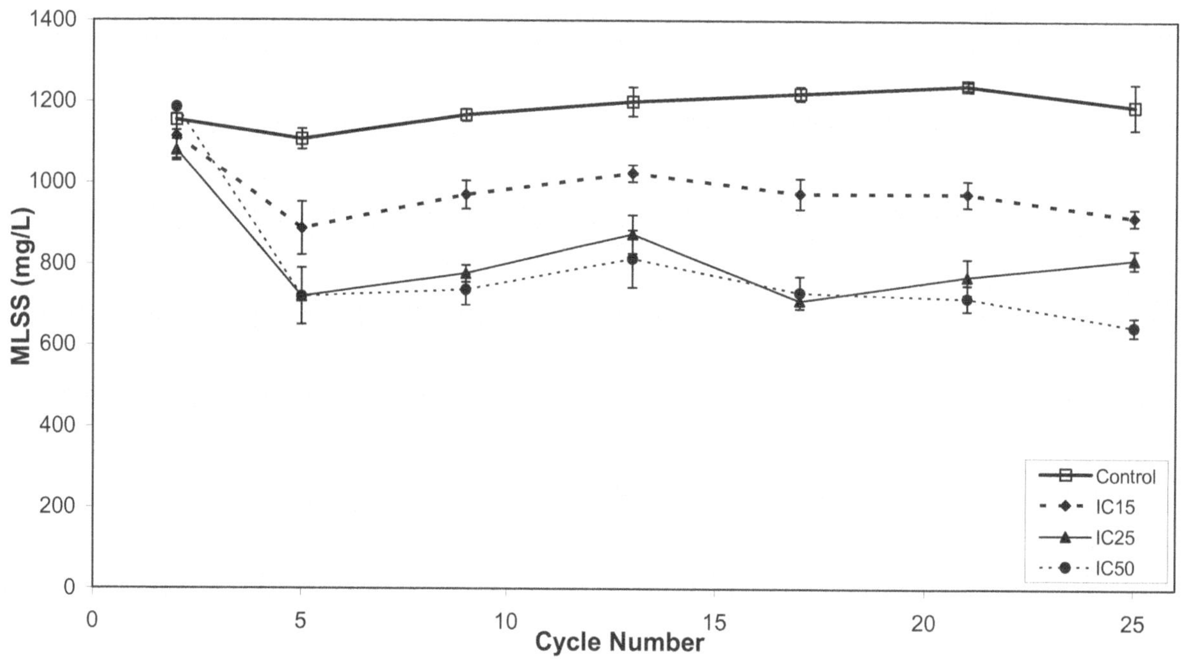

Figure 3-13. MLSS for the 2-Day SRT Reactors Exposed to Cadmium Shock.

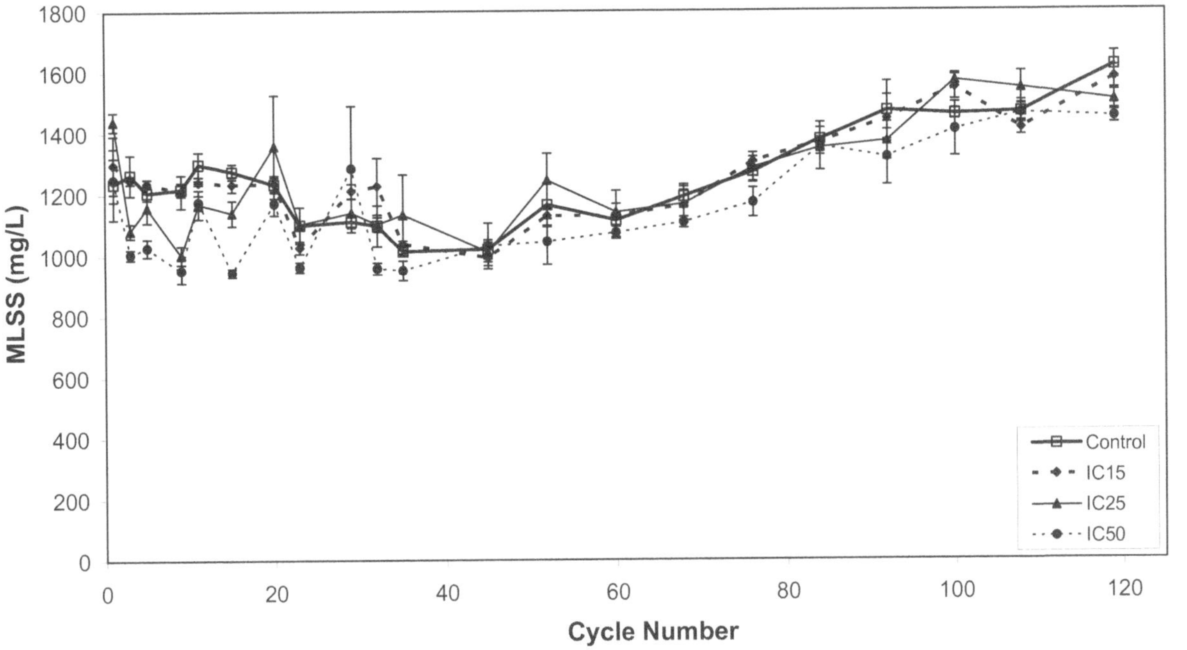

Figure 3-14. MLSS for the 10-Day SRT Reactors Exposed to Cadmium Shock.

3.2.4.3 Specific Oxygen Uptake Rate

SOUR results for the 2- and 10-day SRT reactors show that cadmium severely inhibited respiration, and the degree of inhibition correlated with an increase in cadmium dose (Figures 3-15 and 3-16). For the 2-day SRT reactors, respiration rate recovered to control levels in the IC_{15}-stressed reactor after approximately 4 days (16 cycles) ($\alpha = 0.05$), while SOUR values in the IC_{25} and IC_{50} reactors never recovered to levels measured in the control. Furthermore, the magnitude of the inhibition effect measured after 6 hours (one cycle) of the initial shock was much higher than the expected 15%, 25%, and 50% measured by the batch IC_{XX} assay over 10 minutes. For the reactors, the degree of inhibition was approximately 92%, 96%, and 98% for the IC_{15}-, IC_{25}- and IC_{50}-stressed reactors, respectively. This suggests that either a change occurred in influent or biomass characteristics between the day prior to starting the reactors and the start of the reactors, or that a greater degree of inhibition was incurred due to longer exposure times (6 hours as opposed to 10 minutes).

A similar trend was observed with the 10-day SRT reactors. For this experiment, the SOUR of the stressed reactors recovered to the levels measured in the control reactor by the 17th cycle. This point of recovery corresponded with nitrification recovery in these reactors (Figure 3-8). The magnitude of the inhibition effect measured six hours after the initial shock was much higher than the expected 15%, 25%, and 50% measured by the batch IC_{XX} assay over about 10 minutes. For the reactors, the degree of inhibition was approximately 74%, 84%, and 92% for the IC_{15}-, IC_{25}-, and IC_{50}-stressed reactors, respectively. The SNGR protocol was being resolved during this experiment, so those results are not available for cadmium shock.

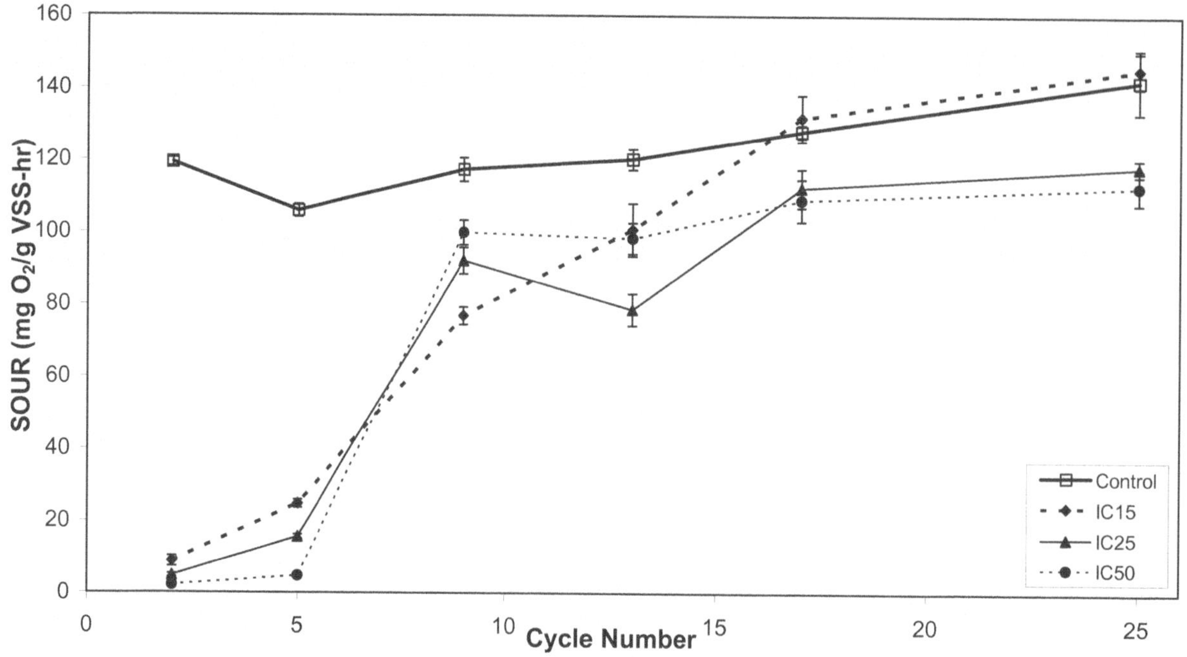

Figure 3-15. SOURs for the 2-Day SRT Reactors Exposed to Cadmium Shock.

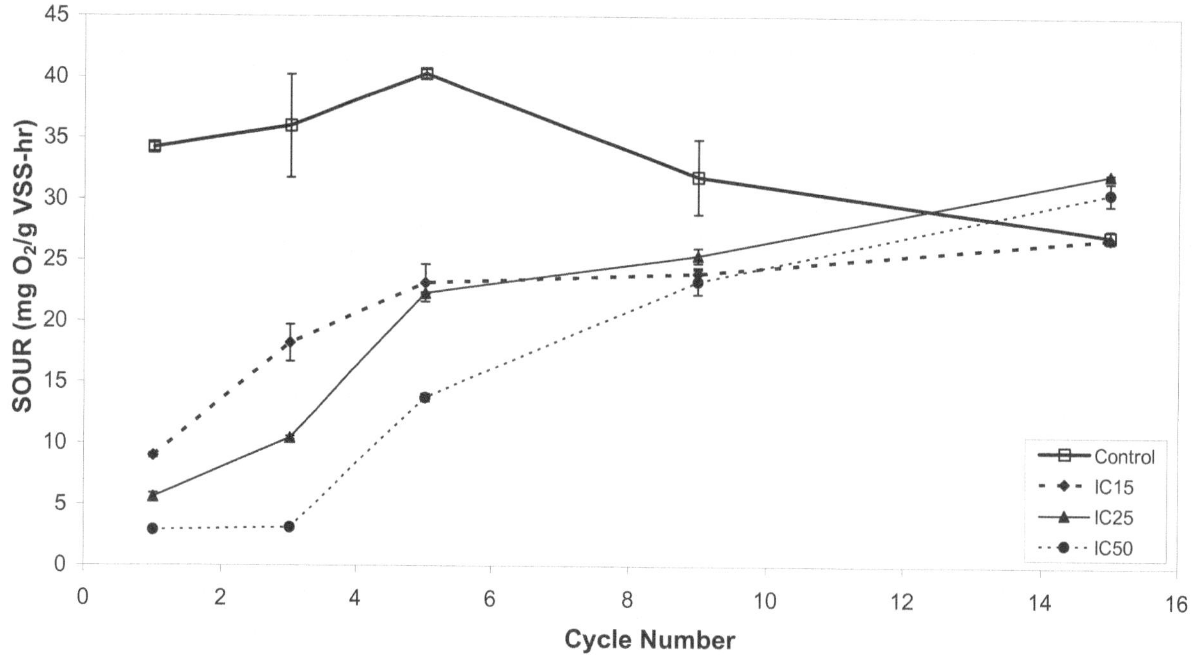

Figure 3-16. SOURs for the 10-Day SRT Reactors Exposed to Cadmium Shock. Data through Cycle 15 Are Shown.

3.3 Discussion

The results from the cadmium source-effect experiment suggest that cadmium is highly toxic to both non-nitrifying, low SRT-activated sludge systems and nitrifying activated sludge systems. Furthermore, cadmium shock affects the treatment process at various levels. The predominant process effects that were observed include deflocculation, respiration inhibition, nitrification inhibition, and COD removal inhibition.

Speciation of cadmium between its soluble and insoluble forms complicates the interpretation of the results presented here. Measurements showed that most of the cadmium existed in an insoluble form and remained in the reactors throughout the experiments. This supports the notion that adsorption to biomass, precipitation, and/or complexation reactions occurred in the system. Cadmium tends to form highly insoluble salts with phosphates (K_{sp} = 2.53 x 10^{-33}), hydroxides (K_{sp} = 7.20 x 10^{-15}), and carbonates (K_{sp} = 1 x 10^{-12}), and forms complexes with these and other anions as well. It is very hard to distinguish between cadmium that precipitated out and adhered to floc materials from cadmium that became entrained into the flocs by sorption, but the alkalinity profiles indicated that otavite formation was probably one of the main processes leading to cadmium immobilization in the 2-day SRT system. Furthermore, the SVI results suggest that cadmium precipitates became integrated into the floc structure for the 2-day SRT biomass. The COD inhibition trends tended to correlate with the presence of measurable soluble cadmium for all reactors. Similarly, nitrification inhibition trends in the 10-day SRT reactors correlated with the presence of soluble cadmium and suggested that the soluble form was the predominant stressor in that system. Finally, nitrification inhibition ceased as soon as the soluble cadmium washed out of the stressed reactors, suggesting that nitrification inhibition in those systems was readily reversible.

Both the 2- and 10-day SRT biomasses experienced severe process disturbances, even at the lowest level of cadmium shock. Therefore, it is apparent that cadmium impacts the performance of both organic matter-degrading heterotrophic bacteria (as detected by deficiencies in sCOD removal, SOUR inhibition in non-nitrifying biomass, and decreased MLSS concentrations) as well as nitrifying bacteria (as directly detected through effluent nitrogen species concentrations). As indicated previously, cadmium complexes phosphate groups. This action may affect metabolic pathways that are regulated through phosphate levels in the cell or that involve the use of inorganic phosphate. For example, ATP production could be impacted, which would decrease the energy content of cells. Heavy metals such as cadmium have also been identified as uncouplers of oxidative phosphorylation (Lewis et al., 1994), which would further decrease intracellular energy levels. Overall, the impact of cadmium on phosphate-regulated pathways of significance in biological treatment systems deserves further study. It is not clear if the mechanism that caused deterioration of COD removal is linked or even the same as the one responsible for deflocculation. Previous work by Bott and Love (2002) suggested that deflocculation occurs as a result of the GGKE mechanism known to be active in Gram-negative heterotrophic bacteria. This mechanism involves an uncoupling of the proton motive force during the K^+ efflux process, but it is not known whether or not the degree of uncoupling explains the loss of COD-removal capacity. Whatever the mechanism, it appears to be reversible.

3.4 Conclusions

- Cadmium primarily associated with the insoluble (mixed liquor) phase during the experiments. Sorption to or integration with biomass was significant (up to 97% of the total cadmium applied to the reactors), resulting in retention of cadmium in the system for long periods of time. Nevertheless, the data suggest that it is the soluble form that was most toxic.

- Significant deflocculation occurred in response to cadmium shock. For both the 2- and 10-day SRT systems, TSS increased significantly after the shock, and the increase in effluent turbidity was proportional to the cadmium concentration applied to the reactors.
- Modest but measurable increases in soluble K^+ were detected in the stressed reactors. This corresponds with previous studies, where deflocculation was linked to the GGKE stress-response mechanism, which is activated when Gram-negative heterotrophic cells are exposed to electrophilic (thiol-reactive) chemicals.
- COD removal efficiency decreased with increasing cadmium concentrations for all the stressed reactors. It is unknown if the mechanism that caused this effect is the same as that responsible for deflocculation, which uncouples oxidative phosphorylation, or if other phenomena (like disruption of catabolic enzymes) are primarily responsible.
- Nitrification in the 10-day SRT reactors was significantly affected by cadmium shock. Ammonia levels increased with increasing cadmium concentrations and a corresponding decrease in nitrate was observed. Effluent pH and alkalinity data were consistent with this. No significant nitrite accumulation was observed, which suggests that cadmium shock targets ammonia oxidizing bacteria. Nitrification inhibition seemed to be readily reversible for the 10-day SRT reactors.
- Cadmium shock did not adversely affect biosolids dewaterability or settleability, as indicated by CST and SVI measurements. In fact, SVIs improved significantly in the 2-day SRT stressed reactors, probably due to an increase in floc density by the precipitated cadmium. In both the 2- and 10-day SRT systems, MLSS concentrations decreased significantly early on in the stressed reactors due to both increased effluent TSS and bacterial growth inhibition.
- Complete recovery was not achieved for the 2-day SRT reactors but was reached in the 10-day stressed bioreactors within 3 SRTs. Recovery in the 10-day SRT systems was linked to washout of soluble cadmium from the reactors. Soluble cadmium in the 2-day SRT reactors was much higher on a mole per cell basis than in the 10-day SRT reactors, and might explain why the 2-day SRT reactors did not recover as promptly.

CHAPTER 4.0

1-Octanol Shock Experiment

4.1 Literature Review

Hydrophobic organic chemicals are used widely in modern industrial practices. They are used as fragrances, food preservatives, and disinfectants, but their most widespread uses are as fuels and industrial solvents. Octanol is a hydrophobic organic chemical that is used as a surfactant, a solvent, and as a tobacco additive in cigarettes. Despite the common use of hydrophobic organic chemicals, few papers were found where the impact of octanol and most hydrophobic organic chemicals on biological wastewater treatment processes was clearly elucidated.

Some industrial organic chemicals are thought to induce toxic effects through narcosis (Bradbury and Lipnick, 1990), a mode of action that slows cellular activity. The structure of 1-octanol gives it both polar properties (due to the alcohol group on the first carbon) and nonpolar properties (due to the long hydrophobic carbon chain). Based on previous work, 1-octanol has been said to function in a manner similar to nonpolar narcotics (Veith et al., 1983; McKim et al., 1987). Based on the chemical properties of nonpolar narcotics, the site of action of these chemicals on organisms is thought to be the cytoplasmic membrane (Sikkema et al., 1995). This is due to the hydrophobic nature of these molecules, which causes the molecules to "seek out" similar molecules. The lipid bilayer that forms the cytoplasmic membrane of all living cells contains hydrophobic centers that are an ideal environment for hydrophobic organic chemicals to locate (Cooper, 2000).

Several effects can result from the insertion of hydrophobic chemicals into the cytoplasmic membrane, one of which is an alteration of the membrane structure. Such an insertion has been shown to cause the membrane to expand, causing deterioration in membrane integrity (Seeman, 1972). Hydrophobic contaminants that migrate into cellular membranes have also been shown to increase the fluidity of the membrane structure. In response to this, some cells are capable of altering the fatty acid content of the cytoplasmic membrane to reduce fluidity and prevent further penetration by hydrophobic contaminants by converting the *cis* fatty acids to the *trans* form (Ingram, 1977; Heipieper et al., 1992).

It has been shown that the membrane thickness, head group hydration, membrane fluidity, and fatty acid composition regulate the activity of both membrane-bound enzymes and cells. Increases in membrane fluidity can result in the passive migration of protons or other ions across the membrane surface, which reduces the membrane potential. Such an action can impact the function of membrane-bound proteins that are controlled by the membrane potential (Sikkema et al., 1995). Membrane expansion may also affect the protein-protein interactions that take place in supramolecular protein complexes, such as those that occur in the electron transport chain. In this case, the protein complexes can become dissociated as hydrophobic molecules accumulate in the cellular membrane, leading to a reduction in adenosine triphosphate (ATP) generation potential (Schneider et al., 1980). There are several cytoplasmic functions, which are critical to successful cellular function that can be disrupted by hydrophobic contaminants. For example, the development of a leaky membrane can cause a cell to lose its ability to regulate internal pH or active transport of chemicals across the cytoplasmic membrane (Poolman et al., 1987). The major result is an overall inhibition of cell growth.

Bearden et al. (1999) showed that exposure to 1-octanol caused a lag in the growth of *Tetrahymena pyriformis*, a freshwater ciliate. After the lag, growth occurred at a rate equal to that of a system of ciliates not exposed to octanol. It was hypothesized that this lag phase was due to the insertion of 1-octanol into the plasma membrane and a loss of unspecified, critical cellular functions that lead to a shutdown in growth. Bearden et al. (1999) conjectured that after the membrane adapted to the presence of the octanol by altering its fluidity, cell function and growth returned to normal.

4.2 Results

Results for the 10-day SRT experiment only show 44 cycles (11 days). Reactor monitoring was ended at cycle 44 for two main reasons: the reactors recovered according to effluent COD and TSS, and a bulking problem occurred at the pilot plant and began to manifest in the source-effect reactors. It was determined that the bulking problem was due to the development of septic conditions in the feed tanks and was possibly exacerbated by low nitrogen concentrations. It is believed that sufficient information was obtained on the effect of octanol prior to cycle 44 to conclude a source-effect relationship from the data obtained.

4.2.1 Inhibitory Concentrations (IC_{XX})

The impact of varying octanol concentrations on the SOUR of the 2- and 10-day SRT biomass was determined using respirometric tests. The IC_{xx} were determined based on a percentage inhibition of the control tests (no contaminant added). Equations describing the IC_{xx} were derived from a best-fit analysis of the data (shown in Appendix C). The results of these calculations are given in Table 4-1, and were the final concentrations added to each of the reactors at the beginning of the first cycle of the source-effect experiment.

When looking only at concentrations, the IC_{XX} concentration values for the 10-day biomass were found to be higher (i.e., less sensitive) than for the 2-day SRT biomass. This trend continued when values were presented on a mass contaminant per mass reactor suspended-solids basis, indicating that the 10-day SRT biomass was less sensitive to octanol shock overall. Although the 10-day SRT biomass was nitrifying when the IC_{XX} experiments were performed, it remained less sensitive than the 2-day SRT biomass, indicating that either octanol did not inhibit nitrifying bacteria or did not inhibit them in the time frame of the assay.

Table 4-1. IC_{15}, IC_{25}, and IC_{50} Values for Octanol in 2- and 10-Day SRT Mixed Liquors.

	2-Day SRT		10-Day SRT	
	mg/L	mg Octanol / g MLVSS	mg/L	mg Octanol / g MLVSS
IC_{15}	40	35	110	124
IC_{25}	76	67	135	157
IC_{50}	167	146	209	252

4.2.2 Influent Characteristics

Average influent characteristics over the course of the octanol source-effect experiments are given in Table 4-2. The influent quality was relatively constant over the time of each experiment with the exception of the suspended solids, which varied significantly as seen by the high standard deviation. Both experiments occurred after influent supplementation was initiated. Accordingly, the influent composition reveals that the wastewater used as the feed for the source-effect experiments with octanol was medium to strong in strength (Metcalf and Eddy, 2003).

Table 4-2. Characteristics of Wastewater Used During Octanol Source-Effect Experiments.

Characteristic	2-Day SRT Biomass Average ± std dev	10-Day SRT Biomass Average ± std dev
Alkalinity (mg/L as $CaCO_3$)	482 ± 83	231 ± 26
Ammonia (mg/L as N)	16.0 ± 3.5	20.0 ± 3.4
COD, soluble (mg/L)	894 ± 25	471 ± 64
COD, total (mg/L)	1043 ± 76	711 ± 118
pH	7.6 ± 0.2	7.2 ± 0.4
TSS (mg/L)	220 ± 95	134 ± 35
VSS (mg/L)	93 ± 36	114 ± 33

4.2.3 Effluent Characteristics

4.2.3.1 Soluble COD

The 2-day SRT effluent soluble COD-removal efficiency was negatively impacted by octanol (Figure 4-1). The first four cycles show that only the IC_{50}-loaded reactor had a significant increase in effluent COD ($\alpha = 0.05$), and after the first few cycles effluent COD levels appeared to be stable and similar for all reactors. The effluent octanol concentrations in all reactors were below detection for all cycles except for the first, when the IC_{50} reactor had an average effluent octanol concentration of 30 mg/L (data not shown). This concentration translates into a theoretical COD of 90 mg/L and accounts for almost all of the additional effluent COD found at the effluent of the first cycle for the IC_{50}-treated reactor. The rapid decrease in octanol from the effluent during this source-effect experiment could be due to a rapid association of the octanol with the hydrophobic regions of the biomass or to degradation.

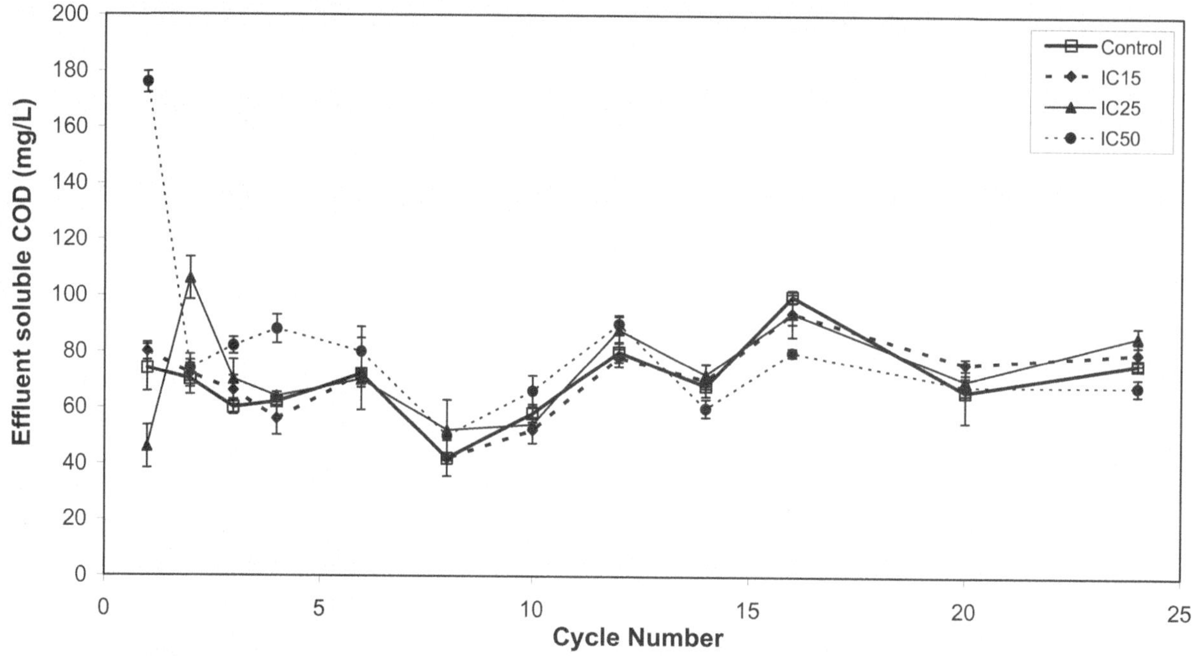

Figure 4-1. Effluent Soluble COD for the 2-Day SRT Reactors Exposed to Octanol Shock.

As was found for the 2-day SRT experiments, the initial cycle for the IC_{50}-shocked 10-day SRT reactor was the only one that showed a significant increase in effluent COD (α=0.05). After the initial cycle, effluent COD levels appeared to be stable and similar for all shocked reactors. The abnormally high effluent COD at cycle 35 for the 10-day IC_{15} reactor was due to a loss of aeration overnight, resulting in incomplete degradation of the waste. The effluent octanol concentration in the 10-day SRT IC_{50} reactor was below detection for all cycles except for the first one, where the average effluent octanol concentration was 12 mg/L. This concentration translates into a theoretical COD of 36 mg/L and accounts for only a fraction of the high residual soluble COD observed in the effluent. As with the 2-day SRT system, the rapid decrease in octanol from the effluent during the 10-day SRT source-effect experiment could be due to a rapid association of the added octanol with the hydrophobic regions of the biomass flocs or to degradation.

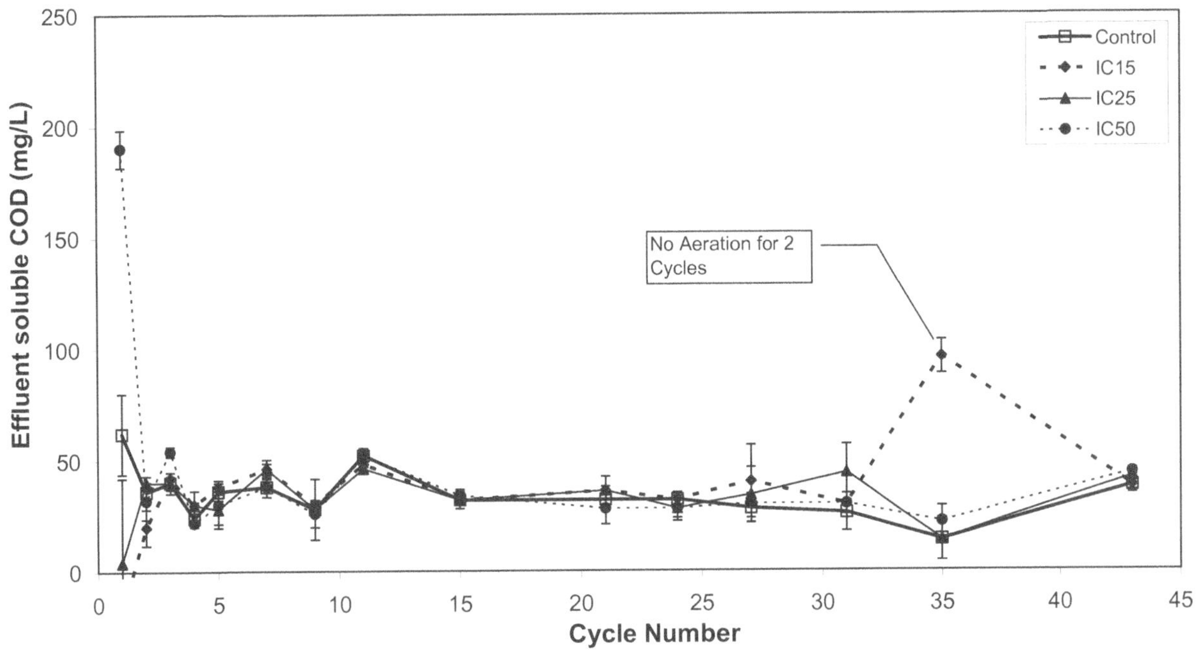

Figure 4-2. Effluent Soluble COD for the 10-Day SRT Reactors Exposed to Octanol Shock.

4.2.3.2 TSS and Potassium Concentrations

Similar performances were seen for the effluent TSS concentrations in both the 2- and 10-day systems, as shown in Figures 4-3 and 4-4. For the 2-day SRT reactors, only the first cycle of the IC_{50} reactor showed a significant increase in suspended solids concentration relative to the other reactors (α=0.05), after which the concentration decreased and was not significantly different from any other reactor. The same phenomenon occurred in the IC_{50} reactor for the 10-day SRT biomass. Furthermore, foaming occurred in both the 2- and 10-day SRT experiments immediately after adding the octanol and persisted for about 2 days (six cycles) in the 2-day SRT system and 3 days (12 cycles) in the 10-day SRT system. The degree of foaming increased with increasing octanol concentration and was more severe in the 10-day system than the 2-day system.

Effluent K^+ concentrations in the 2- and 10-day reactors showed effluent K^+ increases during the first cycle in the IC_{50} reactors (data in Appendix C). Elevated K^+ levels were also seen relative to the control during the first cycle for the IC_{15} and IC_{25} reactors in the 10-day SRT system (α = 0.05). After the second cycle, sustained, elevated K^+ concentrations were not seen in any of the reactors relative to the control. Given that octanol is non-electrophilic, it was not expected to elicit a GGKE potassium efflux response. It is conceivable that octanol's hydrophobic nature caused elevated potassium levels if it inserted into cell membranes and caused them to become leaky and release potassium into the surrounding environment. However, leaky membranes can also release divalent cations in the cytoplasm, thereby reducing the impact of the presence of the monovalent cation by keeping the monovalent to divalent cation concentration in the bulk liquid lower. However, the divalent cation content of bacterial cytoplasms is much lower than that of K^+, so a leak could result in an unfavorable condition for floc structure. The effluent TSS results show an increase in effluent TSS for the IC_{50} reactors, where significant increases in bulk liquid K^+ concentrations occurred. This is similar to what was observed with electrophilic compounds that selectively cause monovalent K^+ efflux and increased M:D cation ratios that correlate with deflocculation events (Bott and Love, 2004). In conclusion, any changes in effluent K^+ are assumed to be due to leaking membranes caused by octanol, and not the GGKE mechanism.

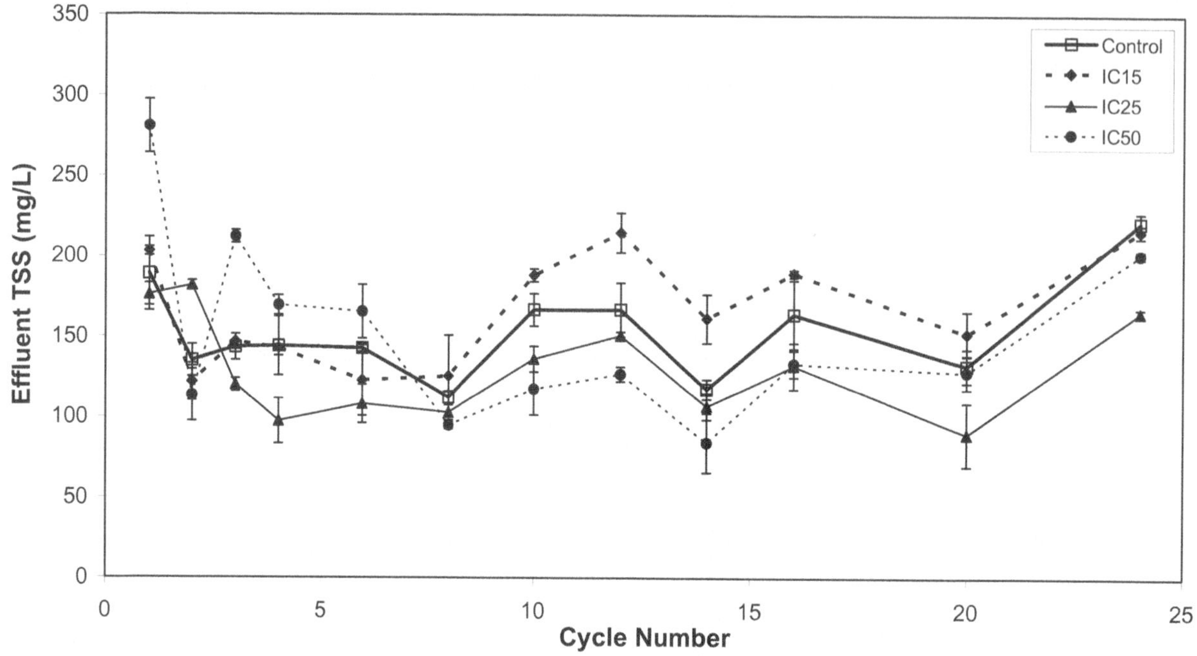

Figure 4-3. Effluent TSS for the 2-Day SRT Reactors Exposed to Octanol Shock.

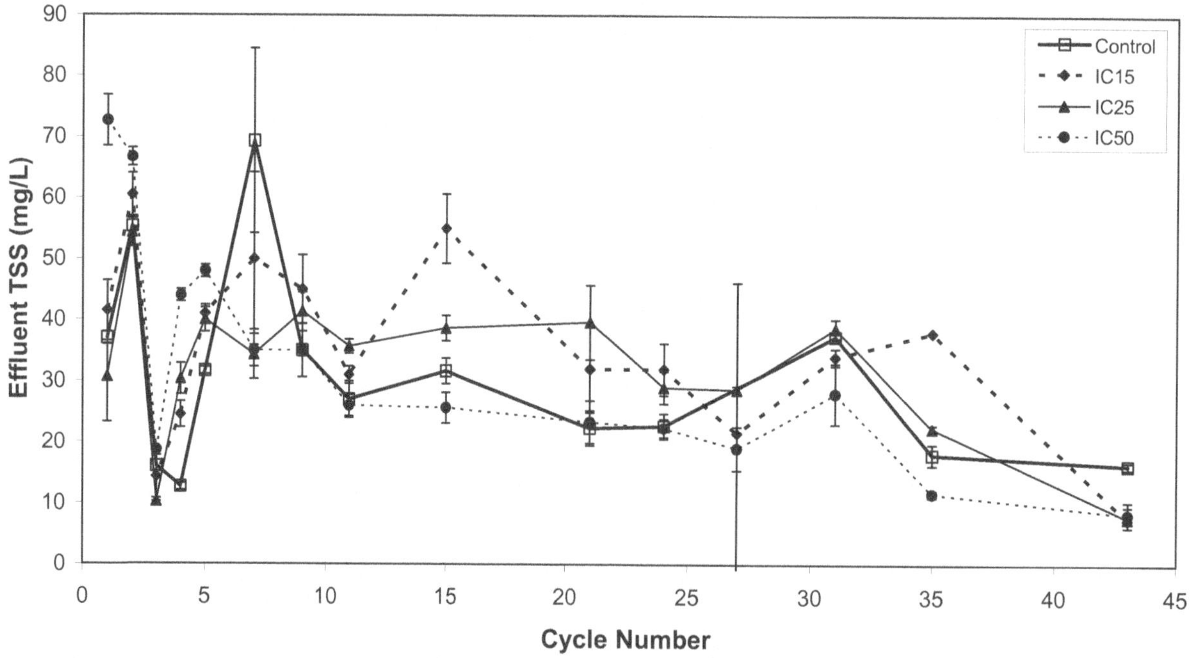

Figure 4-4. Effluent TSS for the 10-Day SRT Reactors Exposed to Octanol Shock.

4.2.3.3 Nitrification Inhibition

Despite complications imposed on data interpretation due to near-limiting nitrogen conditions, the results suggest that octanol had modest to significant impacts on nitrification. Figures 4-5 and 4-6 show effluent ammonia-N and nitrate-N concentrations for the 10-day SRT reactors. The effluent ammonia trends suggest that nitrification was adversely affected in the IC_{50}-treated reactor only during the first cycle, but not thereafter. However, the nitrate data (Figure 4-6) do not correlate entirely with this observation because there was a sustained difference in the nitrate concentrations observed in the control and IC_{50}-treated reactors for the first 25 cycles. In addition, the amount of nitrate generated decreased with increasing octanol dose. SOUR measurements in the mixed liquor (discussed in Section 4.2.4.3) support the notion that growth of the IC_{50} biomass was inhibited during the first cycle, and high ammonia may reflect a reduced demand for ammonia as a nitrogen source during the first cycle. Furthermore, SNGR data (Section 4.2.4.3) support the conclusion that nitrification was inhibited by octanol, which is discussed further below. The cause for the high effluent ammonia concentration observed for the IC_{15} reactor at cycle 35 was a failure in the aeration system for that reactor in the previous two cycles, which inhibited nitrification due to lack of oxygen. Figure 4-6 shows that nitrate concentrations increased with time in all reactors, and that the amount of nitrification occurring correlated with the amount of octanol shock applied to the reactors. Finally, no nitrite was observed in the 10-day SRT effluent (data not shown). These results suggest that octanol inhibited nitrification in the 10-day SRT reactors.

The 2-day SRT pilot plant was operated to prevent nitrification and also provided just enough ammonia to support heterotrophic growth. Effluent ammonia varied in a manner that correlated with variations in MLSS concentration. The data (see Appendix C) show that the 2-day SRT reactors remained devoid of substantial nitrification, as effluent nitrate never exceeded 1 mg/L as N, and nitrite was never detected in the effluent.

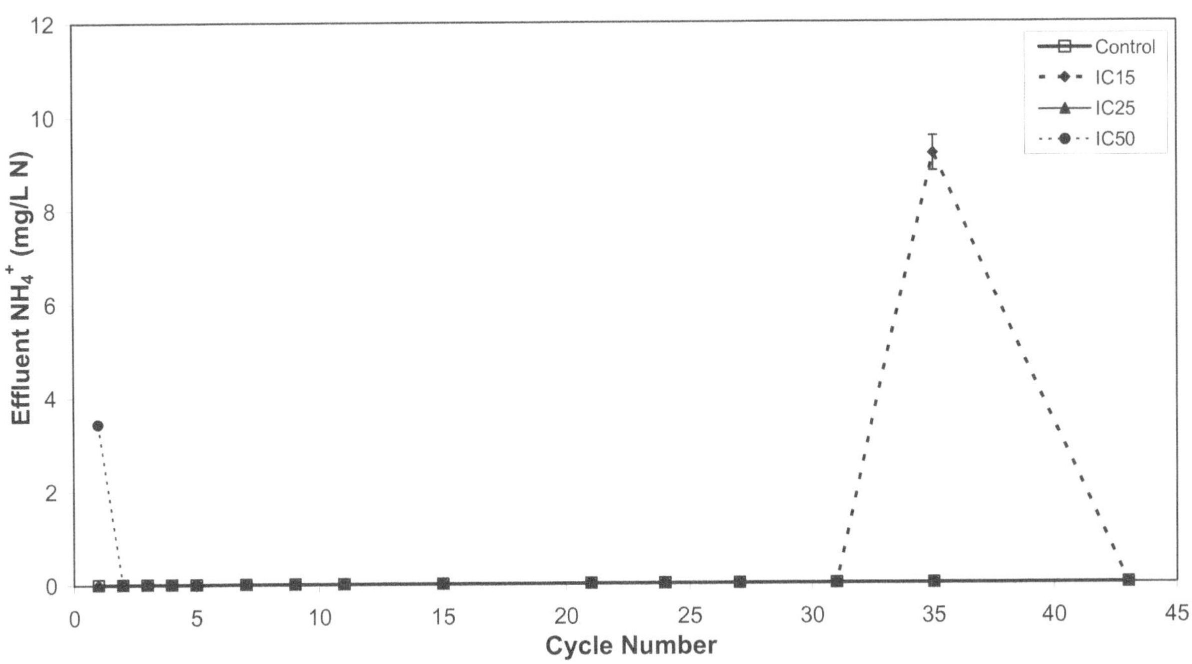

Figure 4-5. Effluent Ammonia-Nitrogen for the 10-Day SRT Reactors Exposed to Octanol Shock.

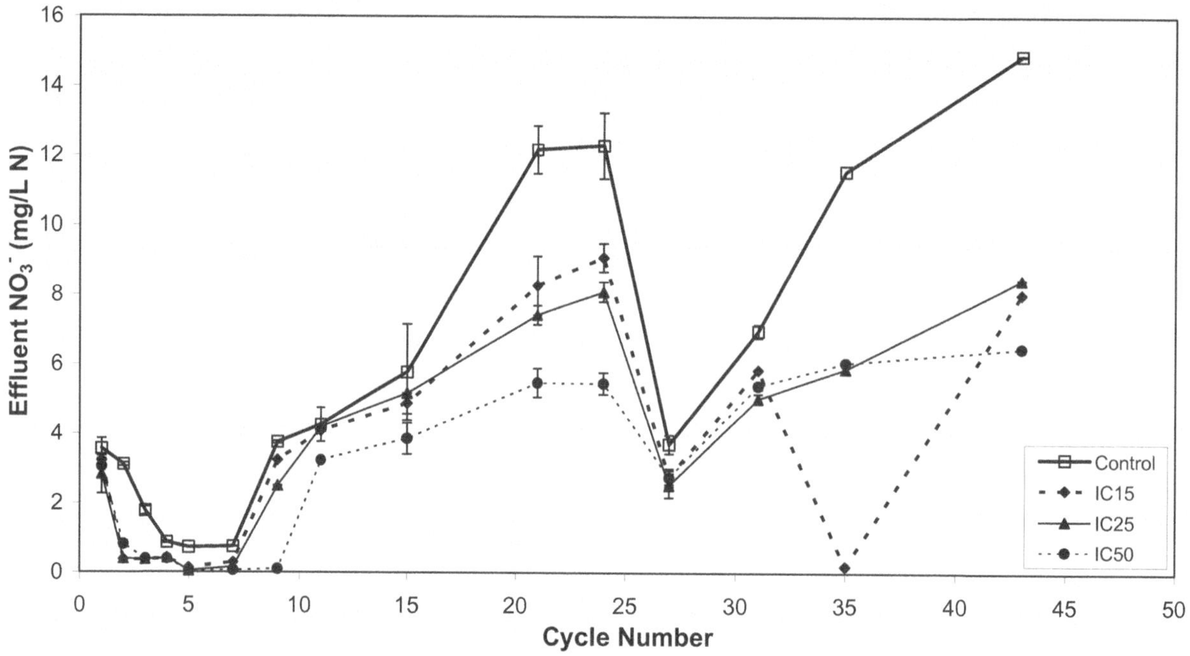

Figure 4-6. Effluent Nitrate-Nitrogen for the 10-Day SRT Reactors Exposed to Octanol Shock.

Alkalinity was monitored because it is consumed during ammonia oxidation and should provide an independent measure that correlates specifically with the extent of ammonia oxidation. In both the 2- and 10-day SRT systems, differences in alkalinity between reactors were small (see Appendix C). For the 10-day SRT system, the control reactor had slightly less alkalinity than the experimental reactors through cycle 44. This was as expected because the 10-day SRT control reactor should have continued to nitrify without inhibition while nitrification inhibition occurred in the octanol-treated systems. Alkalinity concentrations in the 2-day SRT reactors remained constant in all reactors for the duration of the experiment, which is consistent with the observation that the 2-day SRT biomass was not nitrifying with or without octanol. Effluent pH measurements for both the 2- and 10-day SRT biomass were consistent with these nitrification trends.

4.2.4 Mixed Liquor Characteristics

4.2.4.1 Settling and Dewatering Characteristics

Mixed liquor CST measurements were taken for both the 2- and 10-day SRT systems (data in Appendix C). CST measurements for the 2-day SRT biomass were typically over 40 seconds. Some variations occurred between the CST of octanol-shocked biomass and control biomass; however, these variations did not correlate with octanol concentration. The CST measurements for the 10-day SRT reactors did not deviate during the course of the experiments and averaged 14.6 seconds, which is low and indicates that the biosolids were easy to dewater. Collectively, these results suggest that octanol did not adversely impact the dewatering features of the biomass.

The SVI data for the 2- and 10-day SRT reactors showed no negative impact on biomass settling properties due to octanol shock (data in Appendix C). The SVI measurements for the 10-day SRT reactors remained in the good range, below 150 mL/g for the first 20 cycles. Similarly the 2-day SRT biomass SVI values were below 115 mL/g for all cycles. Settling deteriorated in all reactors (including the control) after cycle 20 in the 10-day SRT biomass. Microscopic evaluations suggested that the upward trend in SVI values observed later in the 10-day SRT biomass was due to filamentous bulking that also occurred in the pilot plant systems. This bulking event was determined to be caused by septic conditions and possibly nutrient limitations in the wastewater.

4.2.4.2 Mixed Liquor Solids Trends

MLSS and MLVSS indicate system stability. Throughout the 10-day SRT experiment, the mixed liquor concentrations in all reactors remained consistent over time, and there were no trends indicating a significant deterioration in biomass growth in the octanol-stressed reactors relative to the control. For the 2-day SRT system, the MLSS and MLVSS concentrations in all reactors (including the control) decreased consistently over time; this reduction appears to have been due to a reduction in suspended solids in the influent fed to the laboratory scale reactors relative to what was being fed to the pilot plant. A summary of mixed liquor concentrations is given in Table 4-3, and concentration trends with time are given in Appendix C.

Table 4-3. Mixed Liquor Concentrations in 2- and 10-Day SRT Reactors Exposed to Octanol Shock.

	MLSS (mg/L)	MLVSS (mg/L)	MLVSS:MLSS Ratio
2-Day SRT reactors			
Control	1107 + 213	827 + 136	.75 ± .057
IC_{15}	1127 + 182	837 + 124	.75 ± .030
IC_{25}	1000 + 165	753 + 131	.75 ± .031
IC_{50}	1076 + 130	796 + 90	.74 ± .035
10-Day SRT reactors			
Control	1506 + 165	1344 + 153	.89 ± .021
IC_{15}	1533 + 144	1366 + 127	.89 ± .022
IC_{25}	1537 + 124	1363 + 109	.89 ± .025
IC_{50}	1559 + 138	1389 + 116	.89 ± .021

4.2.4.3 Kinetic Indicators of Inhibition

For the 2-day SRT experiment, SOURs were not significantly lower than the control reactor ($\alpha = 0.05$), but the IC_{15} and IC_{25} reactors were significantly higher during the first cycle (Figure 4-7). For the IC_{50} reactor, the SOUR inhibition relative to the control during the first cycle was only 13% lower, whereas the SOURs for the IC_{15}– and IC_{25}–exposed reactors were 132% and 135% of the control, respectively. Therefore, octanol did not inhibit cellular respiration in biomasses that had 5 hours to adapt to the hydrophobic chemical. Accordingly, lower octanol concentrations that were inhibitory in a short term IC_{XX} test seemed to enhance SOURs in biomass after longer exposure times.

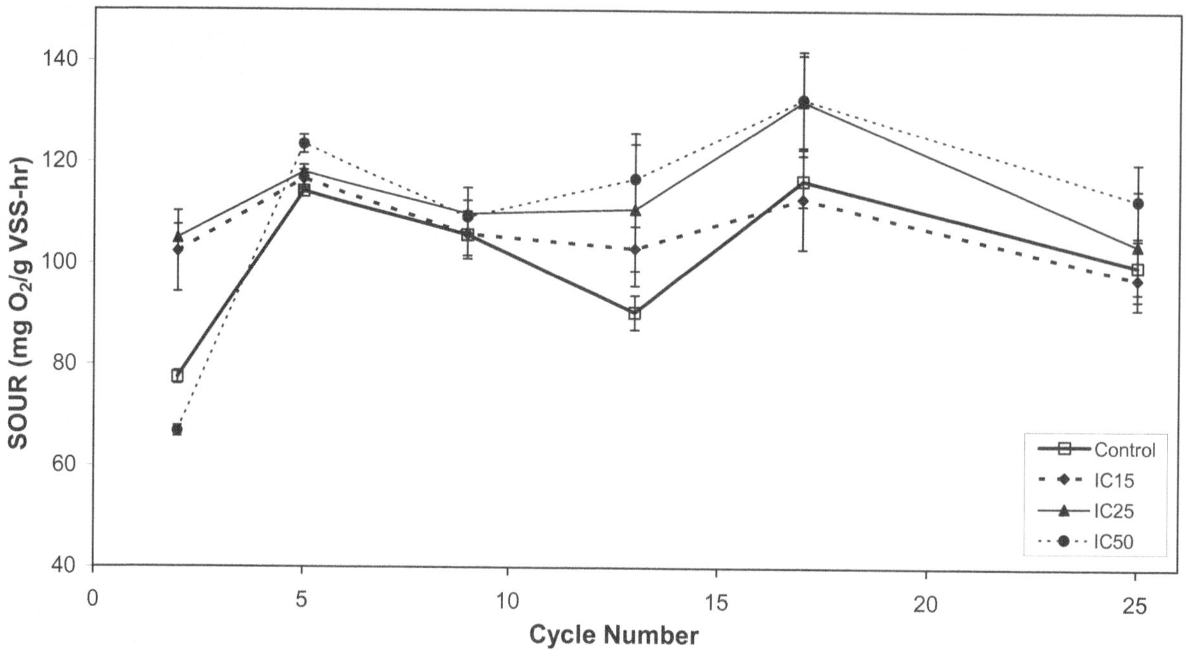

Figure 4-7. SOURs for the 2-Day SRT Reactors Exposed to Octanol Shock.

SOUR results for the first cycle of the 10-day SRT reactors show that octanol inhibited respiration, and the degree of inhibition correlated with an increase in contaminant dose (Figure 4-8). By the second day (cycle eight), however, the SOUR of the stressed reactors was not significantly different from the SOUR of the control reactor ($\alpha = 0.05$). The magnitude of the inhibition effect measured 6 hours after the initial shock was lower than the expected 15%, 25%, and 50% measured by the batch IC_{XX} assay, which occurred over about 10 minutes. Instead, the degree of inhibition was approximately 16%, 21%, and 27% for the IC_{15}-, IC_{25}-, and IC_{50}-stressed reactors, respectively. This suggests that a lesser degree of inhibition occurred due to longer exposure times (6 hours as opposed to 10 minutes), possibly as a result of adaptation of the biomass to the octanol or to degradation of the octanol.

The results from the 2- and 10-day SRT systems are believed to be symptomatic of the degree of nitrification. The relatively higher degree of inhibition observed at the end of the first cycle in the 10-day SRT system could be due to nitrification inhibition. The 2-day SRT system was non-nitrifying and, therefore, any inhibition detected by SOUR would be due to organic carbon-utilizing heterotrophs. The absence of detectable inhibition in the 2-day SRT biomass relative to the control suggests that octanol did not negatively impact heterotrophic mixed liquor populations.

The SNGR test was performed for the nitrifying 10-day SRT system, and decreased in value as octanol concentration increased (Figure 4-9). SNGR for the IC_{15} and IC_{25} reactors were similar throughout, but the IC_{50} response was generally slower than the other experimental reactors and the control. A constant SNGR was not maintained in the control reactor, but it did remain consistently higher than the experimental reactors. Additionally, the degree of inhibition in the stressed reactors appeared to decrease with time. This recovery correlates with the effluent nitrate data, which shows that effluent nitrate concentrations increased with time. These results suggest that octanol inhibits nitrification, and the degree of inhibition tends to increase as the concentration of octanol increases.

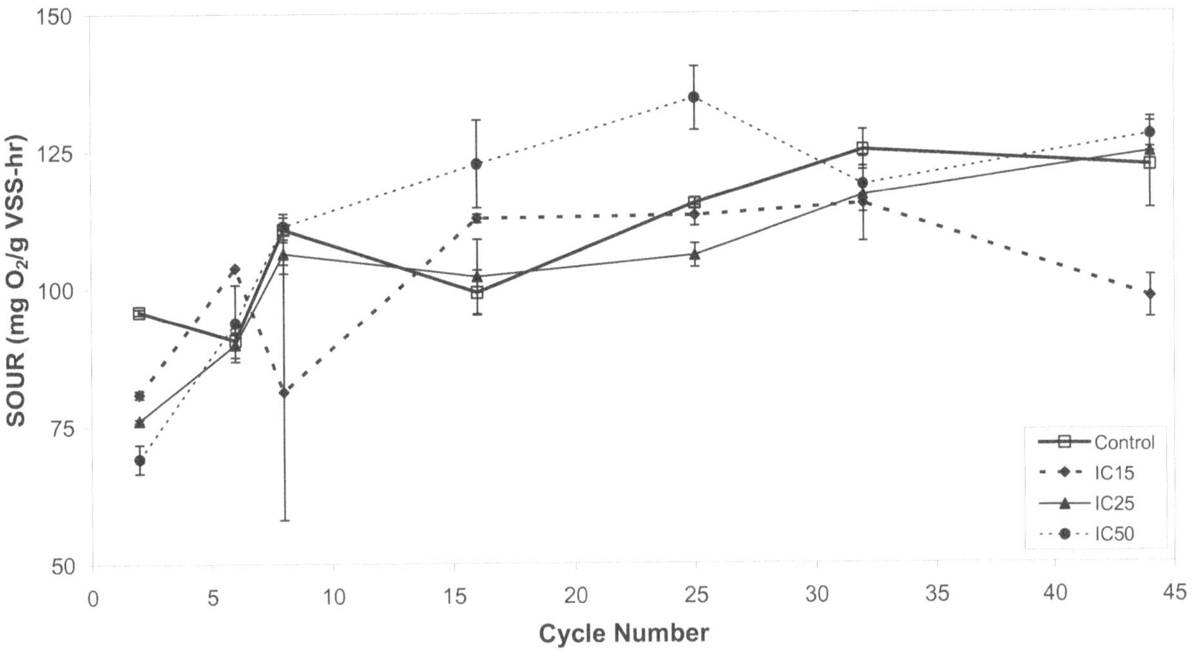

Figure 4-8. SOURs for the 10-Day SRT Reactors Exposed to Octanol Shock.

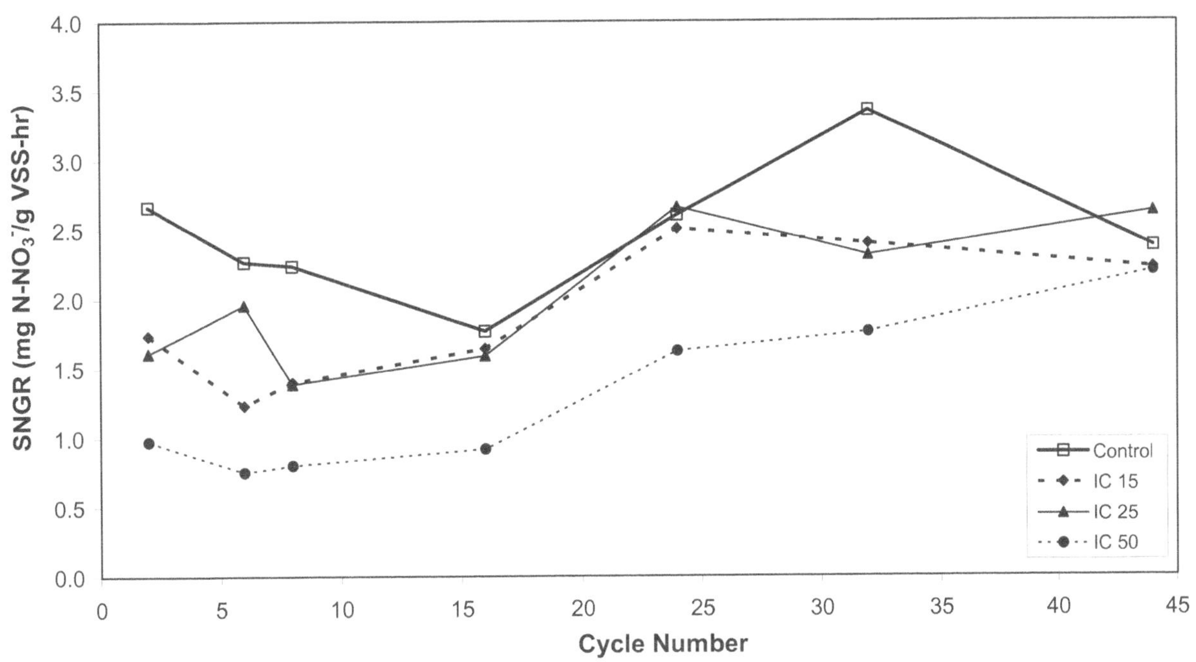

Figure 4-9. SNGRs for the 10-Day SRT Reactors Exposed to Octanol Shock.

4.3 Discussion

Various process effects were observed due to octanol shock and include: deflocculation and loss of solids from the system; a loss of COD-removal capacity; and a loss of nitrification. Octanol represents a model hydrophobic chemical and the results suggest that these chemicals cause a broad range of previously undescribed process effects for activated-sludge populations.

Shock (sublethal) inputs of octanol only caused minor deflocculation to occur, and only at the highest concentration tested. This event was short lived (less than one day) for both the 10-day and 2-day SRT systems. In addition to deflocculation, foaming occurred for about 12 cycles (three days for 2- and 10-day SRT systems) and increased in extent with increasing concentrations of octanol. Octanol is non-electrophilic, and effluent K^+ data suggest that the GGKE mechanism (Bott and Love, 2002) was not activated during this experiment. The slight increase in effluent K^+ noted for the 2-day SRT IC_{50} reactor was likely a result of the influence of octanol on the integrity of the membrane, which may have become leaky. It is thought that hydrophobic, lipophilic chemicals like octanol integrate into the cell membranes of bacteria, causing an increase in membrane fluidity. This increase in membrane fluidity can result in the passive migration of protons or other ions, like K^+, across the membrane surface (Sikkema et al., 1995).

Multiple causal mechanisms may produce a similar process effect. Octanol can function as a surfactant, and both the deflocculation and foaming could occur as a result of these properties. Octanol may disrupt cell-to-cell associations in the bacterial flocs and cause them to break apart in solution, and may increase the surface tension of the water, enabling stable foams to form. These observations are common actions for surfactants, which act as wetting agents and allow for molecules to become more soluble. The deflocculation event was probably short-lived because octanol did not stay in solution very long, as documented by direct octanol analysis. As the biomass grew, it is probable that cell-to-cell interactions were re-established in the absence of octanol, and settleable flocs reformed. Because the deflocculation event was minor, no significant decrease in MLSS was detected in any of the stressed systems. The deterioration in SVI, an indicator of settleability, in the 10-day SRT system was attributed to the development of a filamentous bulking condition in all reactors due to problematic conditions associated with the sewage supplement, and not to the presence of octanol.

The SOUR data show different trends than were expected from the short-term IC_{XX} assays. For the 10-day SRT reactors, only the first cycle showed an inhibition that followed the order of octanol dosage. For the first cycle, the degree of inhibition did not match what was expected based on the initial short-term exposure inhibitory concentration (IC_{XX}) experiments, which were conducted to determine doses for the source-effect experiments. After the first cycle in the 10-day SRT systems, the differences in SOUR between the control and experimental reactors followed no predictable pattern and seemed random. The 2-day SRT reactors did not show significant inhibition as detected by SOUR after 5 hours exposure. These reduced levels of respiratory inhibition after 5-hour exposure in both the 2- and 10-day SRT reactors could be due to adaptation of the biomass to the chemical shock, one that would not be seen over the course of a 15-minute SOUR test.

The inhibition of respiratory activity due to the addition of octanol was expected, as previous work has shown that hydrophobic organic chemicals will inhibit cellular growth and respiration (Sikkema et al., 1995; Bearden et al., 1999). As noted previously, this may be due to integration of the octanol into the cytoplasmic membrane, which makes the membrane more fluid. Enhanced membrane fluidity can cause ions to leak out of the cytoplasm, which reduces the electrochemical potential across the membrane.

A significant loss of COD-removal efficiency was noted during the first cycle of this experiment in the IC_{50} reactors for both the 10- and 2-day SRT systems; however, most of the increase observed in the 2-day SRT effluent could be attributed to residual octanol whereas only 24% of the increase in the 10-day SRT effluent could be attributed to octanol. The SOUR data obtained from these IC_{50} reactors corresponded in that only the first cycle showed any noticeable difference in the SOUR, probably due to nitrification inhibition. The decreased extent of COD removal in the 10-day SRT IC_{50} reactor is consistent with this conclusion. In the 2-day SRT system, the IC_{50}-inhibited biomass saw little impact on SOUR and most of the increase in effluent COD was accounted for as residual octanol. Therefore, little heterotrophic inhibition due to octanol occurred in the shorter SRT biomass. This difference suggests that the sludge age of biomass can be an important factor in defining whether a mixed liquor will be susceptible to inhibition by hydrophobic contaminants. Sludge age defines biomass physiology and whether or not nitrifiers will be present.

Nitrification was clearly impacted by the presence of octanol, as determined through evaluation of effluent nitrogen species and SNGR results from the 10-day SRT system. Furthermore, the effluent data indicate that the ammonia oxidizing bacteria were inhibited primarily, with little to no inhibition of the NOB. When one considers the mechanism of nitrification, some hypotheses for why octanol would inhibit its performance can be developed. To review, nitrification involves the conversion of ammonia to nitrate, typically by AOB and NOB autotrophic bacteria. The reduced nitrogen species serve as the energy source for nitrifying bacteria. AOB convert ammonia to nitrite through several steps. First, ammonia is oxidized to hydroxylamine in the cytoplasm by an inner membrane-bound enzyme called ammonia monooxygenase (AMO). Hydroxylamine is then converted to nitrite by hydroxylamine oxidoreductase (HAO), which is located in the periplasm of the cells (the region between the inner and outer cell membranes in Gram-negative bacteria). These reactions are coupled to several other enzymes, one of which is cytochrome aa_3 oxidase. This cytochrome acts as a proton pump to maintain a proton gradient, which is critical to the ATP-generating (oxidative phosphorylation) process. The nitrite is then used by NOB and is converted to nitrate. This occurs via an inner membrane-bound protein called nitrite oxidase. Again, this reaction is coupled to cytochrome aa_3 oxidase, which functions the same as in ammonia-oxidizers (White, 2000).

4.4 Conclusions

- Minor deflocculation of the mixed liquor occurred due to the addition of octanol, with only the highest inhibitory concentration having an impact. Recovery from this upset event was quick in the reactor systems with both short and long SRTs. The surfactant properties of octanol are a probable causal mechanism for floc disintegration.
- Octanol appears to primarily inhibit nitrifying bacteria over a range of concentrations and, secondarily, heterotrophic microorganisms only at the highest concentrations applied. This action is believed to be due to consequences associated with the insertion of octanol into the cytoplasmic membrane, which disrupts both cytoplasmic functions and the electrochemical gradient associated with energy conservation in cells.

- Octanol appears to have no effect on the dewaterability of mixed liquor, as CST values were not significantly different between chemically-stressed biomass and control biomass.
- Octanol appears to have no effect on the settleability of mixed liquor, as SVI values, when not influenced by filamentous bacteria that proliferated due to other external issues, remained in the excellent range throughout the experiments.

CHAPTER 5.0

AMMONIA SHOCK EXPERIMENTS

5.1 Literature Review

Ammonia is commonly used in a range of industrial processes. Industries that handle protein-rich materials (slaughterhouses, canned food industries) or other nitrogen compounds (e.g., urea in glue production, nitriles and amines in nylon production), tend to generate effluents with high ammonia levels (Campos et al., 1999). The steel industry also produces wastewaters with high ammonia concentrations, particularly from coke ovens, which may contain up to 1,000 mg/L ammonia (Neufeld et al., 1980). Given that free ammonia is extremely toxic for most aquatic ecosystems, ammonia removal through nitrification has become common practice in most wastewater treatment plants. However, ammonia shocks due to industrial discharges or from internal recycle streams originating from biosolids handling facilities may cause problems on the downstream biological treatment. This experiment investigated the effects of ammonia shocks on the activated sludge process and nitrification in particular.

The known mechanisms through which ammonia shocks can cause an impact on nitrification are free ammonia and free nitrous acid inhibition. Neufeld et al., (1980) determined that free ammonia (NH_3 (aq)) inhibits AOBs according to substrate inhibition models. Anthonisen et al., (1976) determined that free ammonia inhibited AOBs at concentrations as low as 10 mg/L N and inhibited NOBs at concentrations as low as 0.1 mg/L N. The relationship between free ammonia and pH is:

$$[NH_3\,(aq)] = \frac{K_a \times [NH_4^+\,(aq)]}{10^{-pH}} \qquad \text{[equation 5-1]}$$

There are limits to the amount of ammonia that can be loaded to a given biological system, and that amount is a function of other constituents in the wastewater. In a study concerning inhibition of activated sludge by high ammonia levels in leachate, Li and Zhao (1999) observed that COD removal from synthetic wastewater (COD of 1,000 mg/L) declined to a maximum of approximately 15% and the dehydrogenase activity of the sludge decreased about 60%, when the ammonia concentrations in the influent increased from 50 to 800 mg/L as N. They suggested that ammonia concentrations in the influent should be lower than 100 mg/L as N in order to avoid significant inhibition of microbial activity. Campos et al., (1999) varied the ammonia loading rate being fed to an activated sludge system by changing the HRT. The system was also fed bicarbonate as a carbon source, which maintained a stable alkalinity. They concluded that the process was extremely efficient and responded well and rapidly to sudden changes in ammonia loading rate. Nitrite accumulation was temporary and rapidly neutralized. Finally, Gilmore et al., (1999) showed that COD loading influenced the location (and, therefore, the degree) of nitrification in an attached growth pilot-scale biological aerated filter system. High COD loads pushed nitrification further along in the filter column where COD concentrations were low. They successfully correlated nitrification activity via nitrate and nitrite concentrations with the activity of AOBs using small subunit ribosomal RNA-targeted DNA probes.

The ammonium ion has been implicated as a possible factor causing activated-sludge settling properties to deteriorate (Novak, 2001). Novak has proposed that loss of nitrification results in the retention of ammonium ion, which increases the M:D in flocs. Previous work has shown that high M:Ds correspond with poor settling characteristics (Higgins and Novak, 1997a). Consequently, Novak suggested that this may be an additional factor to explain why non-nitrifying, low-SRT biomasses settle poorly. Bott (2001) showed that increases in ammonium ion concentration due to nitrification inhibition by electrophilic chemicals were delayed several hours relative to deflocculation and potassium efflux and did not explain deflocculation in chemically inhibited systems.

5.2 Results

5.2.1 Inhibitory Concentrations

Ammonia shock was imposed by elevating the ammonia-N concentration 3X, 9X, or 18X above the average influent concentration (15 mg/L), or 2X, 5X, or 10X above the average influent concentration (39 mg/L) in the 2-day and 10-day SRT shocked reactors, respectively. Ammonium bicarbonate was used to counteract pH effects that could otherwise impose inhibition due to intense nitrification and, therefore, for ammonia/ammonium shock effects to be studied without interferences by dramatic pH swings. Table 5-1 shows a comparison between the target values of the ammonia shocks and the measured ammonia concentration immediately following the shock for the 2-day SRT reactors. Based on the pH of the reactors when they were fed (around 8.3), the highest free ammonia concentration that would have existed in solution was 7.6 mg/L as N in the 18X reactor. A similar analysis was performed from the target values of the ammonia shock concentrations and the measured ammonia concentration immediately following the shock for the 10-day SRT reactors in Table 5-2. For the 10-day SRT systems, Table 5-2 shows that the difference between theoretical and measured ammonia levels in solution increased with increasing concentration, suggesting that part of the total ammonia that was added was rapidly lost in the high concentration reactors. Based on the pH of the reactors when they were fed (around 8.1), the highest free ammonia concentration that would have existed in solution was 2.6 mg/L as N in the 10X reactor. The biomass was aerated with diffused air, and the potential for rapid stripping of ammonia from the bioreactor was great. As free ammonia is stripped, more is formed as the reaction tends toward equilibrium. Consequently, it is assumed that a large fraction of the ammonia that was lost in the first few minutes in the 10X and 5X ammonia-shocked 10-day SRT reactors was due to stripping of free ammonia. This assumption was supported by a noticeable ammonia odor in the laboratory at the beginning of the experiment.

Table 5-1. Theoretical and Measured Concentrations of Total Ammonia in the 2-Day SRT Bioreactors Near Time Zero.

Reactor	Approximate $[NH_4^+]$ + $[NH_3]$ in Influent (mg/L N)[a]	Approximate Initial $[NH_4^+]$ + $[NH_3]$ in Mixed Liquor (mg/L N)[b]	Measured Initial $[NH_4^+]$ in Mixed Liquor (mg/L N)[c]
Control	10[a]	2.6	7.0 (2.6)
3X	40	10	20 (15)
9X	130	33	42 (37)
18X	280	70	98 (94)

[a] Total (ammonia + ammonium) in the influent for the control reactor was measured by ion chromatography. Concentrations for the shock conditions were calculated by adding known amount of supplement to 10.4 mg/L.

[b] The wastewater was diluted 1:4 at the beginning of each cycle.

[c] The values in parentheses correspond to the added ammonia shock concentrations and were obtained by the measured concentrations in each reactor immediately after the shock minus the estimated residual ammonia levels in the mixed liquor prior to shock. A residual ammonia concentration existed in the mixed liquor because the biomass was not nitrifying; this concentration was estimated from the control reactor ammonia values (measured ammonia in control reactor at time zero (7.0) - theoretical ammonia in the control reactor due only to influent addition at time zero (2.6) = 4.4 mg/L initial residual ammonia in the mixed liquor).

Table 5-2. Theoretical and Measured Concentrations of Total Ammonia in the 10-Day SRT Bioreactors Near Time Zero.

Reactor	Approximate $[NH_4^+]$ + $[NH_3]$ in Influent (mg/L N)[a]	Approximate Initial $[NH_4^+]$ + $[NH_3]$ in Mixed Liquor (mg/L N)[b]	Measured Initial $[NH_4^+]$ in mixed liquor (mg/L N)
Control	31[a]	7.8	8.6
2X	71	18	20
5X	191	48	33
10X	391	98	56

[a] Total (ammonia + ammonium) in the influent for the control reactor was measured by ion chromatography. Concentrations for the shock conditions were calculated by adding known amount of supplement to 31 mg/L.

[b] The wastewater was diluted 1:4 at the beginning of each cycle.

5.2.2 Influent Characteristics

The influent characteristics are summarized in Table 5-3. The influent composition reveals that the raw wastewater used as the feed for the source-effect experiments with ammonia was medium strength (Metcalf and Eddy, 2003).

Table 5-3. Characteristics of Wastewater Used During Ammonium Bicarbonate Source-Effect Experiments.

Characteristic	2-Day SRT Average ± std dev	10-Day SRT Average ± std dev
Alkalinity (mg/L as $CaCO_3$)	531 ± 53	258 ± 45
Total Ammonia (mg/L as N)	15.4 ± 4.0	27.3 ± 5.9
COD, soluble (mg/L)	773 ± 75	472 ± 65
COD, total (mg/L)	879 ± 110	676 ± 104
pH	7.5 ± 0.2	7.2 ± 0.4
TSS (mg/L)	141 ± 80	87 ± 29
VSS (mg/L)	52 ± 27	78 ± 26

5.2.3 Effluent Characteristics

5.2.3.1 Soluble COD

Results for effluent soluble COD concentrations show that concentrations did not change significantly between ammonia-stressed reactors relative to the controls (Appendix D). The only exception was the first cycle, where a moderately but statistically higher effluent COD was observed in the 9X and 18X 2-day SRT reactors compared to the control ($\alpha = 0.05$). Although there were variations in the soluble COD over time, these variations were consistent between all reactors. Therefore, ammonia shock did not impact the efficiency of removing the organic fraction of the sewage.

5.2.3.2 TSS and Potassium Concentrations

No significant variations relative to the control were found for effluent TSS in either the 2-day or 10-day SRT reactors spiked with ammonia, suggesting that ammonium as a monovalent cation did not cause detectable deflocculation (data shown in Appendix D). Potassium cation was measured in the effluent of the control and spiked reactors, and results are presented in Figures 5-1 and 5-2. In the 2-day SRT system there was a significant increase in potassium in response to the ammonia shock during the first two and four cycles for the 9X- and 18X-stressed reactors, respectively ($\alpha = 0.05$). For the 10-day SRT reactors, only the 10X-spiked reactor showed increased potassium levels relative to the control, and this effect was only seen during the first cycle. Although these increases seem moderate (about 3–5 mg/L), they are consistent with what has been observed with known potassium effluxing electrophilic chemicals in previous experiments. Two hypotheses are conjectured for the purported efflux observed here: physiological futile cycling mechanisms involving the release of potassium could be activated in response to ammonia shock; or ion exchange mechanisms within the floc may have occurred, through which potassium was released and substituted by NH_4^+ ions at negatively charged sites. The latter statement is supported by literature, which shows that K^+ and NH_4^+ ions compete for active sites in both soils (Chappel and Evangelou, 2000) and zeolite resins used for ammonia removal from natural waters (Choe et al., 1999) or leachates (Liu and Lo, 2001).

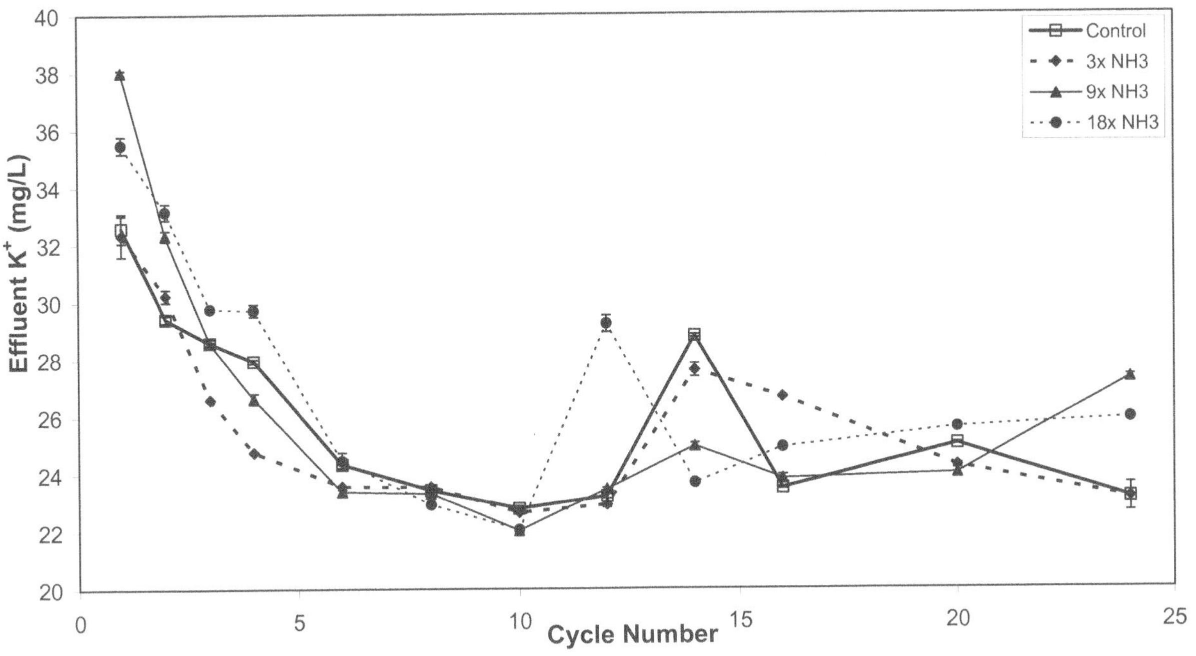

Figure 5-1. Effluent Potassium for the 2-Day SRT Reactors Exposed to Ammonia Shock.

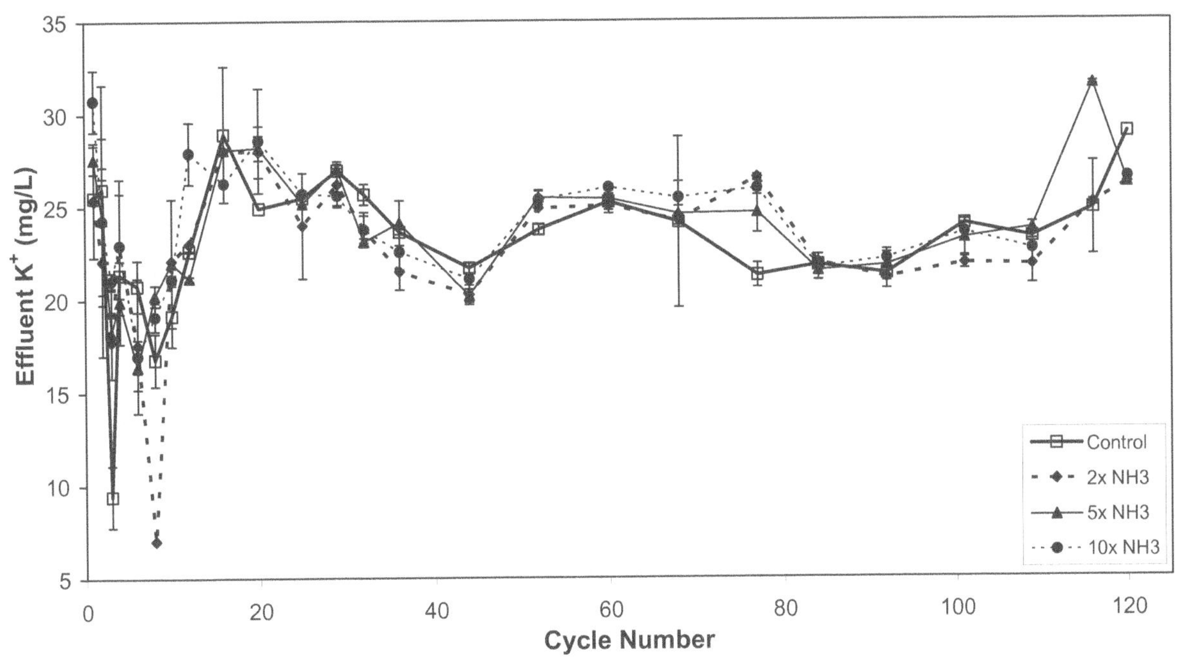

Figure 5-2. Effluent Potassium for the 10-Day SRT Reactors Exposed to Ammonia Shock.

5.2.3.2 Nitrification Inhibition

Effluent (ammonium + ammonia), nitrate and nitrite concentrations throughout the course of the 2-day SRT experiment are presented in Figures 5-3 through 5-5, respectively. The 2-day SRT biomass used for the source-effect experiment was not nitrifying to a significant degree at the beginning of the experiment. Although the total ammonia concentrations in the control reactor were consistently below detection limits during the 6-day period, this is attributed to nitrogen utilization for growth and not to significant nitrification. Additionally, nitrogen was not limiting growth during this period, since the nitrate concentrations in the control effluent were very low (below 2 mg/L as N) but always measurable. Nitrite was below the detection limit during the entire experimental period except for cycle eight when it spiked unexpectedly. Although nitrification was not occurring to a significant extent in the control and the reactors were being operated under temperature-controlled (18°C) conditions to prevent nitrification, the increased ammonia-N load stimulated nitrification in the shocked reactors. This is apparent through increased levels of nitrate (up to 8 mg/L in the 9X-spiked reactor) and a slight accumulation of nitrite, especially in the 9X-and 18X-shocked reactors. With respect to ammonia, it took approximately 12 cycles (three days) and eight cycles (two days) for the effluent total ammonia concentrations to return to background levels in the 18X- and 9X-spiked reactors, respectively. If ammonia loss from the reactors was only due to dilution (Figure 5-3), then it would take approximately 20 and 16 cycles for the effluent ammonia concentration in the 18X- and 9X-spiked reactors to fall below detection, respectively. Clearly, the data show that some fraction of the shocked ammonia was undergoing reaction, but a significant fraction was just washing out. Collectively, the nitrogen results suggest that ammonia shock initiated a higher extent of nitrification by both AOBs and NOBs, but that only partial nitrification was occurring for up to one-third of the oxidized nitrogen.

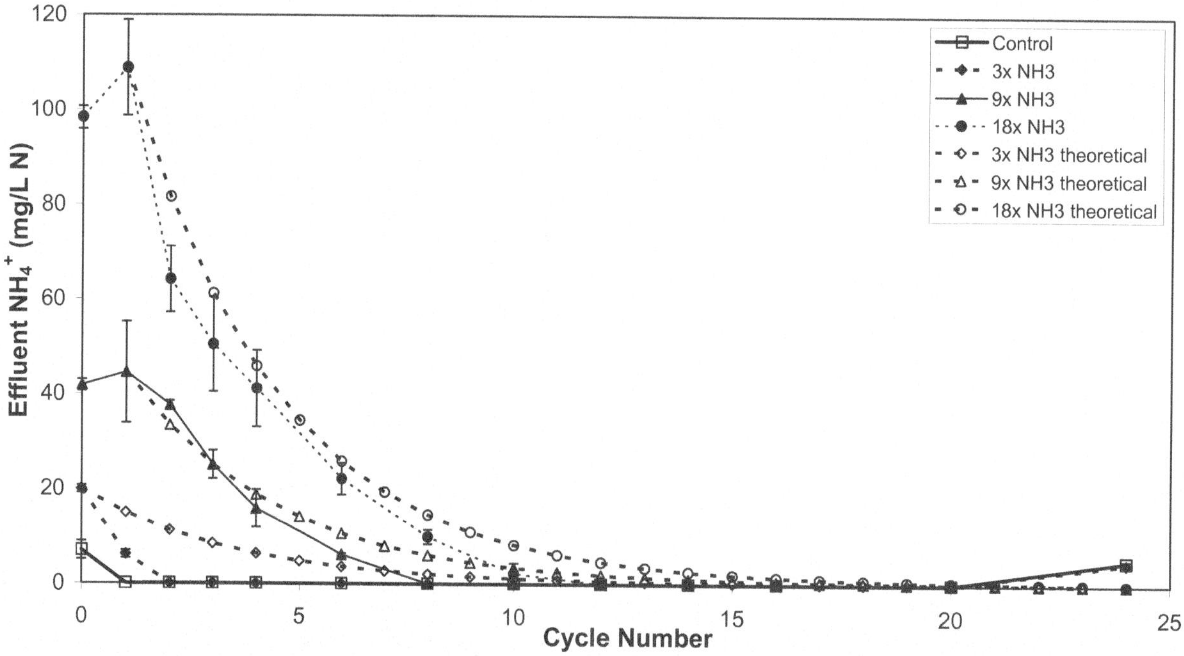

Figure 5-3. Effluent Ammonia-Nitrogen for the 2-Day SRT Reactors Exposed to Ammonia Shock. The First Data Point in Each of the Curves Corresponds to the Initial Ammonia Concentration in the Mixed Liquor Immediately After the Shock. The Dashed Lines Represent Theoretical Dilution Curves.

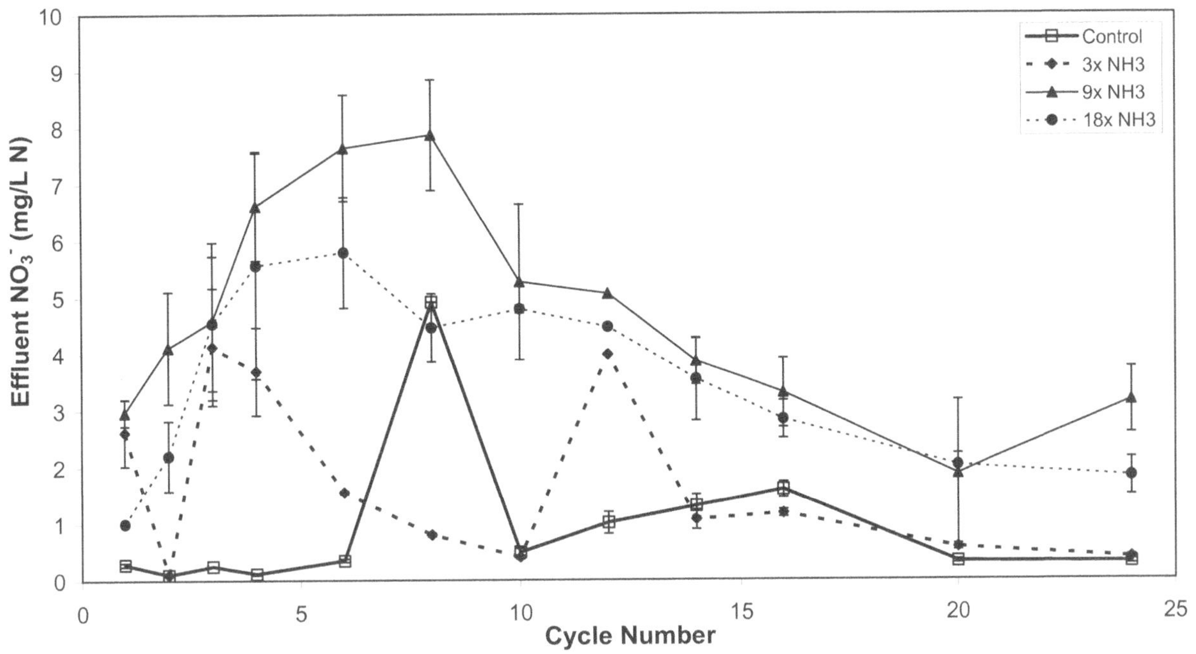

Figure 5-4. Effluent Nitrate-Nitrogen for the 2-Day SRT Reactors Exposed to Ammonia Shock.

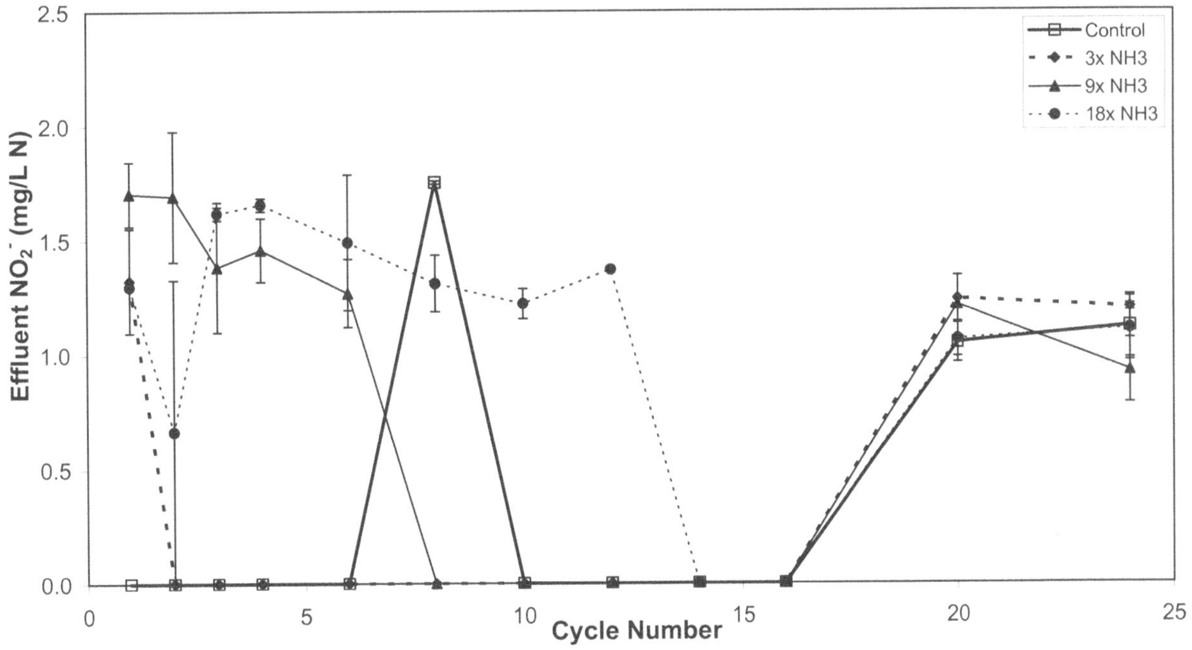

Figure 5-5. Effluent Nitrite-Nitrogen for the 2-Day SRT Reactors Exposed to Ammonia Shock.

Nitrification in the 10-day SRT reactors was disrupted by the ammonia shock at higher concentrations, but recovered rapidly. Figures 5-6, 5-7, and 5-8 show the total ammonia-N, nitrate-N, and nitrite-N concentrations, respectively. The amount of residual ammonia in the effluent during the first few cycles correlated directly with the degree of ammonia shock, with the highest shock concentration yielding the highest effluent concentration. By the end of the third cycle, however, ammonia was below detection in all effluent samples for all reactors and remained there until the end of the experiment. Nitrate concentrations increased with increasing shock concentration during the first cycles and then stabilized to a common value of 7-8 mg/L as N for all reactors. The cause of this apparent decrease in activity is unknown. From cycle 15 on, the nitrate levels remained approximately constant and were not different between reactors relative to the control. Nitrite levels were high in the 5X- and 10X-shocked reactors through the first and third cycles, respectively, but eventually decreased to below the detection limit for the remainder of the experiment for all reactors.

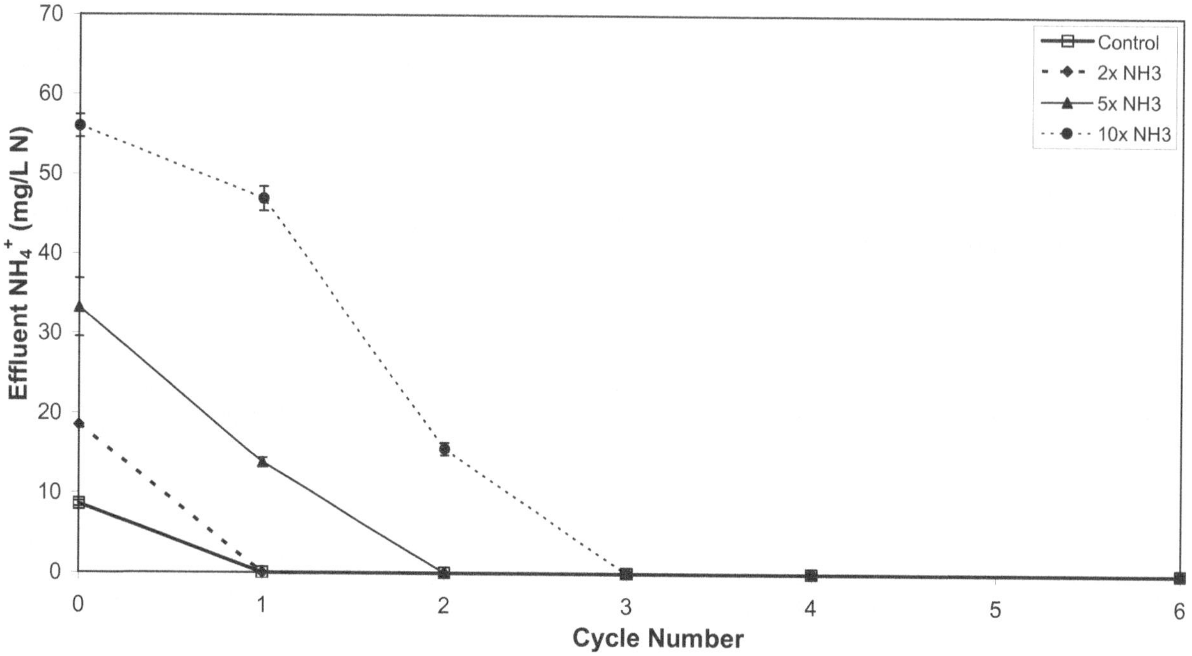

Figure 5-6. Effluent Ammonia-Nitrogen for the 10-Day SRT Reactors Exposed to Ammonia Shock. Data Through Cycle 6 Are Shown.

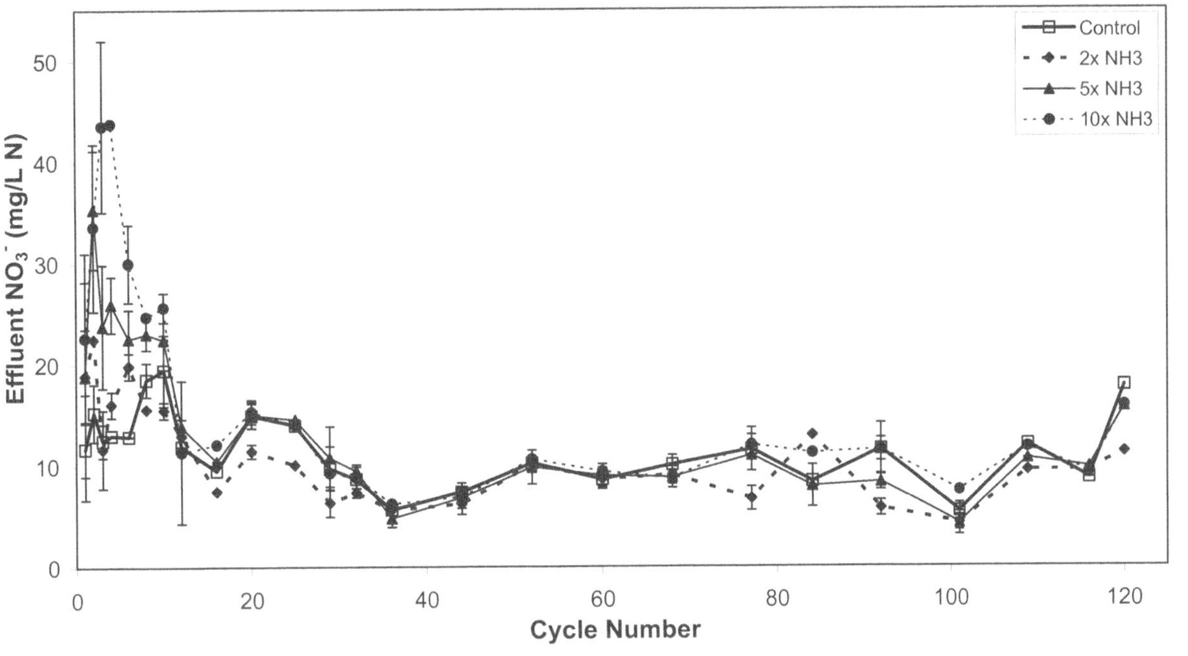

Figure 5-7. Effluent Nitrate-Nitrogen for the 10-Day SRT Reactors Exposed to Ammonia Shock.

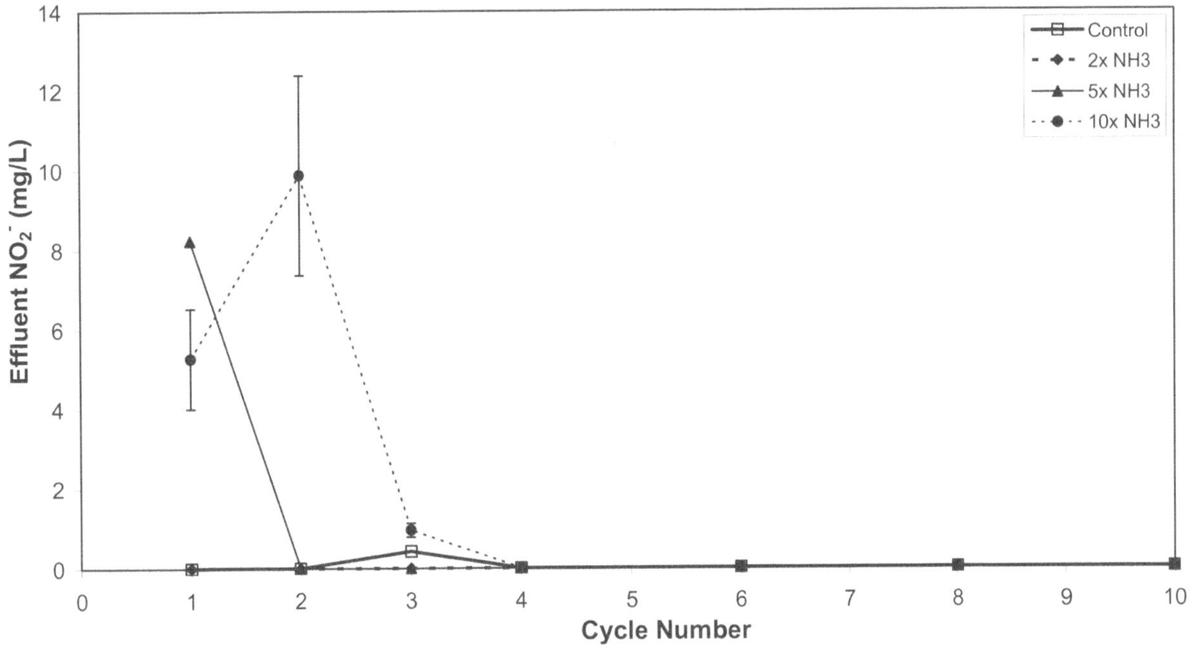

Figure 5-8. Effluent Nitrite-Nitrogen for the 10-Day SRT Reactors Exposed to Ammonia Shock. Data through Cycle 10 Are Shown.

The 2-day SRT reactors showed no significant differences in pH between spiked and control reactors for the duration of the experiment, and pH remained relatively stable (between 8.2 and 8.6) throughout the experiment in all the reactors. There was a significant decrease in pH in the 10X-shocked reactor for the 10-day SRT system shortly after the spike event (see data in Appendix D). This decrease corresponded with high levels of nitrification activity, as shown in

Figures 5-6 through 5-8. The system recovered as soon as the residual ammonia that accumulated after the spike was oxidized to nitrate. Afterwards, the pH tended to remain around 8.1 in all reactors. Consistent with pH trends, the alkalinity values in the effluent correspond with observed pH trends and nitrification activities (see Figures 5-9 and 5-10). In the 10-day SRT system, the 10X-treated biomass experienced a rapid decrease in pH early on, and this occurred as the alkalinity concentration dropped below 50 mg/L as $CaCO_3$. Alkalinity was more stable in the 2-day SRT system. The trend in residual alkalinity levels in the effluent from the first cycle corresponded with the initial ammonium bicarbonate concentrations. During days two and three, the alkalinity dropped slightly below control levels in the 9X- and 18X-spiked reactors, which is consistent with the nitrifying activity in these reactors during this period. Thereafter, the alkalinity stabilized in all reactors at about the same level.

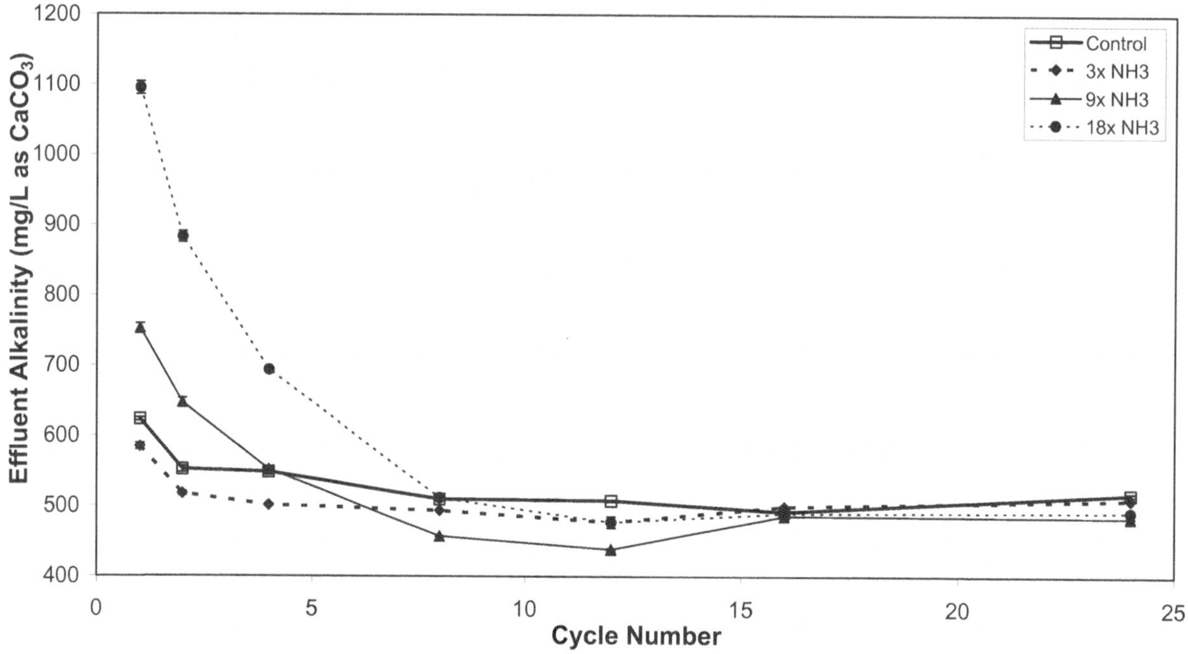

Figure 5-9. Effluent Alkalinity for the 2-Day SRT Reactors Exposed to Ammonia Shock.

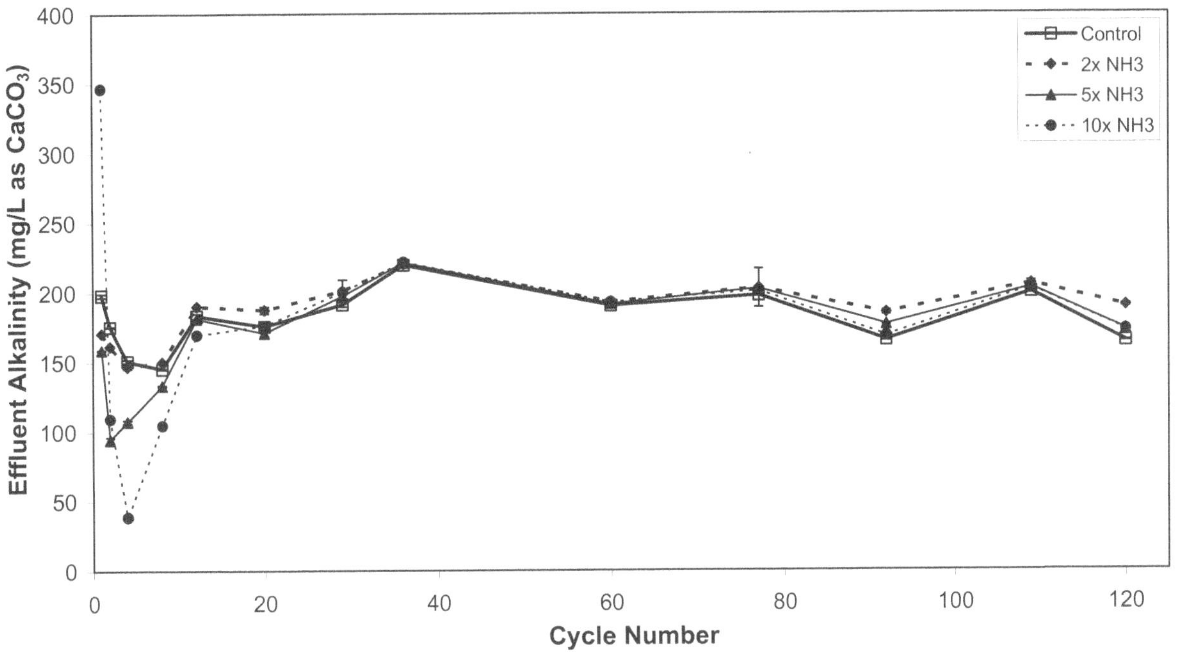

Figure 5-10. Effluent Alkalinity for the 10-Day SRT Reactors Exposed to Ammonia Shock.

5.2.4 Mixed Liquor Characteristics

5.2.4.1 Settling and Dewatering Characteristics

CST was determined for both biomass sources and was relatively stable at all shock levels and in the control for both biomass sources, with the exception of the 2-day SRT control reactor on the last day of monitoring (See Appendix D). The average CST response with the 2-day and 10-day SRT biomass averaged 42 and 13 seconds, respectively, indicating a significant difference between the biomasses. During the first two days of the 2-day SRT experiment, there were statistically higher CSTs for the 9X- and 18X-spiked reactors, compared to the control. However, the CSTs remained lower than 60 seconds and, therefore, the impact of ammonia shock on sludge dewaterability is not considered to be important. In summary, ammonia shock did not significantly affect the dewatering properties of the biomass in either system.

The average SVI values for both systems remained below 150 mL/g for all reactors in both systems. The 2X ammonia-spiked reactor with 10-day SRT biomass actually started to experience high SVIs (up to 300 mL/g) toward the end of the experiment. This increase in SVI did not correlate with any other process effect, and its origin is unknown. For the 2-day SRT reactors, similar to the CST results, the 9X- and 18X-spiked reactors experienced slightly higher SVIs than the control reactor during the entire or some period of the experiment, but the SVIs did not exceed 160 mL/g (data not shown).

5.2.4.2 Mixed Liquor Solids Trends

A summary of the MLSS and MLVSS in each reactor is given in Table 5-4. The suspended solids concentrations in the 2-day SRT reactors were rather variable in all reactors, but followed the same trends among control and shocked reactors. The suspended solids significantly decreased during the first three days of the experiment and slightly recovered towards higher levels during the second part of the experiment (see Appendix D). This variability was probably due to a significant decrease in influent TSS concentrations between the pilot plant and the laboratory reactors. In contrast, the mixed liquor concentrations were rather stable for the 10-day SRT system, and no significant differences could be seen between the control and shocked reactors.

Table 5-4. Mixed Liquor Concentrations in 2- and 10-Day SRT Reactors Exposed to Ammonia Shock.

	MLSS (mg/L)	MLVSS (mg/L)	MLVSS:MLSS Ratio
2-day SRT reactors			
Control	1,180 ± 207	812 ± 138	0.69 ± 0.051
3X NH_3-N	934 ± 280	672 ± 193	0.72 ± 0.055
9X NH_3-N	1,103 ± 257	809 ± 168	0.74 ± 0.061
18X NH_3-N	992 ± 227	732 ± 157	0.74 ± 0.054
10-day SRT reactors			
Control	2,147 ± 167	1,869 ± 150	0.87 ± 0.02
2X NH_3-N	2,220 ± 201	1,923 ± 160	0.87 ± 0.02
5X NH_3-N	2,131 ± 168	1,857 ± 139	0.87 ± 0.02
10X NH_3-N	2,068 ± 136	1,801 ± 107	0.87 ± 0.02

5.2.4.3 Kinetic Indicators of Inhibition

The 2-day SRT biomass, which was not nitrifying at the start of the experiment, started to nitrify in progressively higher degrees as the ammonia shock concentration increased. As noted in Section 5.2.3.1, the 9X and 18X ammonia-shocked reactors experienced moderate but significant inhibition based on poorer COD removal capacity in the first few cycles. This result is consistent with the SOUR response for the 2-day SRT system, shown in Figure 5-11. Statistically lower respiration rates were observed for the 9X- and 18X-spiked reactors during the first two and four days of the experiment, respectively. During the second cycle, approximately 6 hours after the shock, oxygen uptake inhibition in the 9X- and 18X-spiked reactors was as much as 30% and 50%, respectively. These results show that increased ammonia levels caused inhibition to biomass respiration. The causal mechanism for the inhibition is not known, however, but it is not related solely to nitrifier inhibition based on deterioration of COD-removal capacity, which occurred during the same period.

The 10-day SRT biomass did not show any signs of metabolic inhibition due to ammonia shock events based on SOUR (see Appendix D) or SNGR (Figure 5-12). SNGR measurements were highest for the 10X ammonia-spiked reactor and lowest for the control at the end of the first cycle, indicating an increase in nitrifier growth rate with the increased ammonia doses; however, SNGRs were relatively consistent between all reactors after the first cycle. These general trends suggest that nitrification was not inhibited in the 10-day SRT systems and that the observed differences during the first cycle were more likely due to differences in activity of nitrifying biomass.

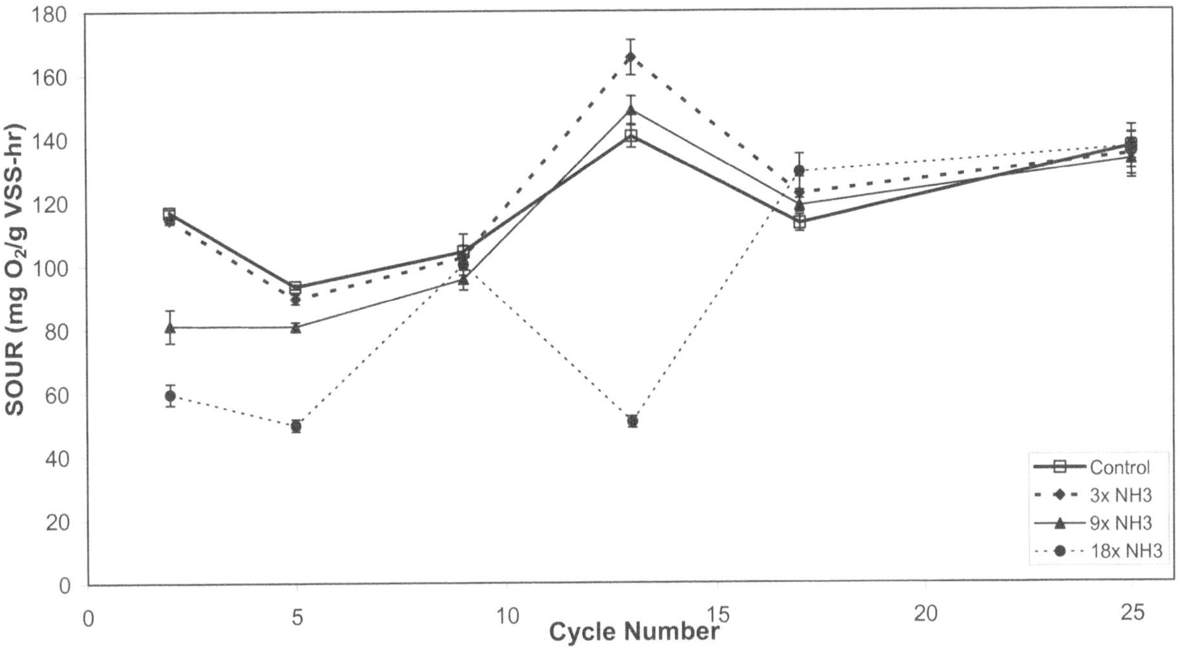

Figure 5-11. SOURs for the 2-Day SRT Reactors Exposed to Ammonia Shock.

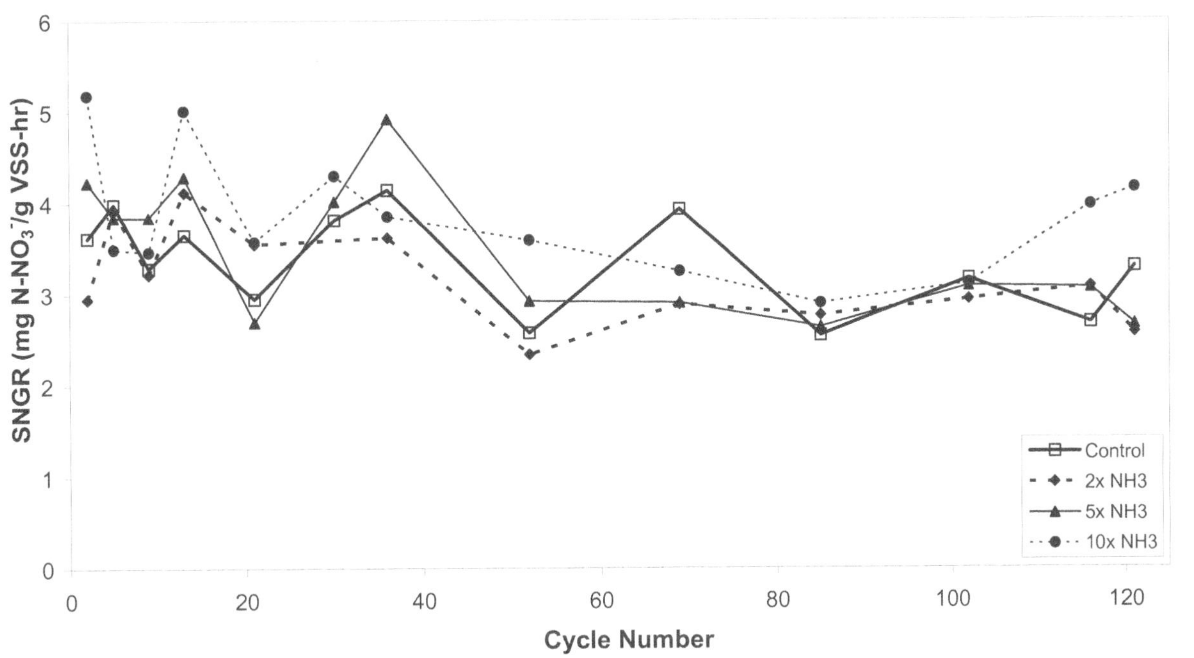

Figure 5-12. SNGRs for the 10-Day SRT Reactors Exposed to Ammonia Shock.

5.3 Discussion

The response to ammonia shock yielded typical results. Most process parameters were not significantly impacted and overall treatment process efficiency was maintained at control levels. Overall, the 2-day SRT biomass, which was initially not nitrifying to a significant degree, seemed to have been more impacted by the ammonia shock than the 10-day SRT nitrifying biomass.

One of the observed effects of ammonia shock in the 2-day SRT system was stimulation of nitrification. Although the biomass was not nitrifying at the start of the experiment, increased ammonia levels in the spiked reactors resulted in increased nitrate levels in the effluent of those reactors. Anthonisen et al., (1976) reported that free ammonia concentrations as low as 0.1 mg/L as N were sufficient to inhibit NOBs. As shown in Table 5-5, the concentrations present in the 2-day SRT reactors were clearly high enough to sustain nitrite inhibition in the 9X- and 18X-spiked reactors for the duration of the experiment. However, effluent nitrite concentrations did not reach high levels in any of the conditions tested, showing that free ammonia did not inhibit nitrite oxidizers to a great extent, even at free ammonia concentrations as high as 8 mg/L as N. The extent to which nitrification occurred was modest, with the highest nitrate concentration, about 7 mg/L, occurring in the 9X reactor. Additionally, the effect lasted only while the ammonia concentrations in the spiked reactors remained elevated and started to decrease on day three. This shows that, although nitrification was stimulated, it was not sustainable after the levels of ammonia returned to control levels in the stressed reactors.

In the 10-day SRT reactors, a brief but significant accumulation of nitrite occurred during the first two cycles in the 5X- and 10X-shocked reactors, which is consistent with the high free ammonia concentrations observed in these reactors during that time period (Table 5-5). Overall, the 10-day SRT nitrifying biomass responded very rapidly to the ammonia shock by nitrifying excessive ammonia. In addition, outside of a brief nitrite buildup, there was no strong evidence of significant metabolic inhibition, which is probably related to the fact that free ammonia concentrations in solution remained very low. Moreover, there is no evidence that heterotrophic populations were impacted during the source-effect experiment. COD removal, SOUR, MLSS, and solids settleability performed approximately the same for all the 10-day SRT reactors.

The effluent TSS concentrations in the 2-day SRT system were similar in the control and stressed reactors, and no deflocculation event was observed. However, the flocculation ability of the biomass, as indicated by the effluent TSS concentrations, was very poor in all the reactors, as effluent suspended solids concentrations reached as high as 220 mg/L. Any further deterioration in effluent TSS would be difficult to see since the starting point was already very unfavorable. On the other hand, settling and dewatering properties were slightly affected by increased NH_4^+ levels and might corroborate Novak's work (2001), which concluded that the ammonium ion affects sludge settleability and reduces floc density in a manner similar to sodium. He also suggested that this effect might be partly responsible for the deterioration in activated sludge settling properties with decreasing SRT due to the loss of nitrification at low SRT. While this theory may be valid, the degree to which ammonia shocks impacted settling and dewatering was quite modest.

The primary evidence that heterotrophic populations were impacted during the source-effect experiments was given by inhibition of oxygen uptake in the 2-day SRT shocked reactors, in comparison to the control. SOUR inhibition was as high as 50% within 6 hours after the shock in the 18X-spiked reactor, revealing that metabolism was affected through some undetermined mechanism. It is possible that the high free ammonia levels may be related to the slow down of metabolism in heterotrophic bacteria, although there is no precedence for this in the literature.

This decrease in SOUR was corroborated by a modest but significant decrease in substrate uptake rate over a 6-hour period, as indicated by high effluent COD concentrations.

Table 5-5. Free Ammonia Concentrations (mg/L as N) Corresponding to Measured ($NH_4^+ + NH_3$) Concentrations at 20°C. Values Were Calculated Using Equation 5-1.

	2-Day SRT System				10-Day SRT System			
Cycle	Control	3X	9X	18X	Control	2X	5X	10X
Initial	0.56	0.56	3.35	7.62	0.42	0.62	0.64	2.26
1	0.00	0.00	3.56	8.43	0.00	0.00	0.26	1.89
2	0.00	0.00	3.87	6.40	0.00	0.00	0.00	0.22
3	0.00	0.00	2.30	4.63				0.00
4	0.00	0.00	1.45	3.86				
6	0.00	0.00	0.54	1.76				
8	0.00	0.00	0.00	0.83				
10	0.00	0.00		0.15				
12	0.00	0.00		0.00				
14	0.00	0.00						
16	0.00	0.00						
20	0.00	0.00						
24	0.39	0.39						

Increased potassium levels were observed in the effluent of the 2-day SRT 9X- and 18X-spiked reactors compared to the control during the first few cycles after the shock and in the 10X-shocked 10-day SRT reactor during the first cycle after the shock. Potassium increase due to high ammonia/ammonium levels could have resulted from physical-chemical ion exchange processes within sludge flocs or other unknown physiological mechanisms.

Activated sludge flocs are composed of bacteria and extracellular polymeric substances that are essentially negatively charged. These sites are occupied by cations such as Ca^{2+}, Mg^{2+}, Na^+, and K^+ and a dynamic equilibrium is established between sorbed and aqueous concentrations of the cations; this equilibrium might be disturbed when sudden alterations of aqueous concentrations of these or other cations occur, creating a potential for ion exchange to occur within the floc matrix. Although no evidence has been gathered that ammonium ions can replace potassium through ion exchange processes in activated sludge, it has been observed in natural waters and soils, as discussed previously. Therefore, ion exchange could be a reason for the observed increased levels in effluent potassium in the spiked reactors.

Physiological mechanisms may also result in increased extracellular potassium. For example, the GGKE stress response causes a significant flow of potassium ions from the inside to the outside of the cells in response to electrophilic shock. This mechanism is not likely to have occurred due to ammonia shock, since neither ammonium nor ammonia has oxidative characteristics. However, other mechanisms could be responsible for the same effect. For example, although at high ammonia concentrations facilitated diffusion seems to be the predominant transport process through which ammonia enters bacterial cells, active uptake of ammonium may also occur (reviewed by Kleiner, 1985). Additionally, some authors have proposed the existence of a K^+/NH_4^+ antiporter in *Escherichia coli* (Barnes and Jayakumar, 1993). This active uptake process would create a potential futile cycle, since the more basic intracellular pH may allow for free diffusion of ammonia from the inside to the outside of the cell (Russel and Cook, 1995), requiring more active transport to take place and more K^+ to be released. Whether or not this process may occur in activated sludge systems is not known. At this point, it is not possible to determine the exact mechanism(s) for the observed effect.

5.4 Conclusions

- The 10-day SRT fully nitrifying biomass was able to assimilate the increased ammonia load within the first four cycles (less than one day). Ammonia accumulated in the system and there were transient spikes of nitrite in the effluent, which was probably due to free ammonia inhibition of the NOB, but complete nitrification was achieved within a day. In short, the 10-day SRT system performed very well in response to the shock. The presence of sufficient alkalinity to offset the potential for pH changes due to the acid production from nitrification was probably key to achieving rapid recovery.
- Ammonia shock loads to a non-nitrifying 2-day SRT biomass at constant pH stimulated nitrification. Despite the presence of free ammonia concentrations that are believed to be quite inhibitory to NOBs, only minimal nitrite accumulation occurred.
- COD-removal efficiency was not affected by up to a 10X shock load of ammonia in the 10-day SRT nitrifying biomass. Soluble COD removal efficiency was only briefly affected by 9X- and 18X-ammonia shock loads in a non-nitrifying 2-day SRT biomass, and that effect was quite moderate.
- SOUR and SNGR measurements indicated that the 10-day SRT biomass was not inhibited by the ammonia spike. Since the biomass population in the 2-day SRT system was predominantly composed of heterotrophs (the reactors were not nitrifying at the beginning of the experiment), the short term SOUR results predominantly represent the impact of ammonia on heterotrophic bacteria. A loss in respiration capacity after the ammonia spike was observed through SOUR measurements, corroborating the notion that the loss in sCOD-removal efficiency was significant, despite being brief.
- The impact of ammonia shock on sludge dewaterability or settleability was modest.
- Elevated potassium levels were observed in the10X (10-day SRT), 9X, and 18X (2-day SRT) spiked reactors after the shock. Since ammonia is not electrophilic in nature, a GGKE basis for potassium efflux is considered unlikely. However, there is precedent for the existence of futile K^+/NH_4^+ antiporter cycles that could have affected cytoplasmic energy levels, thereby explaining the impact of high concentrations of ammonium on heterotrophs.

CHAPTER 6.0

ZINC-COMPLEXED CYANIDE SHOCK EXPERIMENT

6.1 Literature Review

There are three main sources of cyanide that can contribute to wastewater streams: 1) atmospheric deposition; 2) domestic; and 3) commercial/industrial, with industrial being the largest contributor (Wild et al., 1994). Cyanide is produced on an industrial scale of 2 to 3 million tons per year and is found in many different industrial wastewaters (Gijzen et al., 2000). It is prevalent in the wastewaters of manufacturing industries involved in metal plating, pharmaceuticals, iron and steel production, synthetic fibers, plastics, coal gasification, coal coking and re-leaching, petroleum refining, photofinishing, herbicides and insecticides (Lordi, 1980; Luthy, 1981; Gaudy et al., 1982; Wild et al., 1994; Hung and Pavlostathis, 1997; White et al., 1988). The metal plating industry typically produces small volumes of highly concentrated cyanide wastes that originate in plating vats and rinse waters (Lordi et al., 1980). Conversely, the iron and steel industry generally produces large volumes of dilute cyanide wastes (Lordi et al., 1980). Therefore, each industry provides a separate challenge in treating cyanide waste streams. The toxicity and prevalence of cyanide in waste streams make cyanide an important chemical to investigate in relation to biological wastewater treatment systems.

Many different forms of cyanide can exist, both organic and inorganic. Analytically, the term "cyanide" refers to the group of simple or complexed chemical compounds that contain the cyanide ion, CN^-, also known as free cyanide. Simple, or free, cyanide includes the cyanide anion, CN^-, and hydrogen cyanide, HCN. Complexed cyanides are those compounds in which the cyanide anion forms complexes with metals in water. The relationship between simple and complexed cyanides in water is dependent on pH and heavy metal concentrations. Under alkaline conditions, free cyanide is completely ionized and forms stable metal complexes. At neutral and acidic conditions, free cyanide is weakly ionized and formation of volatile hydrogen cyanide is favored (Wild et al., 1994). Due to the volatility of HCN, it has been experimentally shown that up to 66% of cyanide can be stripped from solution by aeration in both water and activated sludge, creating health hazards for people (Raef et al., 1977b; Gaudy et al., 1982).

Cyanide can form both weak and strong metal complexes, depending on the metal present. Strong complexes occur in the presence of iron and cobalt metals and are generally less toxic to microorganisms (Fallon et al., 1991; Zheng and Dzombak, 2003). Typically, cyanide will weakly complex with sodium, potassium, calcium, cadmium, lead, nickel and zinc metals (Wild et al., 1994; Torrens, 2000; Zheng and Dzombak, 2003). Weak cyano-metal complexes, especially zinc (Zn^{2+}) and cadmium (Cd^{2+}), dissociate extensively to the metal ion and CN^- (HCN), creating a greater availability of free cyanide for the biomass and greater potential for biomass and aquatic toxicity (Torrens, 2000). The toxic effects with cyanide compounds are usually attributed to the concentration of hydrogen cyanide in the solution. The toxic effects of the corresponding heavy metals are deemed less significant, but can have additive properties (Blaha, 1976). In wastewater streams, it has been found that most cyanide will exist in the complexed form, but high concentrations of simple cyanide can occur (Blaha, 1976; Lordi et al; 1980; Gijzen et al; 2000). Absorption to biomass has also been documented, but is considered less important (Raef et al., 1977a).

A chemical's toxicity is a key consideration in evaluating its impact on organisms in biological wastewater treatment systems and in the environment (Blum and Speece, 1991). Hydrogen cyanide is considered the most toxic form of cyanide. In humans, the average fatal single dose of cyanide is 50-60 mg, while low doses of cyanide can cause headache, nausea and vomiting (Wild et al., 1994). Concentrations in the range of 0.01-0.1 mg cyanide have been found to kill some aquatic species, especially those in embryonic or early juvenile stages (Blaha, 1976). Sub-lethal concentrations of cyanide also have the potential to impose negative effects on aquatic species, especially when coupled with the toxicity of dissociated heavy metals from complexed cyanides (Blaha, 1976). In general, cyanide complexes with metal species are less toxic than free cyanide. However, those weakly complexed cyanides, such as zinc and cadmium, remain extremely toxic due to the easy dissociation of the compounds in water (Blaha, 1976; Wild et al., 1994). It is these more toxic species and their molecular effects that remain central to the investigation of cyanide in wastewater treatment streams.

The molecular effects of cyanide to both prokaryotic and eukaryotic cells have been extensively studied. Cyanide is a non-specific inhibitor capable of interfering with many different enzymes (Lewandowski, 1984). Specifically, there are three major mechanisms by which cyanide inhibits growth and causes toxicity to cells (Knowles, 1988). The first mechanism is chelation to di- and tri-valent metals in metallic enzymes. Cytochrome oxidases, the terminal enzymes in the electron transport chain, are the most commonly studied metallic enzymes affected by cyanide (Solomonson, 1981; Knowles 1988; Yoshikawa and Caughey, 1990; Arden et al., 1998). Cytochrome oxidase is the cellular component responsible for reducing dimolecular oxygen during respiration, so inhibition of this enzyme leads to inactivation of respiration, or cell suffocation (Solomonson, 1981). Specifically, cyanide prevents the re-oxidation of cytochrome oxidase when the CN^- iron interferes with the heme (iron-containing group) of this enzyme (Solomonson, 1981; Arden et al., 1998). This rapid inhibition of respiration is responsible for the well renowned reputation of cyanide as a deadly poison.

The second mechanism by which cyanide inhibits growth and causes toxicity to cells is by reacting with keto compounds (Knowles, 1988). Reaction of cyanide with these compounds, primarily the α-keto acids pyruvate and α-ketoglutarate, results in the formation of cyanohydrins. This reaction is found to exert a protective effect against CN^- toxicity, by converting the toxic form into the less toxic cyanohydrin form that is easier for bacteria to degrade (Solomonson, 1981). Although some protection is provided through these reactions, the cyanohydrin products are derivatives of enzyme substrates that can also inhibit those enzymes (Knowles, 1988).

The third mechanism by which cyanide inhibits growth and causes toxicity to cells is the reaction of Schiff-base intermediates to form inhibitory nitrile derivatives (Knowles, 1988). All enzymatic reactions of Schiff base intermediates are susceptible to cyanide inhibition (Solomonson, 1981). Susceptible reactions include: 1) inhibition of the enzyme that catalyzes the last step in catabolism of hydroxyproline, an important amino acid found in cartilage, by mammals; and 2) inhibition of clostridial acetoacetate decarboxylase, an enzyme that catalyzes the decarboxylation of acetoacetate in certain anaerobic bacteria (Solomonson, 1981).

Treatment of cyanide containing wastewaters using biological processes is well documented. Industrial wastewaters generally contain levels between 0.01 mg/L and 10 mg/L of total cyanide, but have been known to exceed 10,000 mg/L (Wild, 1994). Since cyanide is a known aquatic pollutant, discharge limits have been imposed on wastewater treatment plants to limit the toxic effects of cyanide on natural waters. In general, cyanide-containing wastewaters arrive at treatment plants predominantly in dissolved, metal-complexed forms (Lordi et al., 1980). If levels of cyanide are not inhibitory, there are four main removal fates by which cyanide can be removed: (1) stripping to the atmosphere as HCN; (2) adsorption to particulates and removal by clarifiers; (3) conversion to another chemical species; and (4) biological degradation (Raef et al., 1977a). Cyanide stripping has been found to account for approximately 15% of total cyanide removed in aerated systems, while adsorption is relatively insignificant because most cyanide is found in the dissociated form (Raef et al., 1977a; Wild et al., 1994). Consequently, most remains in the wastewater as it flows into the bioreactor.

Once in the bioreactor, cyanide can undergo the aforementioned chemical conversion or biological degradation. If the levels of cyanide are too high there is the potential for inhibitory effects and disruption of the biological treatment process (Ludzack and Schaffer, 1962; Raef et al., 1977b; Lewandowski, 1984; Wild et al., 1994; Kholdebarin et al., 1998; Lee and Park, 1998; Gijzen et al., 2000; Torrens, 2000). The commonly accepted mechanism of cyanide toxicity, which is blocking of the respiratory electron transport chain, indicates that mixed aerobic cultures should be more sensitive to cyanide effects than anaerobic cultures (Zintgraff et al., 1969). Within mixed aerobic cultures, the inhibition of nitrification and denitrification are the most commonly observed effects of cyanide toxicity (Zintgraff et al., 1969; Lewandowski, 1984; Daigger and Sadick, 1997; Kholdebarin et al., 1998). This is explained by the fact that concentrations inhibiting autotrophs are often an order of magnitude less than those that inhibit heterotrophs, making it possible for a wastewater treatment plant to remove organic matter while being unable to nitrify (Daigger and Sadick, 1997). Specifically, cyanide is believed to inhibit AOBs, preventing ammonia oxidation to nitrite by interfering with the cytochrome oxidase of the respiratory electron transport system (Kholdebarin et al., 1998). Inhibition of denitrification is caused by inactivation of nitrate and nitrite reductase by cyanide (Solomonson, 1981). Though nitrification is the major inhibition effect, low growth yields have been found in the presence of high cyanide levels as well (Zintgraff et al., 1969). Severe inhibition of bacterial respiration, oxygen uptake rates decreasing with increasing cyanide concentrations, and inhibition of COD removal have also been documented (Ludzack and Schaffer, 1962; Zintgraff et al., 1969).

6.2 Results

6.2.1 Inhibitory Concentrations (IC_{XX})

The impact of varying zinc-cyanide complex concentrations on the SOUR of the 2- and 10-day SRT biomass was determined using respirometric tests. Complexed zinc-cyanide was used to mimic metal-plating industry waste and to minimize volatilization of cyanide. The results of these calculations are given in Table 6-1, and were the final concentrations added to each of the reactors at the beginning of the first cycle of the source-effect experiment. To negate the effects that the dissociated zinc may have on each reactor, an amount equivalent to the uncomplexed zinc in the IC_{50} was added to the control reactor, as computed using the MINEQL+ software program (Environmental Research Software, 1998). These amounts were: 5.41 mg Zn^{2+} in the 2-day SRT control and 8.85 mg Zn^{2+} in the 10-day SRT control.

Table 6-1. IC_{15}, IC_{25} and IC_{50} Values for Cyanide in 2- and 10-Day SRT Mixed Liquors.

	2-Day SRT		10-Day SRT	
	mg/L	mg Cyanide/ g MLVSS	mg/L	mg Cyanide/ g MLVSS
IC_{15}	1.4	1.5	1.7	0.5
IC_{25}	2.3	2.4	3.4	1.1
IC_{50}	5.5	5.7	8.4	2.6

When looking only at the concentrations (mg/L) of cyanide added to the 2- and 10-day SRT reactors, the IC_{XX} concentrations added to the 10-day system were approximately 1.5 times the IC_{XX} concentrations added to the 2-day SRT system. This suggests that the 10-day SRT system is less sensitive to cyanide toxicity than the 2-day SRT system. However, when concentrations are normalized to a mass contaminant per mass reactor suspended solids basis (mg CN^-/g-MLVSS), the 10-day SRT system was found to have concentrations less than half those of the 2-day SRT system. This suggests that the 10-day SRT system is, in fact, more sensitive to the addition of cyanide than the 2-day SRT system. As previously discussed, cyanide is known to have a marked effect on the nitrification process. A significant portion of the oxygen demand in the 10-day SRT system came from nitrifiers, and this may explain why the IC_{XX} for the nitrifying 10-day SRT culture was lower on a mg/g basis than for the 2-day SRT biomass.

6.2.2 Influent Characteristics

Average influent characteristics over the course of the cyanide source-effect experiment are given in Table 6-2. The influent quality was relatively constant over time for each experiment with the exception of the suspended solids and COD, which varied significantly as seen by the high standard deviation. Both experiments occurred after influent supplementation was initiated. Accordingly, the influent composition reveals that the wastewater used as the feed for the source-effect experiments with zinc-complexed cyanide was medium to strong in strength (Metcalf and Eddy, 2003).

Table 6-2. Characteristics of Wastewater Used During Cyanide Source-Effect Experiments.

Characteristic	2-Day SRT Biomass Average ± std dev	10-Day SRT Biomass Average ± std dev
Alkalinity (mg/L as $CaCO_3$)	564 ± 22	356 ± 20
Ammonia (mg/L as N)	24 ± 5.3	11 ± 2.5
COD, soluble (mg/L)	874 ± 80	512 ± 90
COD, total (mg/L)	1,026 ± 158	726 ± 264
pH	7.4 ± 0.2	7.5 ± 0.2
TSS (mg/L)	255 ± 75	151 ± 53
VSS (mg/L)	89 ± 30	72 ± 21

6.2.3 Effluent Characteristics

6.2.3.1 Soluble COD

The effect of metal complexed cyanide on COD removal was moderate to severe, but recovery was rapid in all cases. In the 2-day SRT system, seen in Figure 6-1, COD removal efficiency in the IC_{15} and IC_{25} reactors was unaffected. The IC_{50} reactor experienced a severe loss of COD removal for the first two cycles as compared to the control, but recovered by cycle 3 ($\alpha = 0.05$). In the 10-day SRT system, COD removal efficiency was only briefly affected and moderately so, as seen in Figure 6-2. For the IC_{50} reactor, recovery to control levels occurred by cycle six (1.5 days), while the IC_{15} and IC_{25} reactors were only affected during the first cycle ($\alpha = 0.05$). Because recovery of this and other parameters were apparent before 3X SRT was completed for the 10-day SRT systems, the experiment was discontinued at day 53. These results suggest that cyanide only significantly affected the 2-day SRT IC_{50} reactor, and that recovery was rapid.

6.2.3.2 TSS and Potassium Concentrations

As with effluent COD removal efficiency, effluent TSS was largely unaffected by the addition of zinc-cyanide complex. In the 2-day SRT system the shocked reactors did not exhibit a significant increase in effluent TSS over the control reactor. For the 10-day SRT system, the IC_{25} reactor showed significantly ($\alpha = 0.05$) higher suspended solids relative to the control at cycles two, 10, and 12, while the IC_{50} reactor showed higher suspended solids relative to the control at cycle 1 (shown in Appendix E). These results cannot be attributed to cyanide as no trending effects were seen with increasing cyanide concentration. Overall, effluent TSS seemed unaffected by the addition of zinc-cyanide complex to each reactor.

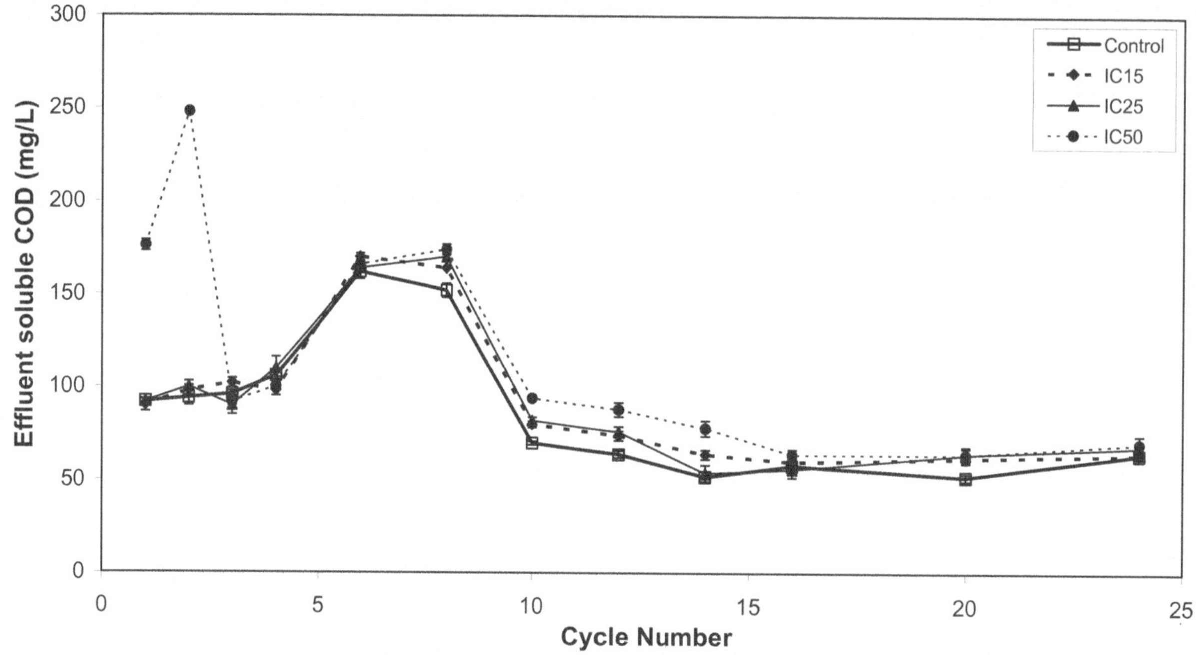

Figure 6-1. Effluent Soluble COD for the 2-Day SRT Reactors Exposed to Cyanide Shock.

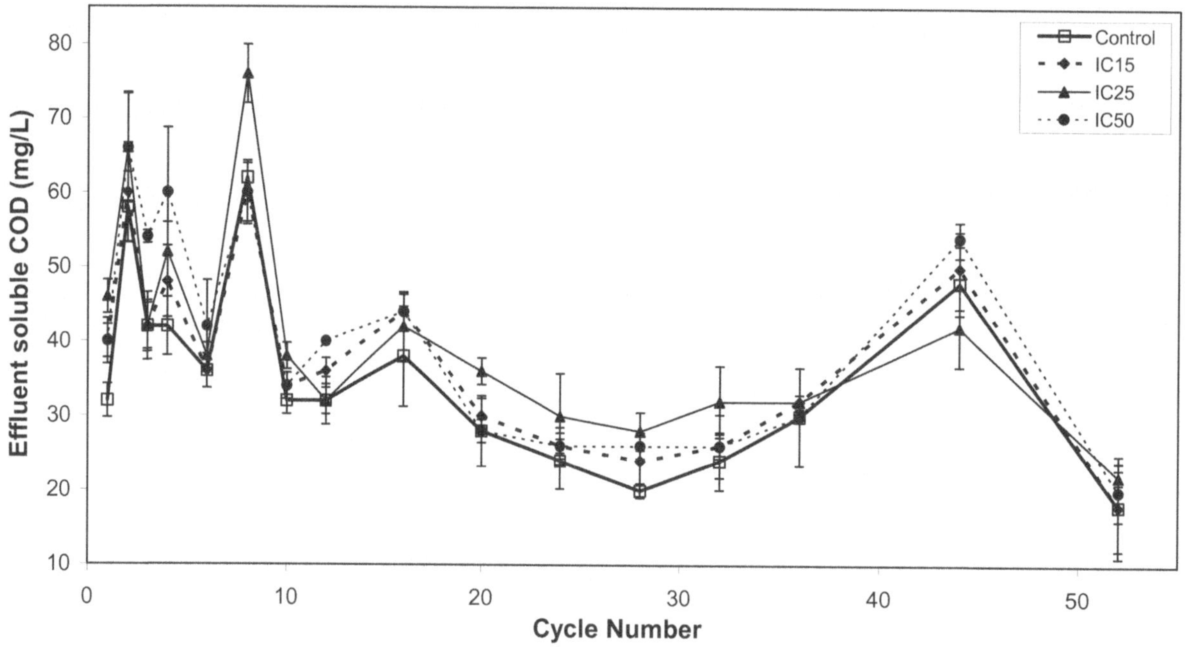

Figure 6-2. Effluent Soluble COD for the 10-Day SRT Reactors Exposed to Cyanide Shock.

Effluent soluble potassium levels did not show significant increases in either the 2- or 10-day SRT reactors for any concentration (data in Appendix E). There was a decreasing potassium concentration trend for all reactors in the 10-day SRT systems, which followed the decrease in MLSS and MLVSS in the same system. Since this decrease occurred in all reactors, including the unshocked control, it is not attributed to the cyanide shock. Therefore, the overall decrease in soluble potassium levels was probably due to the decreasing biomass concentrations and not due to the activation of any specific stress response mechanism. These data also suggest that the addition of zinc, a thiol reactive heavy metal, was not present in its free and reactive form in sufficient quantity to activate a GGKE response, as discussed previously for cadmium.

6.2.3.3 Nitrification Inhibition

As previously discussed, cyanide is a known inhibitor of nitrification and severe disruption was expected. The 2-day SRT pilot plant reactor was being operated to prevent nitrification and was not nitrifying the day before the experiment began. Furthermore, effluent ammonia, nitrite and nitrate results confirm that no significant level of nitrification was occurring during the 2-day SRT biomass shock experiments (data shown in Appendix E). In contrast to the 2-day SRT system results and consistent with our expectations, the nitrifying 10-day SRT biomass shows nitrification inhibition patterns, which the systems recovered from relatively quickly (shown in Figures 6-3 through 6-5). All 10-day SRT reactors were nitrifying prior to cyanide addition. After cyanide addition, a dramatic increase in effluent ammonia was observed in the IC_{50} reactor, illustrating an inhibition of AOBs at high concentrations of cyanide (Figure 6-3). The ammonia levels in this reactor recovered to control levels by cycle four (1 day) ($\alpha = 0.05$). With the decrease in ammonia, a sharp increase in nitrite was observed (Figure 6-4), suggesting a recovery of the AOBs. The increase in nitrite was most likely due to the NOBs being unable to deal with the rapid influx of nitrite, causing a delay in nitrite conversion; however, some inhibition of NOBs may have occurred. There were moderate levels of nitrite found in the IC_{15} and IC_{25} reactors, suggesting slight inhibition of nitrite oxidation that corresponded with the level of zinc-cyanide complex introduced to each reactor. Figure 6-5 demonstrates lagging trends in nitrate conversion consistent with the increasing inhibitory concentrations in the shocked reactors. Overall, there was clear inhibition of nitrification in the 10-day SRT system with the addition of increasing concentrations of zinc-cyanide complex, but the systems all recovered to within control performance levels by cycle 10 (2.5 days) ($\alpha = 0.05$). This suggests that weak metal-complexed cyanide inhibits nitrification, but that this inhibition is rapidly reversed.

Alkalinity was monitored because it is consumed during nitrification and should provide an independent measure of the degree of nitrification inhibition. Alkalinity levels in the 2-day SRT system were consistent with a non-nitrifying system and further reinforce the notion that the 2-day SRT system was not nitrifying during this experiment (data shown in Appendix E). In the 10-day SRT system, the control reactor had significantly lower alkalinity concentrations than the three shocked reactors (Figure 6-6). This was expected because nitrification was not interrupted in the control reactor, so more alkalinity should have been consumed. There were increasing levels of alkalinity corresponding to increasing inhibitory concentrations added to the reactors during the first five cycles. This correlates well with the brief inhibitory effects that were observed among the nitrifying 10-day SRT shocked reactors.

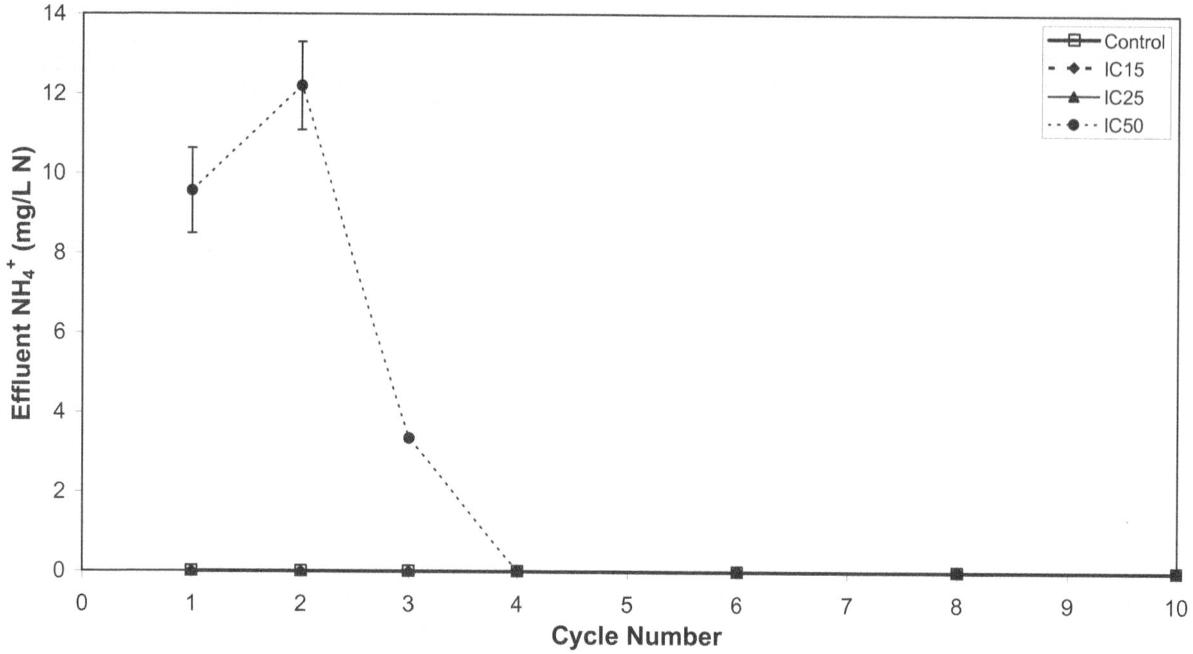

Figure 6-3. Effluent Ammonia-Nitrogen for the 10-Day SRT Reactors Exposed to Cyanide Shock. Data through Cycle 10 Are Shown.

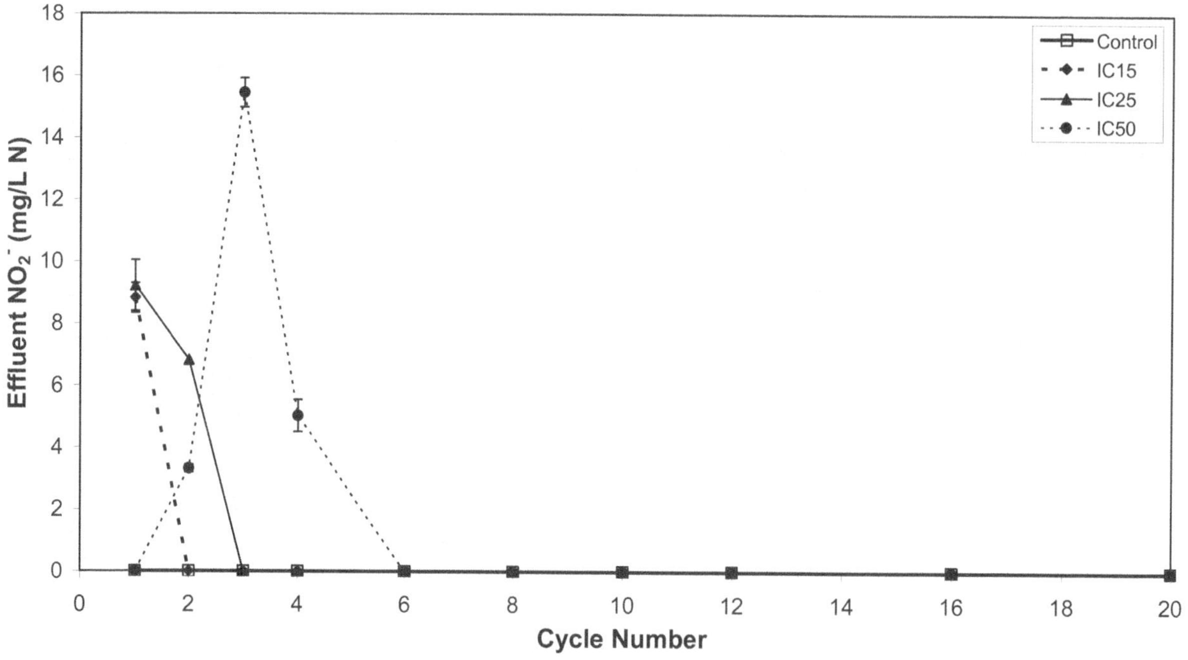

Figure 6-4. Effluent Nitrite-Nitrogen for the 10-Day SRT Reactors Exposed to Cyanide Shock. Data through Cycle 20 Are Shown.

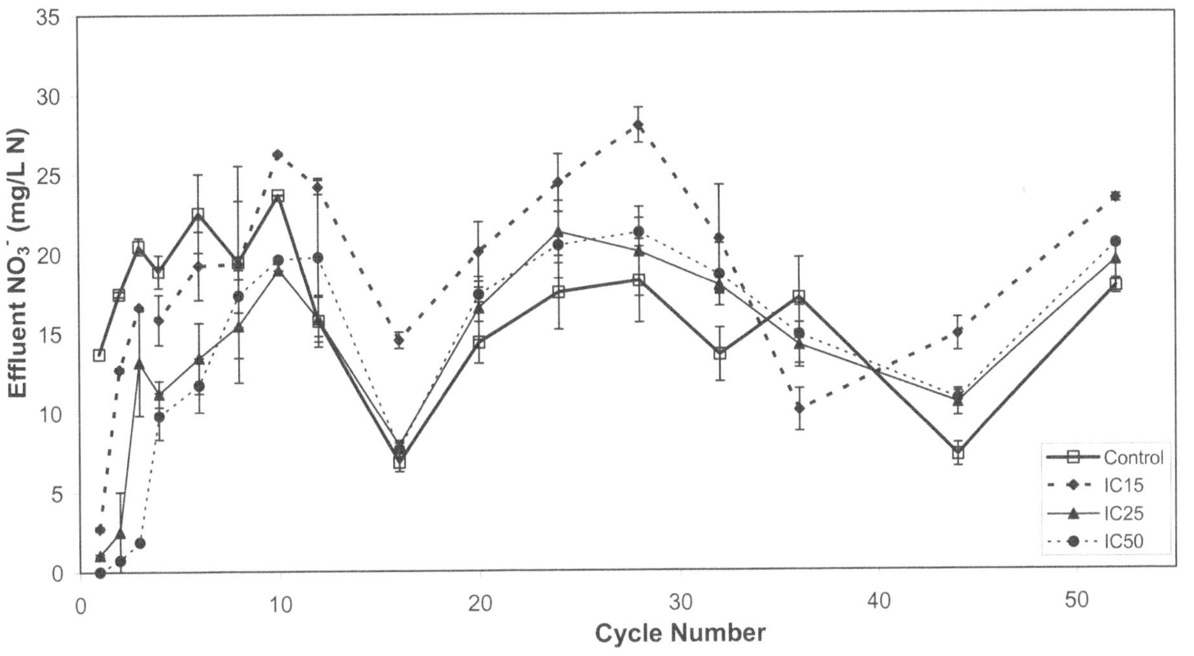

Figure 6-5. Effluent Nitrate-Nitrogen for the 10-Day SRT Reactors Exposed to Cyanide Shock.

Effluent pH levels in the 2- and 10-day SRT systems were consistent with what was determined from the alkalinity data, and are provided in Appendix E. For the 2-day SRT shocked reactors, pH was slightly elevated above the control for most of the experiment; however, the difference from the control reactor was small and there were no significant differences between reactors for the duration of the experiment. Similarly, effluent pH values for the 10-day SRT system were higher than the control during cycle one in the IC_{50} reactor, which correlates with the alkalinity increase that occurred due to nitrification inhibition discussed previously. There were some unexplained pH outliers as well, indicating that pH is influenced by a number of factors and is helpful in corroborating nitrification inhibition that is detected by independent methods, but does not confirm nitrification inhibition alone. This suggests that pH-stat bioassays for detecting nitrification inhibition may have difficulty distinguishing nitrification inhibition from other pH-sensitive responses that may occur in full-scale activated sludge systems.

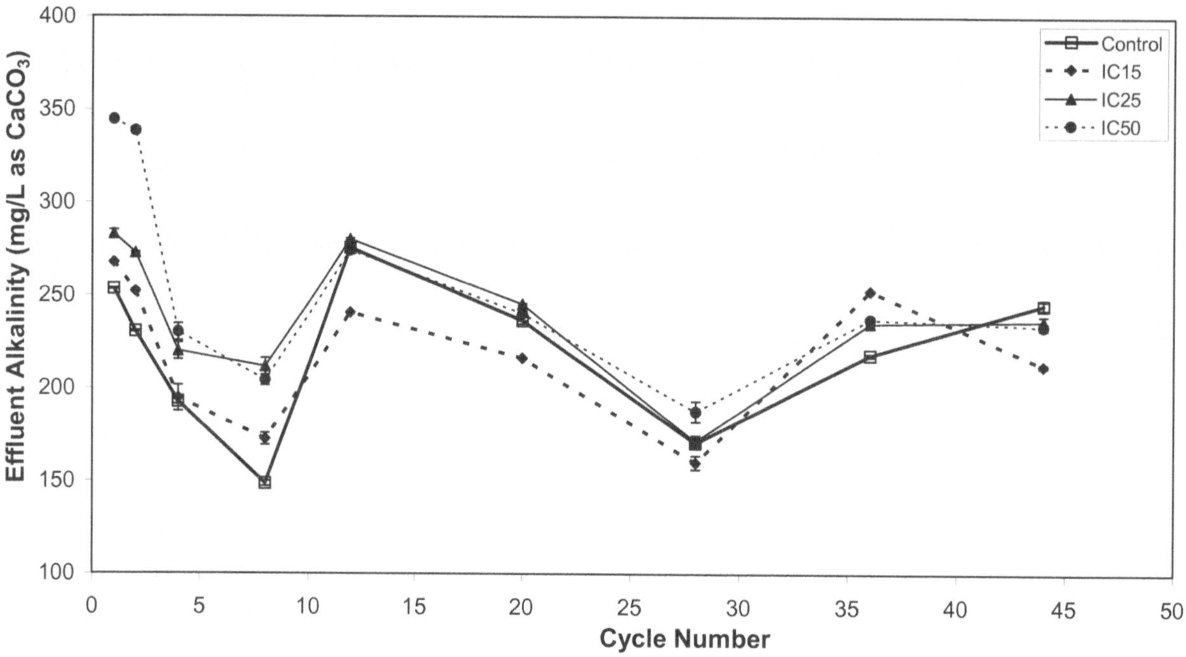

Figure 6-6. Effluent Alkalinity for the 10-Day SRT Reactors Exposed to Cyanide Shock.

6.2.4 Mixed Liquor Characteristics

6.2.4.1 Settling and Dewatering Characteristics

CST measurements were taken for the mixed liquor of both the 2- and 10-day SRT systems. The values were relatively constant over time for both systems, with the 2-day SRT reactors averaging 19.8 ± 0.4 seconds CST and the 10-day SRT reactors averaging 16.5 ± 0.3 seconds CST. These low values indicate that cyanide had little impact on the dewaterability of activated sludge.

The SVI data for the 2- and 10-day SRT reactors showed minimal to temporary impact on SVI that could be correlated with the addition of zinc-cyanide complex. Beginning at cycle 14 (day four) the SVI in all the 2-day SRT reactors started to trend upwards with the control increasing at the slowest rate, but the cause is unknown and cannot be explicitly connected to the addition of zinc-cyanide complex (data in Appendix E). Only the 10-day SRT IC_{50} shocked reactor showed any direct impact after the initial chemical shock, with an elevated SVI relative to all other reactors during cycle two (the first cycle when SVI was measured) (Figure 6-7). Despite the distinct change in the IC_{50} reactor, the SVI remained below 120 mL/g and returned quickly to control reactor levels. This observation correlates with the loss of ammonia oxidation in the IC_{50} reactor, and is consistent with Novak's hypothesis that the ammonium cation may cause weak floc structure (Novak, 2001). Together, these results indicate minor to no impact of cyanide on the settleability of the biomass.

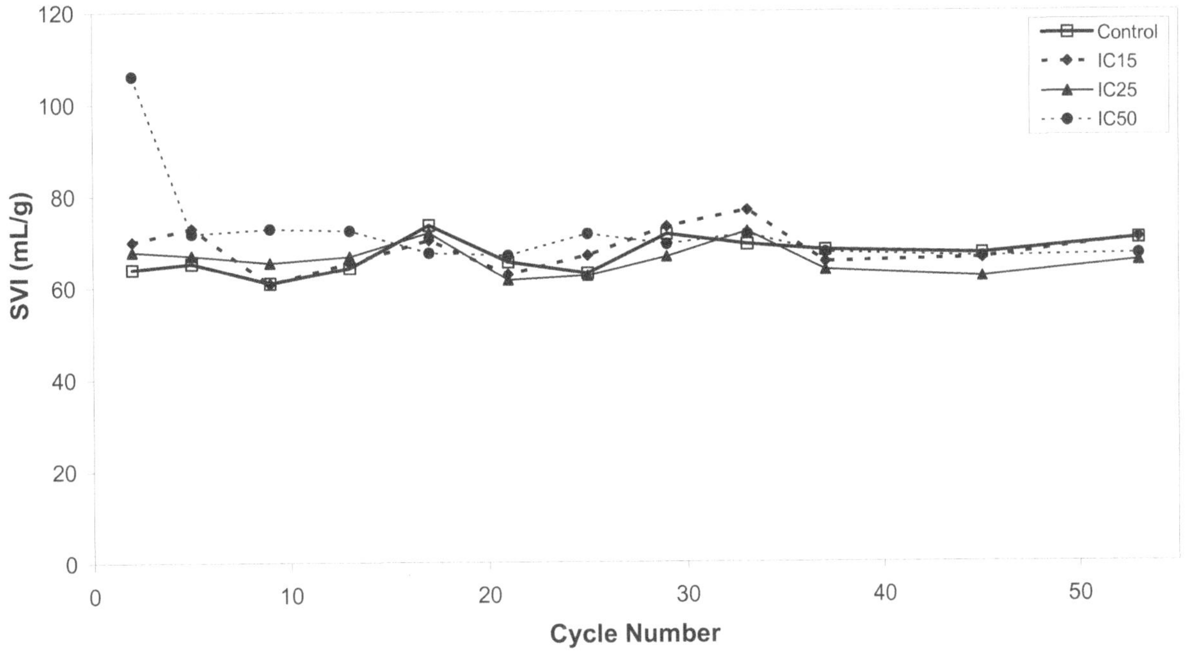

Figure 6-7. SVI for the 10-Day SRT Reactors Exposed to Cyanide Shock.

6.2.4.2 Mixed Liquor Solids Trends

There was no significant difference between MLSS and MLVSS concentrations due to chemical shock in either the 2-day or 10-day SRT systems. This is consistent with the fact that inhibition in the 10-day SRT shocked reactors was predominantly directed toward the nitrifying bacteria, which make up a small percentage of the biomass. A summary of mixed liquor concentrations is given in Table 6-3. Throughout the 10-day SRT zinc-cyanide complex source-effect experiment, the mixed liquor concentrations in all reactors decreased consistently over time (data in Appendix E). The decrease continued until a quasi-steady state was reached at around 3,000 mg/L. This effect was seen in other source-effect experiments discussed in this report and is believed to be due to higher suspended solids content in pilot plant influent versus the lab-reactor influent, which led to a higher MLSS concentration of the biomass that was brought into the laboratory at the start of the experiment. The MLSS measurements in the 2-day SRT system remained relative constant at around 1,075 mg/L (data in Appendix E).

6.2.4.3 Kinetic Indicators of Inhibition

SOUR results showed that heterotrophic inhibition occurred during zinc-cyanide shock in the IC_{50} reactor containing the non-nitrifying 2-day SRT biomass, as shown in Figure 6-8. This response is consistent with the sCOD effluent results shown previously for this reactor. The reactor remained inhibited relative to the control reactor (a=0.05) until the 13th cycle. The magnitude of the inhibition effect (78%) measured six hours after the initial shock was higher than the expected 50% measured by the batch IC_{XX} assay, which was conducted over about 10 minutes. These results suggest that a greater degree of inhibition occurred due to longer exposure times (six hours as opposed to 10 minutes), and indicate that the zinc-cyanide complex can have a significant effect on the respiration rate of heterotrophic biomass.

Table 6-3. Average Mixed Liquor Concentrations in 2- and 10-Day SRT Reactors Exposed to Cyanide.

	MLSS (mg/L)	MLVSS (mg/L)	MLVSS:MLSS Ratio
2-day SRT reactors			
Control	1,128 ± 182	792 ± 90	0.71 ± 0.05
IC_{15}	1,084 ± 180	765 ± 90	0.71 ± 0.05
IC_{25}	1,086 ± 171	771 ± 86	0.72 ± 0.05
IC_{50}	1,000 ± 255	705 ± 168	0.71 ± 0.05
10-day SRT reactors			
Control	3,966 ± 669	2,612 ± 423	0.66 ± 0.02
IC_{15}	3,964 ± 809	2,607 ± 512	0.66 ± 0.02
IC_{25}	4,071 ± 763	2,666 ± 468	0.66 ± 0.01
IC_{50}	4,140 ± 812	2,708 ± 503	0.66 ± 0.01

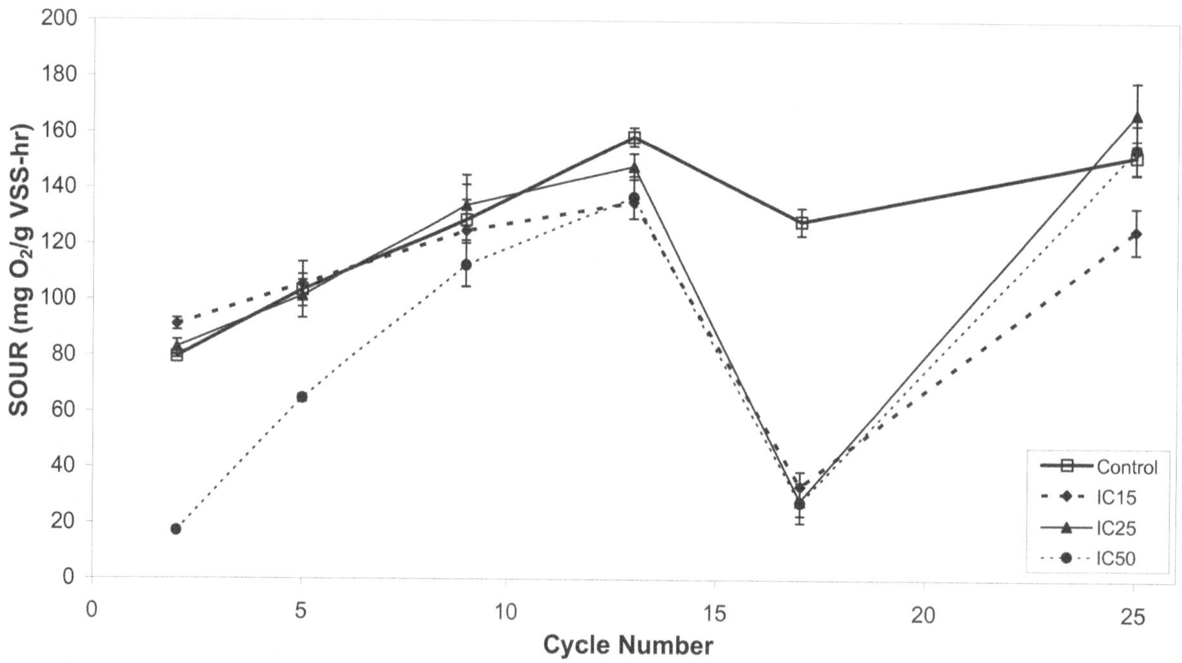

Figure 6-8. SOURs for the 2-Day SRT Reactors Exposed to Cyanide Shock.

SOUR results for the first cycle of the 10-day SRT reactors show that, with the exception of the IC_{15} reactor, cyanide inhibited respiration and the degree of inhibition correlated with an increase in contaminant dose (Figure 6-9). Recovery of the IC_{25} reactor occurred by cycle 10 and recovery of the IC_{50} reactor occurred by cycle 30 ($\alpha = 0.05$). Since sCOD removal was only moderately inhibited in this experiment relative to the more severe effect seen with the 2-day SRT biomass, it appears that nitrification inhibition contributed significantly to the observed SOUR results. Also, the degree of inhibition in the IC_{15}, IC_{25} and IC_{50} reactors 6 hours after inhibition was approximately –3%, 27%, and 65%, respectively. Therefore, the magnitude of the inhibition effect in the IC_{25} and IC_{50} reactors was similar to the extent of inhibition expected based on the 10 minute batch IC_{XX} assay.

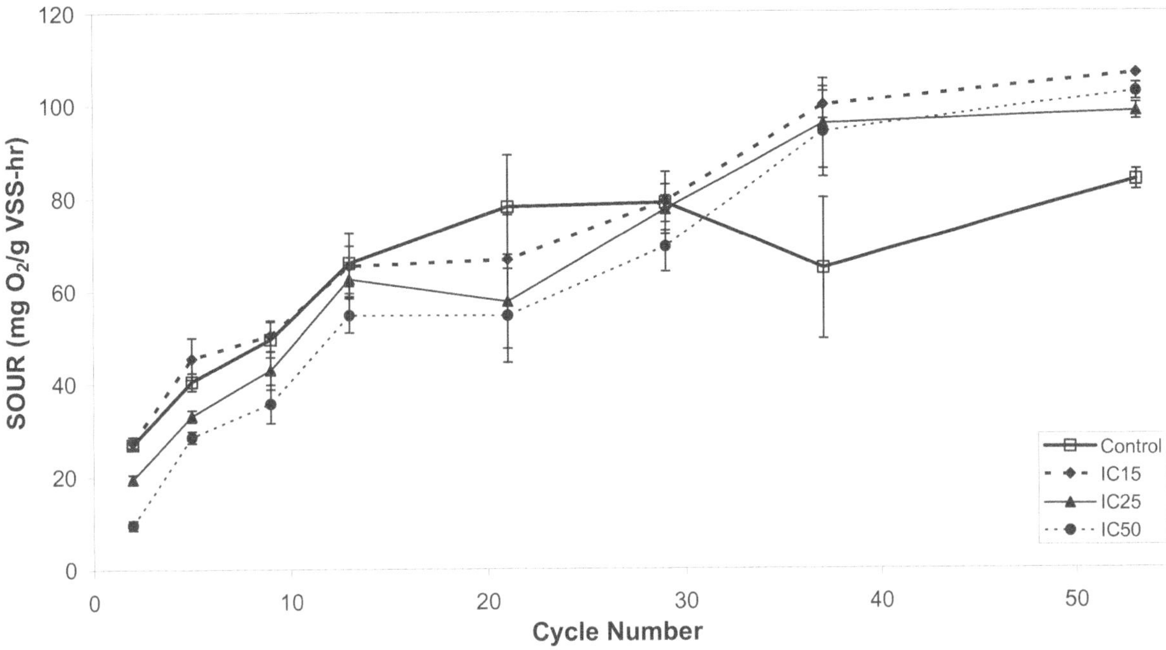

Figure 6-9. SOURs for the 10-Day SRT Reactors Exposed to Cyanide Shock.

Figure 6-10 shows that the SNGRs for the 10-day SRT shocked reactors were significantly (α=0.05) less than the control for the first 30 cycles. It is also shown that the SNGR response corresponded with the level of zinc-cyanide addition. All shocked reactors recovered by cycle 37, whereas recovery of nitrification (according to effluent nitrogen species measurements) for all shocked reactors occurred around cycle 10 (Figure 6-5). Since the SNGR reflects biomass rate of metabolism, it is a more sensitive indicator of nitrification inhibition than effluent performance. The effluent nitrogen species reflect performance after 5 hours of reaction, which may be occurring slower due to inhibition but would not be detected if the effluent decreased to control levels within the reaction time frame. Overall, the SNRG assay results correspond with other water quality measurements that indicated nitrification inhibition, as discussed earlier.

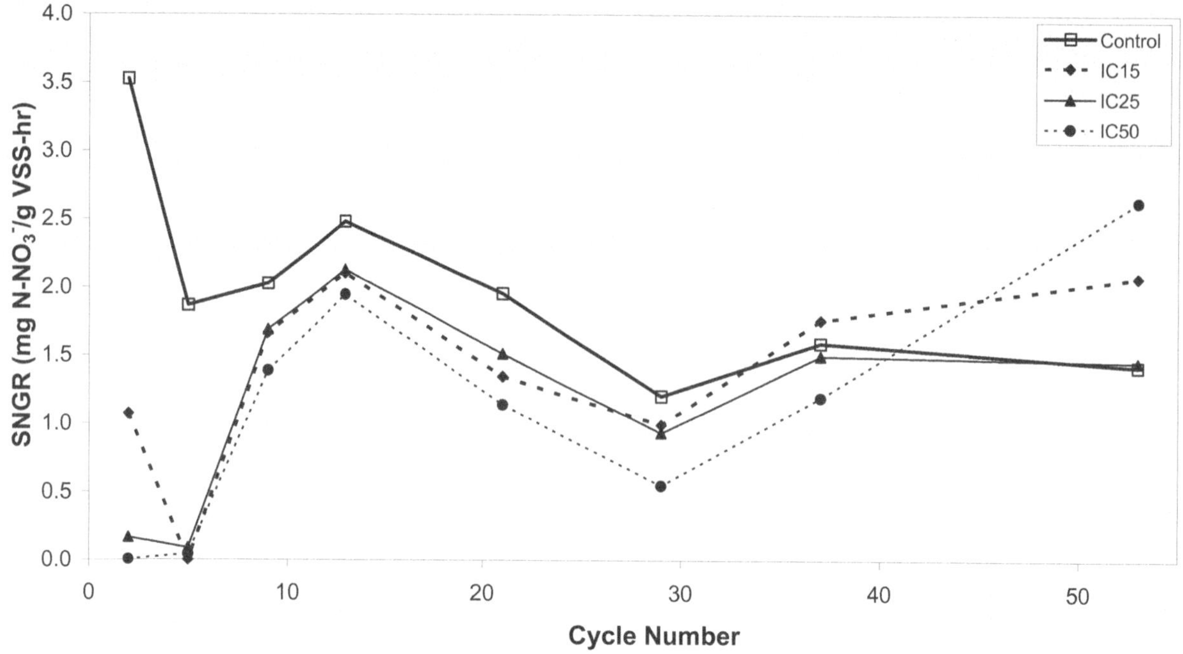

Figure 6-10. SNGRs for the 10-Day SRT Reactors Exposed to Cyanide Shock.

6.3 Discussion

The results from this experiment suggest that a shock load of zinc-cyanide complex impacted COD removal efficiency, settleability and nitrification to varying degrees but did not appear to cause deflocculation or impact dewaterability. Furthermore, the results suggested that recovery was relatively rapid in terms of effluent performance, even though the biomass remained inhibited for up to 30 reaction cycles.

Deflocculation of the biomass was not observed when sublethal shock inputs of zinc-cyanide complex occurred. Although several reactors showed elevated effluent TSS levels relative to the control reactors in both systems, no discernable pattern for these deviations could be seen. Therefore, the presence of cyanide could not be attributed to the noted differences in effluent TSS, suggesting that cyanide does not cause deflocculation at the concentrations that were tested. In addition, measured soluble potassium levels for all reactors revealed no discernable patterns, suggesting that cyanide does not activate the GGKE mechanism. These results are consistent with previous studies on eukaryotic mitochondria, which showed that cyanide did not elicit a potassium efflux response (Zhang et al., 2001).

The SOUR data coupled with the SNGR and reactor effluent data revealed that the zinc-cyanide complex inhibited both heterotrophic and nitrifying metabolisms, given the different responses observed between the 2-day and 10-day SRT experiments. The 2-day SRT IC_{50} reactor showed a low SOUR that took 1.6xSRT to recover. The spike in effluent sCOD at the same time and the lack of detectable nitrification based on effluent nitrate and nitrite measurements suggests that this inhibition effect was directed toward heterotrophic bacteria. In contrast, the 10-day SRT biomass experimental systems were nitrifying and showed significant nitrification inhibition at the higher shock doses. The SNGR and SOUR data showed, however, that the inhibition lasted longer than could be inferred from effluent nitrate and nitrite data. For example, the effluent nitrate data suggested recovery by cycle ten for the IC_{50}-shocked reactor, whereas SOUR and SNGR data both suggest that it actually took 30 cycles. Furthermore, since no anomalies in effluent sCOD were observed in the 10-day SRT systems, the SOUR inhibition was most likely primarily due to nitrification inhibition. The mechanism for the respiratory inhibition noted here is likely direct inhibition of the cytochrome oxidases required for cells to utilize oxygen, and are used during respiration by heterotrophic bacteria and both AOBs and NOBs. As discussed in the literature review, cyanide binds to cytochromes and prevents them from using oxygen as the terminal electron accepter; a phenomenon termed "cell suffocation" (Solomonson, 1981; Patel et al., 1992). Given the universal nature of this inhibition, it is quite reasonable that it explains both the sCOD removal and nitrification inhibitions observed.

A loss of COD removal efficiency was noted during the first cycle of this experiment in the stressed reactors for both the 10- and 2-day SRT systems, but the response was much more severe for the IC_{50}-shocked 2-day SRT system. When examining the cyanide shock loads on a mass cyanide to mass MLVSS basis, the 2-day SRT IC_{50} reactor received the highest shock load among all cyanide source-effect experiments, but not the highest concentration (see Table 6-1). This result suggests that using mass cyanide load per mass MLVSS may be a better predictor of impact on sCOD removal efficiency for weak acid cyanide shock loads. Although the exact mechanism for the loss of COD removal is not known, it probably relates to interferences with enzymes responsible for carbon utilization and respiration, as discussed previously. Furthermore, the mechanism of inhibition appears to be rapidly reversible.

The cyanide-induced nitrification inhibition appeared to strongly impact the AOBs and may have impacted the NOBs as well, although the latter is less clear. The AOBs were inhibited first, as shown by the sudden increase in ammonia in the IC_{50} reactor. This inhibition recovered rather quickly, however (0.25 x SRT). The nitrite data spiked but recovered quickly also, so it is not clear if the spikes observed were due to the sudden influx of nitrite or inhibition (or some combination of both). Interestingly, when examining a theoretical dilution curve for the cyanide concentrations added to each of the shock reactors (assuming no reaction or absorption), inferences can be made regarding the free cyanide concentrations required to inhibit nitrification. For the IC_{50} reactor, ammonia oxidation was inhibited for the first day of the experiment (four cycles). At cycle four, when ammonia oxidation resumed, theoretical levels of free cyanide in the IC_{50} reactor were approximately 3 mg/L as CN^- (Figure 6-11). Since no other reactor experienced this concentration of cyanide and also showed much lower degrees of inhibition, it may be that dilution was a significant factor in relieving the system of its inhibition (as opposed to adaptation over time). Overall, it appears that free cyanide concentrations above 3 mg/L are detrimental to ammonia oxidation.

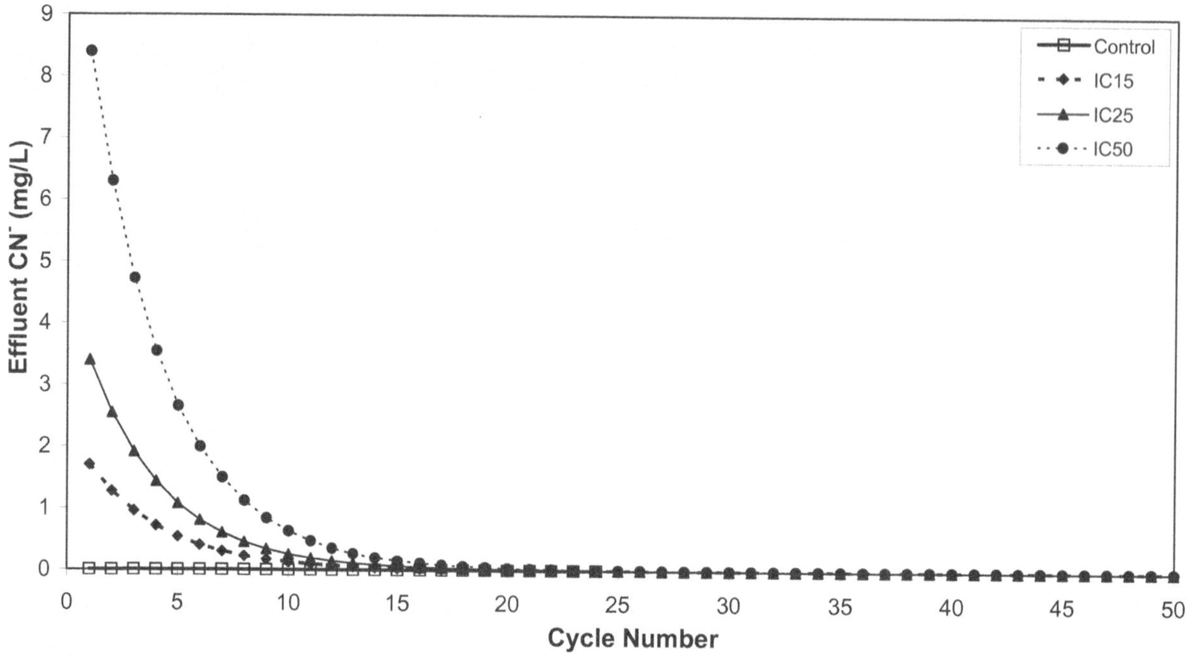

Figure 6-11. Theoretical Dilution of Free Cyanide for the 10-Day SRT Reactors Exposed to Zinc-Cyanide Complex Shock.

6.4 Conclusions

- Zinc-cyanide complex inhibited both heterotrophic and autotrophic nitrifying bacteria. The nitrifying bacteria were affected at all concentrations tested and the degree of inhibition corresponded with cyanide dose. The heterotrophic bacteria responded most severely to the IC_{50} concentration imposed upon the 2-day SRT biomass, which received the highest mass of cyanide per g MLVSS of all experiments. Because both groups of bacteria rely upon cytochrome oxidases to respire and because these biomolecules are known to be inhibited by cyanide, it is assumed the basis for the inhibition observed is primarily related to inhibition of these oxidases.
- The inhibition patterns were, in all cases, reversible. From the standpoint of effluent performance, the recovery was more rapid than for other toxins tested in this study.
- Zinc-cyanide complex definitely inhibited AOBs, which are the bacteria responsible for the first step in nitrification. The degree to which the zinc-cyanide shocks affected NOBs is unclear from this study and deserves further attention.

CHAPTER 7.0

1-CHLORO-2,4-DINITROBENZENE SHOCK EXPERIMENTS

7.1 Literature Review

1-Chloro-2,4-dinitrobenzene (CDNB) is an uncharged organic electrophile with numerous industrial and pharmaceutical applications. Electrophilic chemicals are those that act as oxidizing agents by accepting electrons from other compounds. CDNB is considered a high volume chemical, with production in the United States exceeding a million pounds per year (EPA, 2004). It is a prevalent industrial chemical with the major use being in the manufacture of azo and sulfur dyes (Sax and Lewis, 1987). It is also used as a reagent for the detection and determination of nicotinic acid, nicotinamide, and other pyridine compounds (Budavari, 1989); as an algicide in coolant water in air conditioning systems (Clayton and Clayton, 1982); and in the manufacture of fungicides, rubber chemicals and explosives (Environmental Defense, 2004). CDNB also has many applications as a therapeutic drug. The major pharmaceutical uses are as an immunostimulant in the treatment of leprosy and some forms of cancer, and as a treatment for alopecia and warts (Reynolds and Prasad, 1982). Because of its prevalence in industrial and medical applications, CDNB has been detected at low levels in the environment and in the discharges from azo and sulfur dye manufacturing operations. The environmental prevalence of this compound makes it an important one to study in relation to its effects on different environmental systems, specifically the wastewater treatment process.

Despite the apparent widespread use of CDNB and other uncharged organic electrophilic chemicals in industry, very few studies have been performed that look at its effects on activated sludge treatment. One study showed that CDNB causes a rapid deflocculation in activated sludge (Bott and Love, 2002). The same was seen for N-ethylmaleimide (NEM), which is another organic chemical with similar electrophilic properties. It was proposed that this deflocculation event was caused by a rapid efflux of potassium which reduced floc strength by increasing the intrafloc monovalent:divalent (M:D) cation ratio (Bott and Love, 2002). This ratio has been previously found to be an important factor in controlling the strength of activated sludge flocs (Higgins and Novak, 1997a; Higgins and Novak, 1997b; Novak et al., 1998) and a rapid increase in this ratio likely weakened the floc strength, which resulted in release of small non-settleable particles from the floc surface. A comprehensive literature search revealed that no other treatment process effects have been examined for CDNB. However, Weber and Sherrard (1980) studied cadmium, a charged electrophilic metal, and showed a slight decrease in COD removal efficiency and a decrease in nitrification rate as the influent cadmium concentration increased. These observations are consistent with those presented in this report in Chapter 3.0. Collectively, these results suggest the possibility that other electrophilic chemicals, such as CDNB, may induce the same upset effects.

After a thorough search of current and past literature on CDNB and other uncharged organic electrophiles such as NEM and methylglyoxal, it was found that much of the work that examines the effects of this class of chemicals has been performed using pure cultures of Gram-negative bacteria, like *Escherichia coli*. Although *E. coli* is not a prevalent organism in activated sludge cultures, it is Gram-negative and most organisms that have been phylogenetically characterized in activated sludge systems are Gram-negative (Wagner and Loy, 2002). Therefore, it is expected that highly conserved functions found in Gram-negative bacteria will occur in activated sludge communities as well; consequently, results from pure culture studies may be used to draw inferences about effects on the activated sludge process.

From pure culture testing, organic electrophiles have been found to damage proteins and DNA (Apontoweil and Berends, 1975a; Apontoweil and Berends, 1975b; Ferguson et al., 1995). It has been found that the mode of action of organic electrophilic agents is against the thiol bonds found in proteins (Ferguson et al., 1995; Ferguson et al., 1997; McLaggan et al., 2000). Specifically, it has been shown that methylglyoxal reacts with the side chains of the amino acids arginine, lysine and cystine. Methylglyoxal has also been found to react with DNA at the bases guanine, adenine and cytosine (Ferguson and Booth, 1998). It has been speculated that, due to its chemical structure, CDNB acts on proteins and DNA in the same manner (Ness et al., 1997). Supporting its action on DNA, Summer and Göggelmann (1980) showed that CDNB and 1-fluoro-2,4-dinitrobenzene (FDNB), a chemical homologue, are mutagens in *Salmonella typhimurium*, a Gram-negative model organism. It is hypothesized that the mutagenicity of these compounds is due to the action of the halogen group and the degree of mutagenicity increases with increased electronegativity of the halogen (Gupta et al., 1997). From this information, it is important to note that CDNB is a known mutagen and likely damages proteins and DNA at the molecular level.

From the literature, it appears that an important cell protection mechanism induced by CDNB and other organic electrophiles is a glutathione gated potassium efflux (GGKE) system. Glutathione (GSH) is a tripeptide that occurs in both eukaryotic and prokaryotic cells and appears to be a major cellular defense against electrophilic compounds (Apontoweil and Berends, 1975a; Ferguson and Booth, 1998; Ferguson et al., 1998; He et al., 1998; Summer and Göggelmann, 1980). Glutathione contains a reduced thiol bond which sacrifices itself to prevent electrophiles from acting upon proteins and DNA in order to prevent damage (Bott and Love, 2002). The glutathione is then either oxidized by the electrophile (oxidized glutathione: GSSG) or forms a glutathione-S-conjugate with the compound (glutathione conjugate: GSX). In the case of CDNB, it forms a conjugate called 2,4-dinitrobenzyl-S-glutathione (DNBSG), which is likely exported from the cell and degraded by a transpeptidase (Ness et al., 1997; Vuilleumier, 1997). Booth and coworkers have shown on numerous occasions that in addition to aiding in detoxification, the oxidized glutathione and glutathione-S-conjugates that form from reactions with electrophiles activate KefB and KefC potassium efflux channels in *E. coli* (Ferguson et al., 1995; Ferguson and Booth, 1998; Ferguson et al., 1998). Booth's group also showed that the efflux of potassium from cells is coupled with an influx of H^+, leading to cytoplasmic acidification and an increased tolerance to CDNB and other electrophilic compounds (Ferguson et al. 1997; Ferguson et al., 1998; Ness et al., 1997). It has been proposed that the increased local potassium levels seen in our labs when activated sludge is exposed to electrophilic chemicals is due to the GGKE protection mechanism previously described (Bott and Love, 2002).

Other electrophile protection mechanisms have been seen in bacteria that more specifically protect the DNA of the organism. One of these mechanisms is a protein known as Dps, which binds to DNA non-specifically. This protein has been shown by Ferguson and co-workers (1998) to increase cell viability in the presence of NEM, although whether this is due to protection afforded to the DNA through physical binding or activation of genes that control DNA repair mechanisms is not known. Tse-Dinh (2000) found that another enzyme known as topoisomerase I is found to have a significant protective effect against electrophile attack, specifically against hydrogen peroxide and NEM. This enzyme has two possible actions that would afford protection of the DNA. One is regulation of the supercoiling of DNA and the other is the need for topoisomerase I to prevent R-loop formation at sites with high transcription rates while the cell is mounting a defense against the stress (Tse-Dinh, 2000). The last protection mechanism found in the literature is the activity of catalases that protect cells against peroxides. Vattanaviboon et al. (2001) found that cells exposed to NEM induce a 10-fold increase in each of the different catalase enzyme levels and that this afforded the cells an additional level of protection from the electrophilic agent.

7.2 Results

7.2.1 Inhibitory Concentrations (IC_{XX})

The impact of varying CDNB concentrations on the respiratory activity of the 2- and 10-day SRT biomass is shown in Table 7-1. This table also shows the concentrations added to each of the reactors at the beginning of the first cycle for the source-effect experiment.

Table 7-1. IC_{15}, IC_{25} and IC_{50} Values for CDNB in 2- and 10-Day SRT Mixed Liquors.

	2-Day SRT		10-Day SRT	
	mg/L	mg CDNB/g MLVSS	mg/L	mg CDNB/g MLVSS
IC_{15}	1.6	2.5	8.6	5.1
IC_{25}	4.1	6.3	14	8.4
IC_{50}	43	65	50	29

When looking only at concentrations, the IC_{XX} concentrations for the 10-day biomass were found to be higher (i.e., less sensitive) than for the 2-day SRT biomass. However, when concentrations are normalized to a mass contaminant per mass reactor solids basis, the 10-day specific IC_{XX} concentrations were found to be substantially lower for the IC_{50} value. When these experiments were conducted, only the 10-day SRT system was nitrifying. It may be that the relative impact of SOUR inhibition in the 10-day SRT system was greater per unit biomass due to a larger nitrifier population. Although no literature has been found showing the impact of organic electrophiles on nitrifying populations, nitrifiers have been found to be more sensitive to shock loads of electrophilic metal toxins than heterotrophs (Bagby and Sherrard, 1981). This same pattern was reported for cadmium in Chapter 3.0 of this report. These results suggest that the inhibition concentration for nitrifiers should be much lower than that for heterotrophs. Since a significant portion of the oxygen demand in the 10-day system is due to nitrifiers, this may explain why the IC_{50} for the nitrifying 10-day SRT culture is lower on a mg/g basis than for the 2-day SRT biomass.

7.2.2 Influent Characteristics

Average influent characteristics over the course of the CDNB source-effect experiments are given in Table 7-2. The influent quality was relatively constant over the time of each experiment with the exception of the suspended solids, which varied somewhat as seen by the standard deviation. Both experiments occurred after influent supplementation was initiated. Accordingly, the influent composition revealed that the wastewater used as the feed for the source-effect experiments with CDNB was medium to strong in strength (Metcalf and Eddy, 2003).

Table 7-2. Characteristics of Wastewater Used During CDNB Source-Effect Experiments.

Characteristic	2-Day SRT Biomass Average ± std dev	10-Day SRT Biomass Average ± std dev
Alkalinity (mg/L as $CaCO_3$)	584 ± 8	271 ± 23
Ammonia (mg/L as N)	17.6 ± 3.0	30.8 ± 4.6
COD, soluble (mg/L)	835 ± 62	558 ± 93
COD, total (mg/L)	1023 ± 91	744 ± 138
pH	7.6 ± 0.2	7.3 ± 0.3
TSS (mg/L)	241 ± 85	110 ± 40
VSS (mg/L)	105 ± 38	87 ± 30

7.2.3 Effluent Characteristics

7.2.3.1 Soluble COD

Effluent sCOD data show that removal efficiency was impacted by CDNB in both the 2- and 10-day SRT systems, as seen in Figures 7-1 and 7-2, respectively. The 2-day SRT effluent sCOD removal efficiency was negatively impacted by CDNB (Figure 7-1). Soluble COD removals in the control, IC_{15} and IC_{25} reactors remained similar for the first three cycles, before the latter two began to diverge from the control reactor levels. The elevated levels for these two reactors, though significant, were very slight (less than 20 mg/L sCOD) and no distinct pattern was observed during the experiment. The IC_{50} reactor showed elevated effluent sCOD with respect to the control reactor for the first eight cycles (2 days) before recovering to control levels by cycle 10. The elevated levels in the IC_{50} experiment peaked at cycle two at a concentration well above the sCOD equivalent of the CDNB concentration added before decreasing back to control levels and recovering from the shock event. After cycle 12, all the effluent sCOD concentration in the shocked reactors and the control were not statistically different (α=0.05).

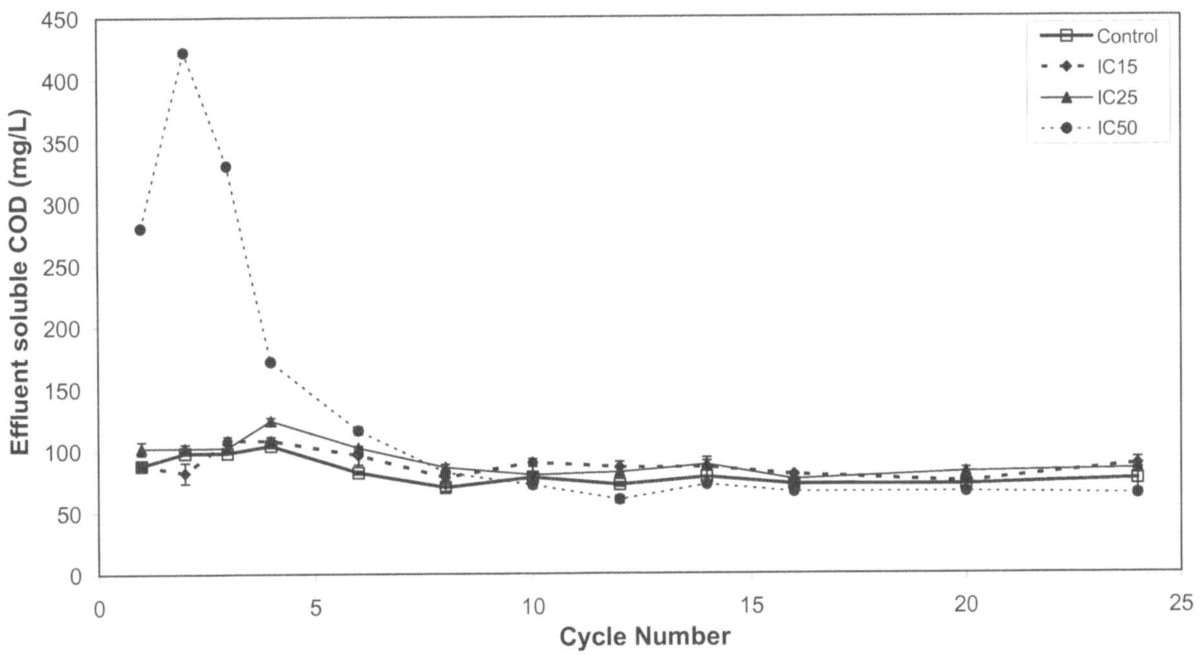

Figure 7-1. Effluent Soluble COD for the 2-Day SRT Reactors Exposed to CDNB Shock.

The effluent sCOD in the 10-day SRT systems showed similar patterns to that observed in the 2-day SRT systems. Effluent sCOD remained elevated in the IC_{15} and IC_{25} stressed reactors relative to the control until cycle 20 and 24, respectively (at the α=0.05 level). For the IC_{50} stressed reactor, the elevated sCOD levels peaked at cycle two and remained elevated above the control until the 32nd cycle (α=0.05). After the 32nd cycle, the performance of all reactors was similar. Inhibition of sCOD removal efficiency was highest for the IC_{50} shocked reactor, but the inhibition of the IC_{15} and IC_{25} reactors did not significantly differ (α=0.05) from each other.

Soluble CDNB concentrations were measured in both the 2- and 10-day SRT biomass for the duration of the experiment (data in Appendix F for 10-day SRT reactors). For the 2-day SRT biomass, no detectable level of soluble CDNB was found for any of the cycles. For the 10-day SRT biomass, soluble CDNB concentrations were detectable in the IC_{50} and IC_{25} reactors through the fourth and first cycles, respectively. CDNB was not detectable in the IC_{15} or control reactors. In addition, CDNB appeared to exit the system faster than it would theoretically be diluted out (shown in appendix F). These observations are consistent with the notion that CDNB complexes with glutathione in cellular cytoplasms (Ness et al., 1997; Vuilleumier, 1997). If it forms a complex, the chromatographic retention time or the ideal detection wavelength for measurement via HPLC would change and it would be undetected.

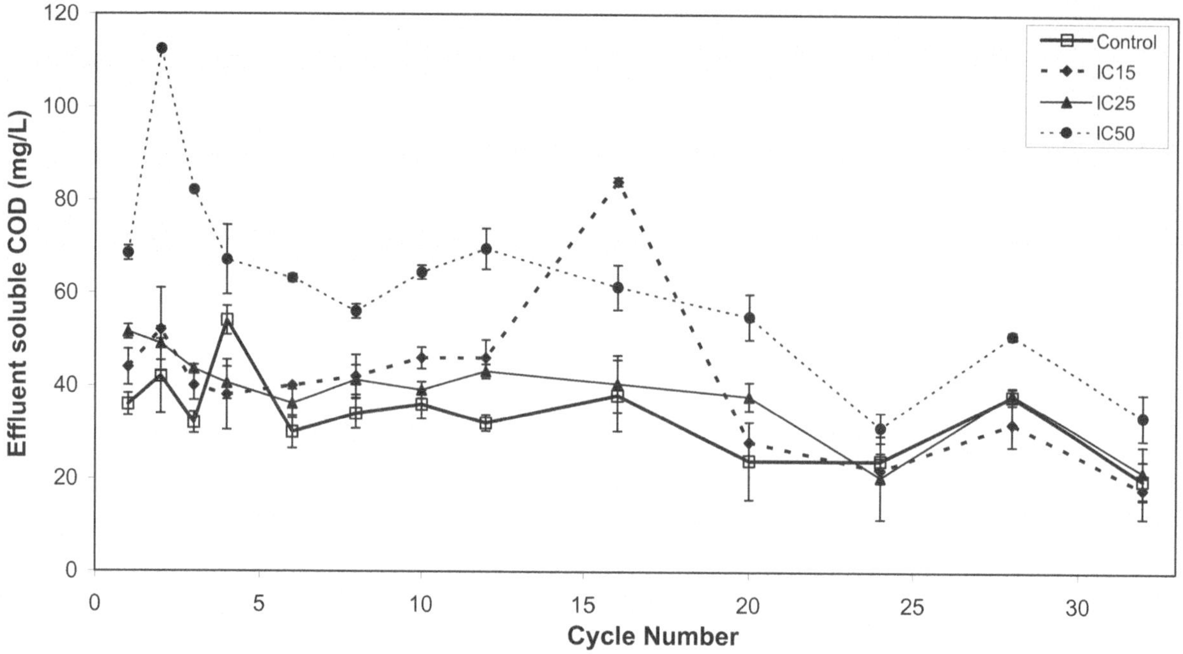

Figure 7-2. Effluent Soluble COD for the 10-Day SRT Reactors Exposed to CDNB Shock. Data Through Cycle 32 Are Shown.

7.2.3.2 TSS and Potassium Concentrations

Effluent TSS was significantly impacted by CDNB shock, as expected based on previous studies (Bott and Love, 2001). In all cases, the increases in effluent TSS were likely due to high turbidity caused by deflocculation. Deflocculation was observed during the first cycle after shock in both the 2- and 10-day SRT reactors and for all CDNB concentrations, resulting in average effluent TSS concentration increases that were 33 to 57% higher than the values observed for the control reactors (Figures 7-3 and 7-4). The effluent TSS in the 2-day SRT shocked reactors increased immediately after shock for only the highest CDNB concentration tested, although all shocked reactors were visually more turbid when compared with the control. The effluent TSS concentration for the IC_{50} reactor was significantly higher than the control and the other reactors during the first three cycles ($\alpha = 0.05$), after which the levels returned to those of the control (Figure 7-3). Despite the apparent increase in turbidity for the IC_{15} and IC_{25} reactors, the effluent TSS was not significantly different from the control when Dunnett's test was applied. The reasons for the increase in effluent TSS seen for the control reactor after cycle 8 and the IC_{50} shocked reactor after cycle 16 are unknown.

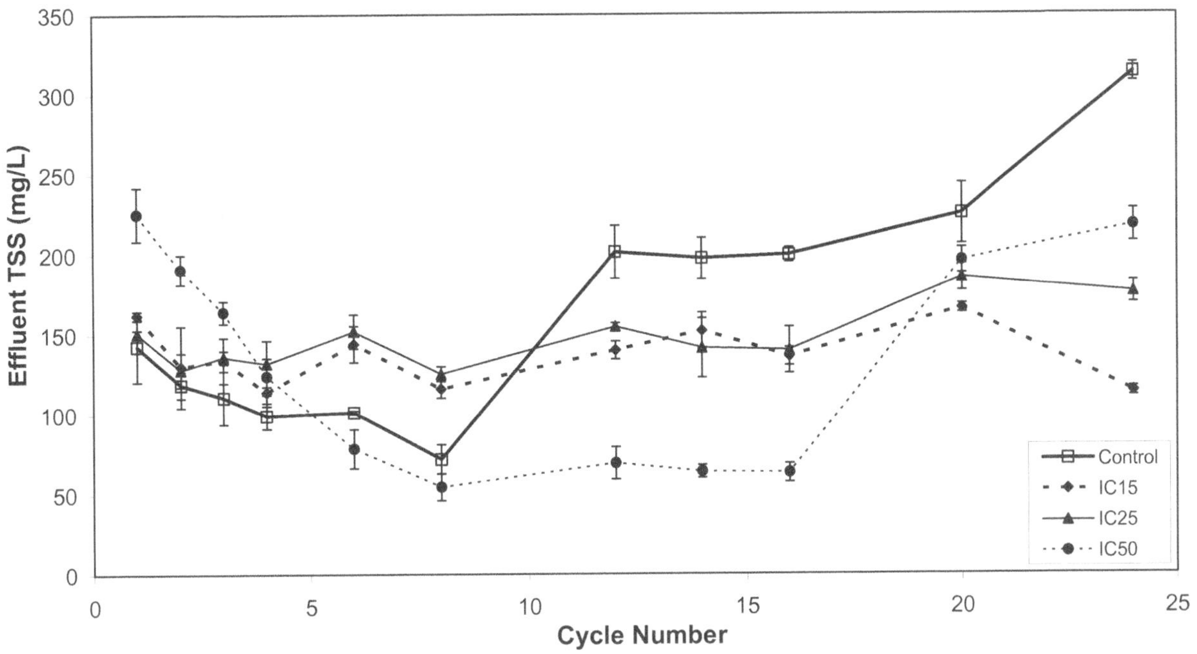

Figure 7-3. Effluent TSS for the 2-Day SRT Reactors Exposed to CDNB Shock.

For the 10-day SRT system, increased concentrations of CDNB corresponded with increased effluent turbidity during the first cycle (at the $\alpha = 0.05$ level) for the IC_{25} and IC_{50} reactors, which remained elevated relative to control levels until cycles 8 and 12, respectively ($\alpha = 0.05$). The IC_{15} reactor showed no significant difference from the control reactor (α=0.05). Elevated effluent TSS levels seen in the control reactor during cycles 16 and 20 were due to a filamentous bulking problem that manifested in the control reactor prior to the 5th cycle and cleared up by cycle 25. Although not shown, the IC_{50} shocked reactor experienced a filamentous bulking event that caused the effluent TSS levels to rise starting with cycle 68. The increase in filamentous content in the mixed liquor started around cycle 40 and continually worsened until the activated sludge blanket would not settle below the decant tube used to collect effluent samples. The same problem began to occur at cycle 100 for the IC_{25} reactor. A reason for the bulking observed in these reactors is unknown.

Average effluent potassium concentrations significantly increased during the first cycle of the 2- and 10-day SRT reactors (α=0.05) (shown in Appendix F). For the 2-day SRT reactors, the effluent potassium concentrations in the IC_{15} and IC_{50} reactors were determined to be significantly higher than those of the control (α=0.05) (shown in Appendix F). The change in the liquid phase potassium concentrations was around 4.5 mg/L for the IC_{50}-stressed biomass and 1 mg/L for the IC_{15}-stressed reactors. An apparent return of effluent soluble potassium levels to those of the control occurred by the fourth cycle (one day) for both the IC_{15} and IC_{50} reactors. For the 10-day SRT system, the change in liquid phase potassium concentrations was around 9, 7, and 4 mg/L for the IC_{50}-, IC_{25}-, and IC_{15}-stressed reactors, respectively. Effluent potassium concentrations returned to control levels for all reactors by the fourth cycle (one day). The smaller level of potassium efflux is consistent with the fact that the biomass concentration in the 2-day SRT bioreactors was considerably smaller than in the 10-day SRT system.

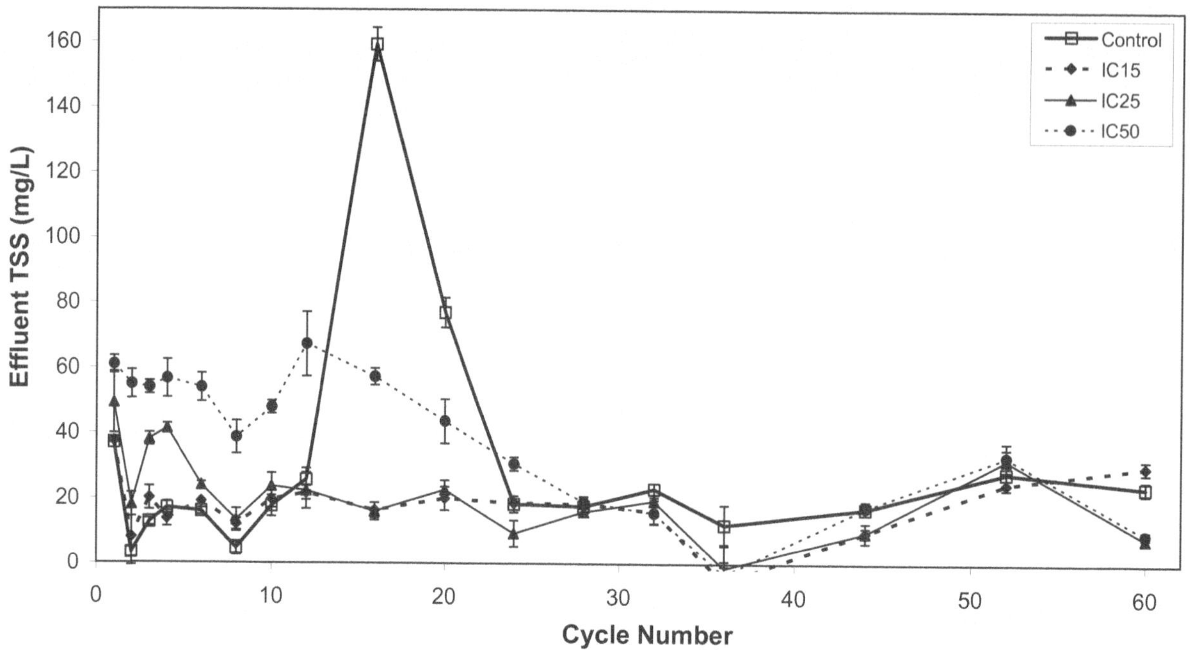

Figure 7-4. Effluent TSS for the 10-Day SRT Reactors Exposed to CDNB Shock. Data Through Cycle 60 Are Shown.

This rapid, yet brief increase in potassium in response to CDNB is consistent with previous observations seen by members of the research team (Bott and Love, 2002). As stated previously for cadmium, electrophilic chemicals like CDNB and cadmium are believed to activate a glutathione-gated potassium efflux protective stress response in activated sludge (Bott and Love, 2002), and the increases in potassium concentration observed here are consistent with that theory. Furthermore, these authors have previously shown that this increase is significant and rapid after an electrophilic stress event, and they hypothesized that it is high intrafloc concentrations that lead to such severe deflocculation in response to electrophilic contaminants (Bott and Love, 2004). The effluent TSS remained elevated for several days for the 10-day SRT IC_{50} stressed reactor, despite the rapid return of bulk liquid potassium levels to those seen in the control within one day. This observation is consistent with the notion that deflocculated sludges need to "regrow" into flocculant sludges after cation-induced deflocculation events (Novak et al., 1999).

7.2.3.3 Nitrification Inhibition

The impact of CDNB shock was, perhaps, most significantly apparent with nitrification performance. The effluent ammonia, nitrite and nitrate data for the 10-day SRT systems are shown in Figures 7-5 through 7-7. In these reactors, which were nitrifying prior to CDNB addition, effluent ammonia values increased with increasing CDNB concentrations beginning with the first cycle (Figure 7-5), indicating that CDNB inhibited AOBs. Ammonia oxidation recovered in the 10-day SRT reactors after 24 and 68 cycles for the IC_{25} and IC_{50} stressed systems, respectively. Recovery from the loss of ammonia oxidation took 2.5 times longer than recovery from the deflocculation event. The IC_{15} reactor showed no inhibition based on effluent ammonia concentrations. Variations in nitrite concentration corresponded with apparent recovery in ammonia removal for the IC_{25} and IC_{50} reactors (Figure 7-6), indicating that nitrification inhibition was also targeted against the NOB. In addition, a significant increase in nitrite in the

IC_{15} shocked reactor indicated that NOB were, in fact, inhibited in that system as well. The lag phases in the buildup of nitrite in the IC_{25} and IC_{50} stressed reactors (12 and 52 cycles, respectively) provide further evidence that the AOB were inhibited at the higher concentrations. They may have been inhibited in the IC_{15} reactor, but this would not be detected through effluent ammonia concentration if the duration of the reaction phase was sufficient for ammonia oxidation to occur completely, even if at a slower nitrification rate. Effluent nitrite levels in the IC_{15}, IC_{25} and IC_{50} reactors returned to control levels by cycles 20, 32 and 84, respectively, suggesting that the NOBs recovered. Trends seen through increases in effluent nitrate concentrations correlate with trends for nitrite (Figure 7-7), and indicate that nitrification recovery occurred by cycle 68 for the IC15-shocked reactor, and cycle 76 for the IC25- and IC50-shocked reactors. The latter two reactors diverged again from the control and IC15-shocked reactor at the end of the experiment, and the reason for this is unknown.

The 2-day pilot plant reactor was being operated to prevent nitrification and was not nitrifying the day before the experiment began. Effluent ammonia and nitrate results confirm that no significant level of nitrification was occurring during the 2-day SRT shock experiment (shown in Appendix F).

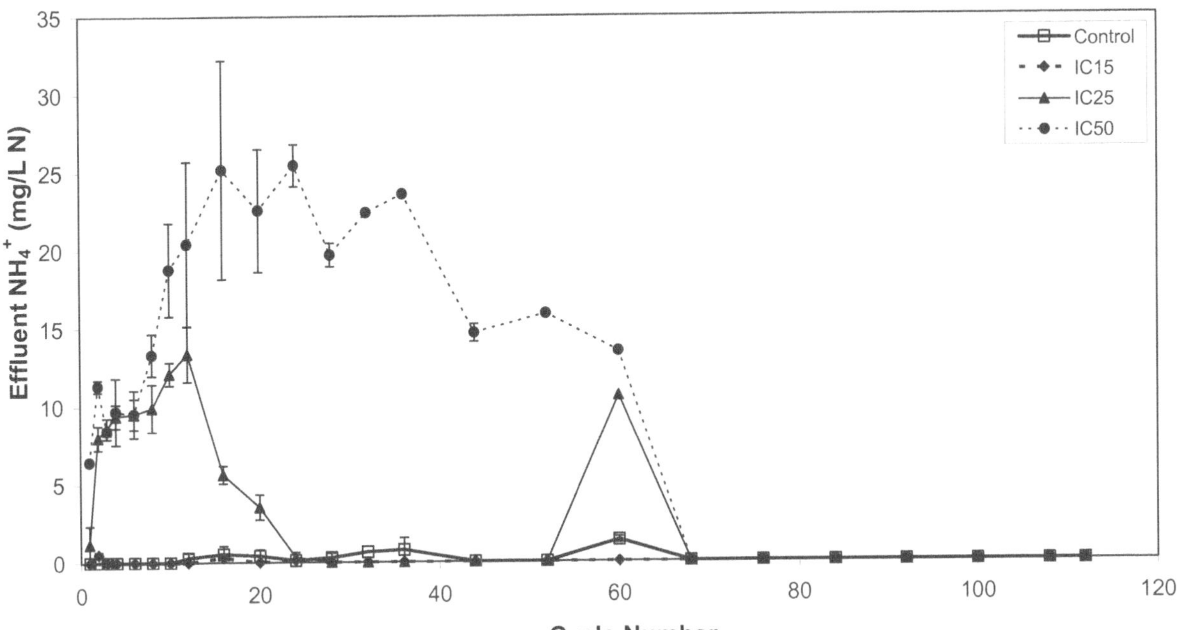

Figure 7-5. Effluent Ammonia-Nitrogen for the 10-Day SRT Reactors Exposed to CDNB Shock.

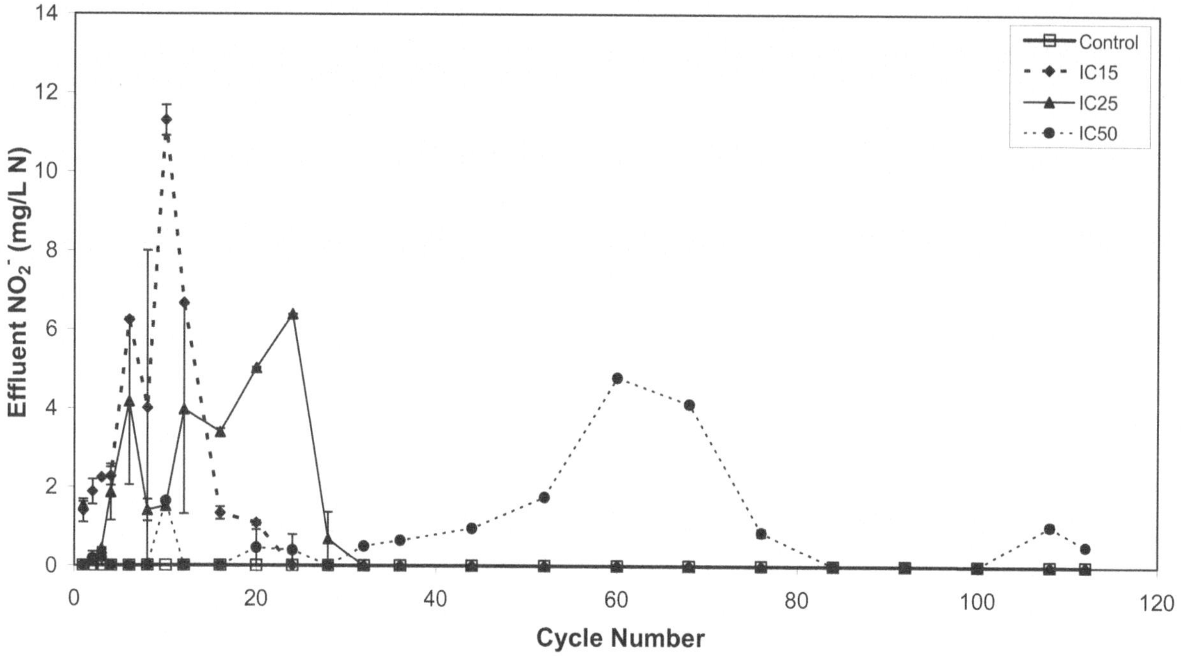

Figure 7-6. Effluent Nitrite-Nitrogen for the 10-Day SRT Reactors Exposed to CDNB Shock.

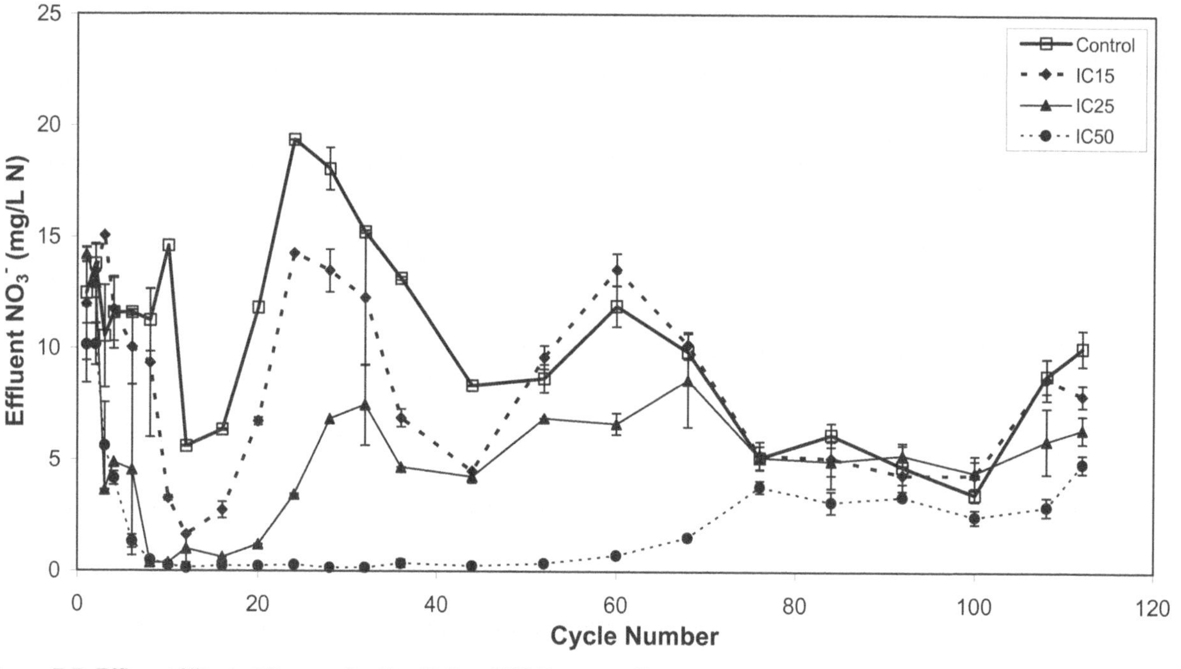

Figure 7-7. Effluent Nitrate-Nitrogen for the 10-Day SRT Reactors Exposed to CDNB Shock.

Alkalinity measurements corroborate the nitrification inhibition trends that were observed with effluent nitrogen species. As shown in Figure 7-8 for the 10-day SRT biomass, the control reactor had significantly lower alkalinity than the IC_{25} and IC_{50} experimental reactors (α=0.05) and alkalinity increased with increasing CDNB exposure. AOB metabolism of ammonia is the major point in nitrification when acid is formed and alkalinity consumed. The difference in alkalinity levels in the IC_{25} and IC_{50} reactors relative to the control is consistent with the degree of AOB inhibition over time in those same reactors, as shown by effluent ammonia concentrations in Figure 7-5. Alkalinity in the IC_{25}-shocked reactors returned to control levels around cycle 76, but recovery to control levels did not occur for the IC_{50} shocked reactor. This result suggests that the apparent recovery of ammonia oxidation took longer than can be inferred from effluent ammonia data. The reason for the spike in alkalinity at cycle 12 is unknown, but was consistent for all reactors.

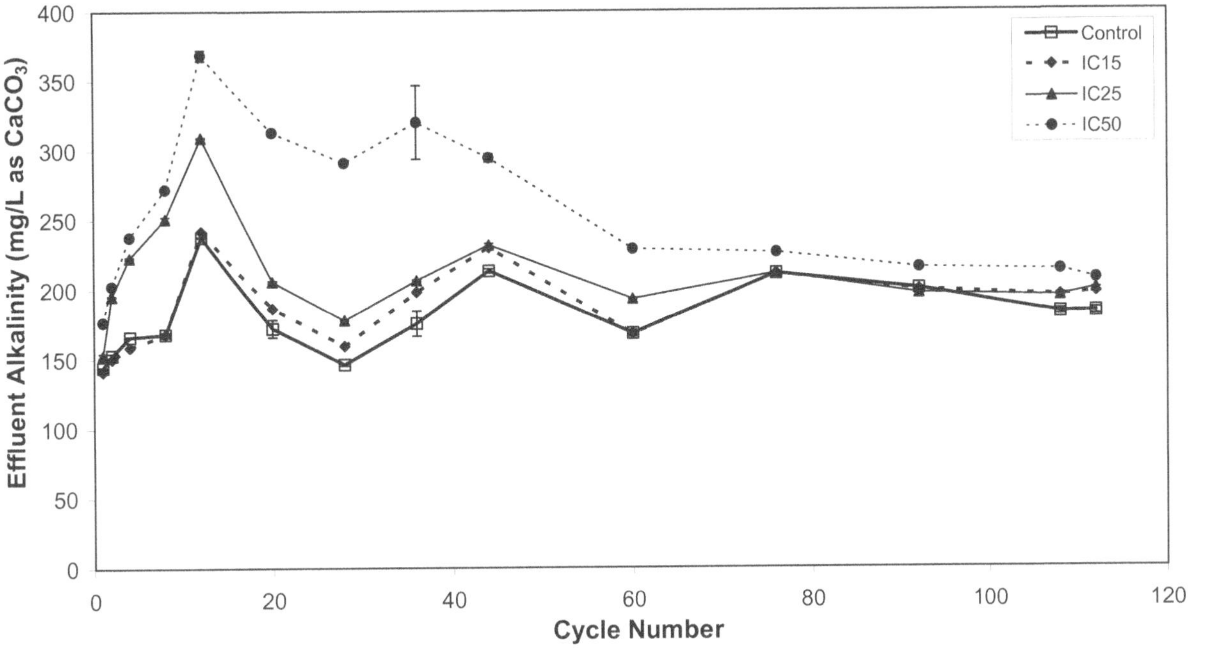

Figure 7-8. Effluent Alkalinity for the 10-Day SRT Reactors Exposed to CDNB Shock.

Alkalinity concentrations in the 2-day SRT reactors showed that the control reactor had moderately (but significantly, $\alpha = 0.05$) lower alkalinity than the IC_{15} experimental reactor during the first two cycles only. In contrast, the alkalinity in the IC_{50} reactor was found to be substantially lower for the first cycle only (shown in Appendix F). It is speculated that CDNB inhibition may have influenced alkalinity levels, but in an unknown way. No other reactors were significantly different from the control for alkalinity during this experiment, which is consistent with the absence of substantial nitrification in these reactors.

Effluent pH trends corresponded with effluent alkalinity, and were elevated relative to the control for the IC_{25} and IC_{50} shocked reactors for the 10-day system beginning with the first cycle (Figure 7-9). The pH recovered to control levels for these two reactors after 44 cycles (11 days) and for the IC_{15} reactor after 24 cycles (6 days). The 2-day SRT system had no significant pH differences between reactors for the duration of the experiment (Appendix F).

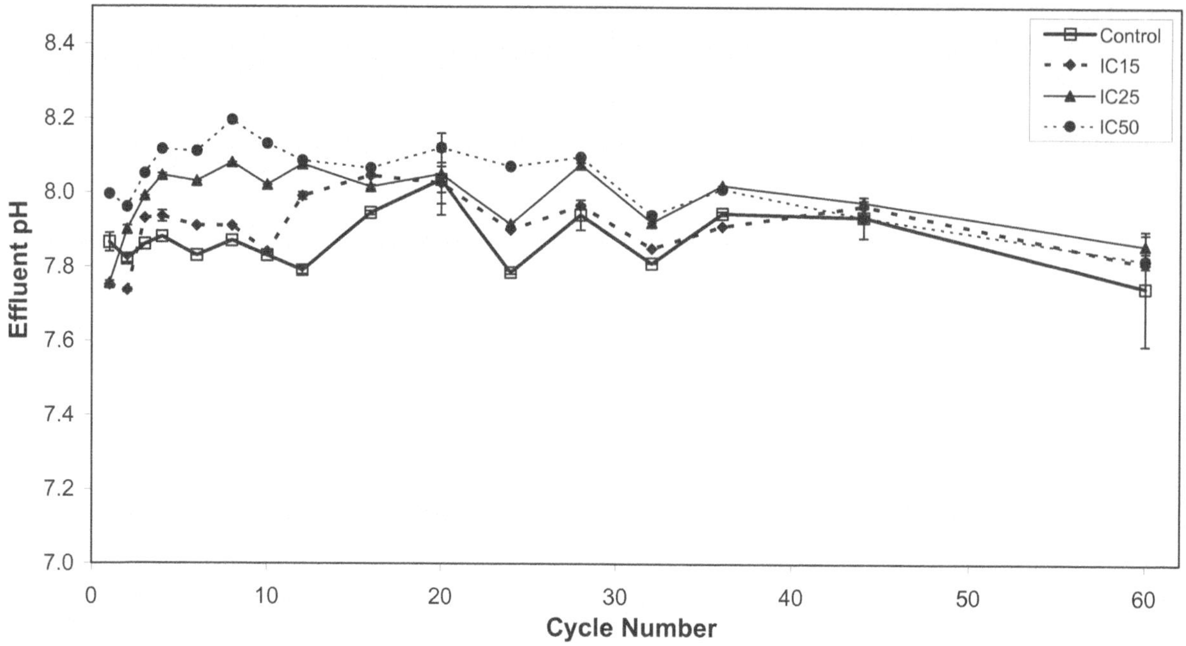

Figure 7-9. Effluent pH for the 10-Day SRT Reactors Exposed to CDNB Shock. Data Through Cycle 60 Are Shown.

7.2.4 Mixed Liquor Characteristics

7.2.4.1 Settling and Dewatering Characteristics

CST measurements were taken for the mixed liquor of both the 2- and 10-day SRT systems. CST values were significantly different from control levels for the IC_{50} reactor of the 2-day SRT biomass (see Appendix F). For the first cycle, the level was significantly higher than the control and for the second and third days, the levels were significantly lower than those of the control ($\alpha = 0.05$). By the last cycle, the IC_{50} reactor was not different from the control, but the IC_{15} and IC_{25} reactors were significantly lower and the values did not change over the course of the experiment. Furthermore, the CST values measured at the end of the experiment in all the 2-day SRT reactors (including the control) were unusually high (more than 100 sec) relative to measurements for this biomass for other shock experiments, and suggests that an unrelated factor that affected CST was activated. Contrary to the 2-day SRT results, the CST values for the 10-day SRT biomass did not change significantly over the course of the experiment, with the average value being 10.8 seconds for all reactors. The values were very low and typical of biosolids that are easy to dewater. Overall, no definitive trends were observed relating CST to the CDNB concentration, so no conclusions can be drawn as to the impact of CDNB on biosolids dewatering capacity.

SVI measurements for the 2- and 10-day SRT reactors were not negatively impacted due to the addition of CDNB (shown in Appendix F). The SVI for the 2-day SRT reactors remained below 180 mL/g in all cases, indicating acceptable settling qualities. Although the IC_{50} reactor SVI was significantly lower than that for the other reactors for most of the test, the shocked reactor SVIs were never significantly higher than those of the control. In contrast, the SVI measurements for the 10-day SRT reactors did increase with time. These increases corresponded with the filamentous bulking events discussed earlier and could not be explained or correlated to the presence of CDNB.

7.2.4.2 Mixed Liquor Solids Trends

MLSS concentration trends in the 2- and 10-day SRT CDNB-stressed reactors did not indicate a significant CDNB-initiated deterioration relative to the controls, despite the visually apparent turbidity in all the stressed reactors (Table 7-3 and in Appendix F). MLSS concentrations decreased over time in the 2-day SRT systems, and increased over time for the 10-day SRT systems. The MLSS did decrease in the IC_{50} reactor for the 10-day SRT system after cycle 68 in response to the filamentous bulking event, but that response could not be clearly correlated with CDNB stress. These patterns seem to reflect changes in the suspended solids composition of the influent wastewater over time during the experiments (which were performed at different times).

Table 7-3. Mixed Liquor Concentrations in 2- and 10-Day SRT Reactors Exposed to CDNB Shock.

	MLSS (mg/L)	MLVSS (mg/L)	MLVSS:MLSS Ratio
2-Day SRT reactors			
Control	1,372 ± 215	1,016 ± 134	0.74 ± 0.04
IC_{15}	1,176 ± 253	867 ± 156	0.74 ± 0.04
IC_{25}	1,259 ± 203	930 ± 129	0.74 ± 0.04
IC_{50}	1,389 ± 175	1,010 ± 117	0.73 ± 0.04
10-Day SRT reactors			
Control	1,822 ± 355	1,567 ± 293	0.86 ± 0.03
IC_{15}	2,066 ± 491	1,764 ± 401	0.86 ± 0.03
IC_{25}	2,134 ± 579	1,831 ± 473	0.86 ± 0.03
IC_{50}	1,763± 626*	1,492 ± 492*	0.85 ± 0.03*

*Values from cycle 78 until the end were not included for the 10-day SRT IC_{50} reactor due to biomass loss from filamentous bulking.

A microscopic analysis was also performed on both the 2- and 10-day SRT systems to compare variations in microbial stability indicators for the shocked reactors relative to the controls. In both sets of experiments, immediate effects were observed on floc morphology and microbial diversity. Floc density visually decreased with increasing CDNB concentration, and corresponded with an increase in the number of free cells (not floc associated). There was a decrease in the number of ciliates, especially stalked and crawling ciliates, that corresponded with increased CDNB concentration in all reactors. In the 10-day SRT shocked reactors, the number of shelled amoebae was markedly lower in the IC_{25} and IC_{50} reactors when compared with the control, indicating higher levels of ammonia as shelled amoebae are sensitive to ammonia. These indicators were not observed in any of the 2-day SRT reactors, probably because the non-nitrifying biomass retained a high ammonia residual. The metazoan population (i.e. rotifers, worms, etc.) and diversity was also lower in the 10-day SRT and not observed in the 2-day SRT reactors, possibly due to washout from the short SRT. For all 10-day SRT shocked reactors except the IC_{50} reactor, microscopic parameters were consistent with parameters for the control reactor by cycle 20 (day 5). The recovery point for the 2-day SRT system was cycle five (day 2) except for the IC_{50} reactor, which did not recover before the end of the experiment.

7.2.4.3 Kinetic Indicators of Inhibition

SOUR results for both the 2- and 10-day SRT systems showed that increased CDNB dose resulted in increased inhibition of SOUR. In the 2-day SRT system, only the IC_{50} reactor was statistically lower than the control ($\alpha = 0.05$) (Figure 7-10). By the fifth cycle, the SOURs of all stressed reactors were not significantly different from the SOUR of the control reactor. Despite the fact that statistical differences were not found, apparent recovery of SOUR values for the IC_{25} and IC_{50} reactors was only achieved after cycle 10. The magnitude of the inhibition effect measured five hours (end of cycle one, reaction phase) after the initial shock was higher than the expected 15%, 25%, and 50% measured by the batch IC_{XX} assay over about 10 minutes the day before the experiment was started. For the reactors, the degree of inhibition was approximately 22%, 31%, and 86% for the IC_{15}, IC_{25}, and IC_{50} stressed reactors, respectively. A similar trend was seen with the 10-day SRT biomass, although the inhibition of SOUR during the first cycle was statistically significant relative to the control for all stressed reactors (Figure 7-11). The IC_{15} reactor SOUR values returned to control levels by cycle 5, while for the IC_{25} and IC_{50} reactors recovery was observed only by cycle 9 ($\alpha = 0.05$). As with the 2-day SRT system, the magnitude of the inhibition effect measured 6 hours after the initial shock was much higher than the expected 15%, 25%, and 50% measured by the 10-minute batch IC_{XX} assay. The degree of inhibition was approximately 36%, 63%, and 88% for the IC_{15}, IC_{25}, and IC_{50} stressed reactors, respectively. Collectively, these results for the 2- and 10-day SRT shocked systems indicate that a greater degree of inhibition occurred due to longer exposure times (6 hours as opposed to 10 minutes). These results suggest that CDNB has an effect on the respiration rate of biomass, but this effect tends to be short lived.

The SNGR test confirms the nitrification inhibition patterns observed by other measurements for the 10-day SRT shocked reactors. Figure 7-12 shows that the SNGR tended to decrease with increasing CDNB concentration. The control reactor did not maintain a constant SNGR, but it did remain consistently higher than the experimental reactors for all but cycle 85 and the last cycle. Additionally, the stressed reactors recovered from inhibition with time, and only the IC_{50} reactor did not recover by cycle 113 (day 28). This observed recovery correlates with the effluent nitrate, nitrite, and ammonia data, which shows decreasing levels of ammonia and nitrite and increasing levels of nitrate with time. The inability of the IC_{50} reactor to recover according to the SNGR tests does not correspond with the effluent nitrogen species data, as recovery was noted at 2X SRT. This suggests that the rate of nitrification remained slow in the IC_{50} reactor throughout the experiment, but that the 5-hour reaction phase was long enough to allow effluent ammonia, nitrate, and nitrite concentrations to trend toward recovery by the end of the experiment. These results support the conclusion that CDNB inhibits nitrification, and the degree of inhibition increases as the concentration of CDNB increases.

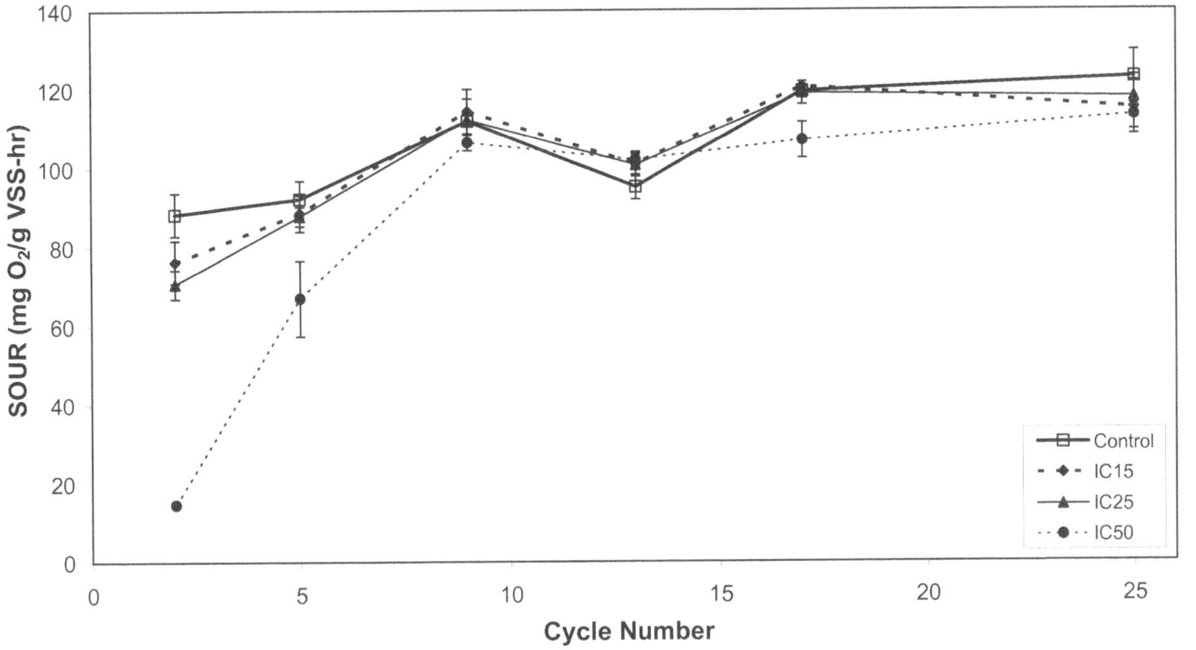

Figure 7-10. SOURs for the 2-Day SRT Reactors Exposed to CDNB Shock.

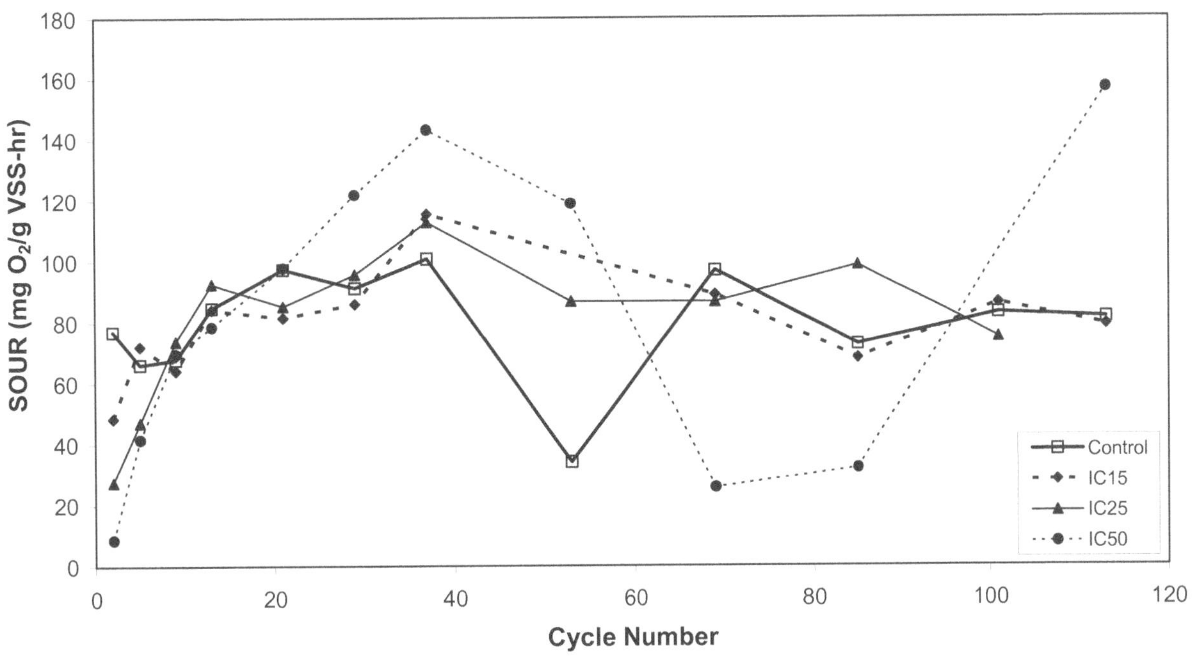

Figure 7-11. SOURs for the 10-Day SRT Reactors Exposed to CDNB Shock.

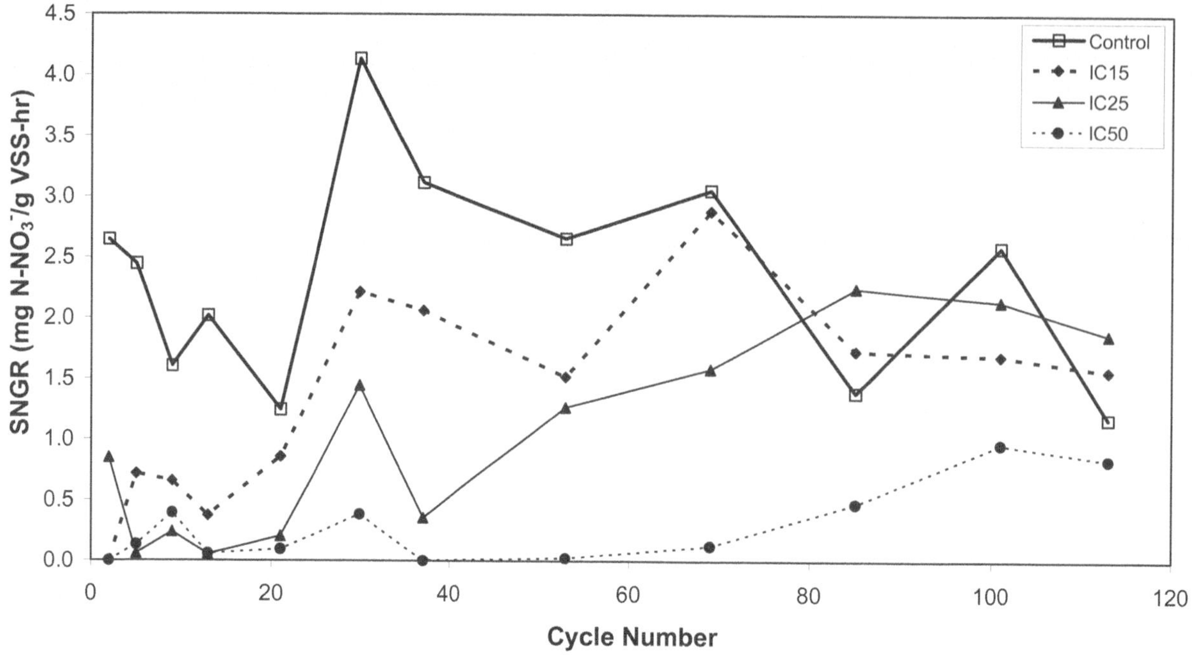

Figure 7-12. SNGRs for the 10-Day SRT Reactors Exposed to CDNB Shock.

7.3 Discussion

CDNB proved to be a strong toxin that yielded significant process deterioration at multiple concentrations. The results suggest that CDNB impacted nitrification most severely. COD removal efficiency and settleability were also affected to varying degrees.

Deflocculation of the biomass was observed when sublethal shock inputs of CDNB occurred. In the 2-day SRT system this event was short lived (0.5X SRT) for the IC_{50} reactor but no significant differences were noted for the IC_{25} and IC_{15} reactors. Similarly, deflocculation was short lived (0.6X SRT) for the IC_{50} reactor with the 10-day SRT biomass, and recovery occurred even sooner for the IC_{25} and IC_{15} reactors (0.2X and less than 0.1X SRT, respectively). In conjunction with this data, dramatic increases in effluent potassium levels were seen for the first cycle in all stressed reactors for both 2- and 10-day SRT systems, and correlated with increasing CDNB concentration. When the deflocculation event is considered in conjunction with the increased effluent potassium levels, it appears that the GGKE mechanism (Bott and Love, 2002) was activated by CDNB shock during this experiment. As CDNB is an uncharged organic electrophile, this supports previous research performed with this chemical on activated sludge cultures by this research team (Bott and Love, 2002; Bott and Love, 2004) and on pure cultures by others (Ferguson et al. 1997; 1998b; Ness et al., 1997).

The SOUR data show different trends than were expected from the short-term IC_{XX} assays. For both the 2- and 10-day SRT reactors, inhibition levels measured approximately five hours after shock were substantially greater than those expected from the short-term IC_{XX} assays (36%, 63%, and 88% for the 10-day SRT and 22%, 31%, and 86% for the 2-day SRT). After the first cycles, the differences in SOUR for the reactors followed no predictable pattern. Overall, the inhibition of respiratory activity due to the addition of CDNB was expected, as previous work has shown that organic electrophilic chemicals, including CDNB, inhibit cellular growth and viability (Ferguson and Booth, 1998; Ferguson et al., 1995; 1997; 1998; Ness et al., 1997; Tse-Dinh, 2000; Vattanaviboon et al., 2001). This may be due to the influx of H^+ that accompanies the efflux of potassium that was seen. This influx of H^+ would temporarily dissipate the proton motive force, which is necessary for the generation of energy (White, 2000). If dissipation of the proton gradient did occur, this would decrease the rate of oxygen consumption for energy generation by the cells. Re-establishment of the proton gradient with an uptake of potassium would possibly return the cells to normal respiration rates. As the elevated effluent potassium levels were only noted for the first day or less, rapid recovery from SOUR inhibition corresponds well with this theory.

A significant loss of sCOD removal efficiency was noted during the first cycle of this experiment in the stressed reactors for both the 2- and 10-day SRT systems. For the 2-day SRT system, a significant loss of sCOD removal efficiency was noted during the first six cycles of the experiment (0.75X SRT) in the IC_{50} stressed reactor, but only for the first cycle for the IC_{25} reactor. For the 10-day SRT system, these levels remained elevated until 0.9x SRT for the IC_{50} stressed reactor and until 0.6x SRT for the IC_{15} and IC_{25} stressed reactors. For both systems, the time frame for loss of sCOD removal efficiency did not correlate with the decrease in SOUR or with the efflux of potassium from the cells and also remained longer than the deflocculation event. As the soluble CDNB amounts were below detection levels for the majority of the experiment, it suggests that the inhibition of sCOD removal was not due to continued inhibition by CDNB but rather to extended damage done by CDNB on the process. The exact mechanism for this loss of sCOD removal is not known and an extensive literature search did not reveal any known mechanisms. It is, however, conceivable that the dissipation of the proton motive force discussed above could account for the loss of sCOD removal efficiency. As H^+ ion movement is often coupled with solute transport (White, 2000), a disruption of the proton motive force would impede bacterial uptake of soluble substrate (sCOD). Another possibility is that proteins required for the uptake and metabolism of substrate in the wastewater were oxidatively damaged by CDNB and repair or replacement of these proteins was required before control levels of COD removal could be attained. Although the exact mechanism of inhibition is unknown at this time, it appears to be reversible.

Finally, nitrification was clearly negatively impacted by the presence of CDNB. The process of nitrification involves the conversion of ammonia to nitrate by autotrophic bacteria, where the reduced nitrogen species serve as the energy source for nitrifying bacteria. The full process of nitrification involves two different groups of bacteria, the AOB and NOB. For both groups, the reactions are coupled to several enzymes, one of which is cytochrome aa_3 oxidase which is the terminal electron acceptor for both nitrifying groups (Madigan et al., 2003) and is a critical component of a sustained proton motive force. This proton motive force, as previously stated, is critical to the ATP-generating (oxidative phosphorylation) process (White, 2000).

The CDNB-induced nitrification inhibition appeared to most strongly impact the NOB rather than the AOB. This can be seen from the effluent ammonia, nitrite, and nitrate data as well as from the SNGR data obtained for the 10-day SRT reactors. From the effluent ammonia and nitrite data for the IC_{25} and IC_{50} reactors, ammonia oxidation was clearly inhibited and recovered more quickly in the IC_{25} system (0.6X SRT vs. 1.7X SRT for IC_{50}). This observation is supported by the alkalinity data. Coupled with an apparent lack of inhibition in the IC_{15} stressed reactor, these results indicate that ammonia oxidation was inhibited more with increasing CDNB concentrations. Recovery of ammonia oxidation was more rapid than recovery of nitrite oxidation. A clear buildup of nitrite was seen in all three stressed reactors and was coupled to initiation of rapid ammonia oxidation, based on the effluent ammonia data. This buildup was first seen in the IC_{15}, which did not exhibit ammonia oxidation inhibition based on effluent measurements. When ammonia oxidation recovered in both the IC_{25} and IC_{50} reactors, a buildup of nitrite was also seen initially. A decline in nitrite levels was coupled with an increase in nitrate levels for the stressed reactors, indicating a recovery of nitrite oxidation and thus a recovery of complete nitrification. As recovery of nitrite oxidation appeared to take longer than ammonia oxidation, it can be concluded that NOBs were more sensitive than AOBs to CDNB shock loads, although more carefully controlled studies should be done to confirm this. As nitrification took longer to return to control levels than any of the other monitored parameters, it is concluded that nitrification was the most sensitive process parameter to the CDNB shock.

It is not clear if the mechanism that caused deterioration in nitrification is linked to the one responsible for deflocculation or loss of sCOD removal efficiency. A loss in efficiency of the proton motive force due in part to potassium efflux could link a causal mechanism for deflocculation or loss of sCOD removal efficiency. However, analysis of the recently sequenced genome for *Nitrosomonas europaea* (Chain et al., 2003) suggests that AOBs do not have all the necessary genes needed for an active GGKE system, suggesting that the proton motive force theory linked to GGKE activation does not explain the nitrification inhibition observed. It is conceivable that oxidative damage to critical proteins could have caused process deterioration. Whatever the mechanism(s) for these effects, they appear to be reversible.

7.4 Conclusions

- Deflocculation of mixed liquor occurred upon addition of CDNB. For both the 2- and 10-day SRT systems, TSS increased significantly after the shock, and the increase in effluent turbidity was proportional to the concentration added to the reactors. All systems recovered, indicating that the deflocculation mechanism was reversible in the time period of the study.
- Increases in soluble potassium were detected in the stressed reactors within 5 hours of the shock event. This corresponds with previous studies, where deflocculation was linked to the GGKE stress response mechanism, which is activated when Gram-negative cells are exposed to electrophilic chemicals. The potassium efflux was brief and recovered within one day of the shock, whereas recovery from deflocculation took two to three times longer in the 10-day SRT system, suggesting that GGKE-induced deflocculation can only be reversed if new biomass is grown.
- CDNB inhibited mixed liquor respiration rates as seen by SOUR. This inhibition may have occurred because GGKE disrupts the cytoplasmic proton gradient, which is needed for cellular function. Both effluent potassium and SOUR values recovered to control levels at the same time, which corroborates the hypothesis that these events are linked.
- COD removal efficiency decreased with increasing CDNB concentrations for all the stressed reactors. The causal mechanism for this effect is unknown, but two hypotheses are provided: (i) disruption of the proton gradient through the GGKE mechanism, or (ii) oxidative damage of solute transport or catabolic proteins.
- The increase in mixed liquor SVI corresponded with filamentous bulking events. The cause of this bulking and apparent recovery in the control reactors is not known. The

failure for bulking to occur in the shocked reactors during one of the filamentous bulking events may indicate that deflocculated biomass has increased susceptibility to toxins because of a larger exposed surface area.

- CDNB had no observed effect on the dewaterability of mixed liquor as CST values were not significantly different between chemically stressed biomass and control biomass.
- Nitrification was severely inhibited by CDNB, as was seen from both SNGR data as well as effluent ammonia, nitrate, nitrite, and alkalinity concentrations. NOBs were more susceptible to CDNB than AOBs, but all were affected to some extent. Time to recovery corresponded with increasing CDNB concentration. Nitrification appeared to be recoverable, albeit more slowly than recovery from loss of COD removal or deflocculation, which suggested a different causal mechanism at work.

CHAPTER 8.0

2,4-DINITROPHENOL SHOCK EXPERIMENTS

8.1 Literature Review

2,4-Dinitrophenol (DNP) is the most commercially important dinitrophenol. It is used in many industrial processes, such as in the manufacture of dyes, picric acid (trinitrophenol), picramic acid, wood preservatives, photographic developers, explosives and pesticides (acaricides, herbicides, and insecticides) (Agency for Toxic Substances and Disease Registry, 1995). DNP is also used as a pH indicator (Agency for Toxic Substances and Disease Registry, 1995) and in the detection of potassium and ammonium ions (Budavari, 1989). In addition to releases in the environment from manufacturing units, other sources of DNP may include pesticide runoff, automobile exhaust, release from DNP-containing wastes that are disposed off in landfills, accidental spills during transport, and leaks from storage containers (Agency for Toxic Substances and Disease Registry, 1995). Data from the U.S. EPA indicates that from 1996 to 2000, approximately 310,000 pounds of DNP were released by industry into the environment, with more than 50% of this value corresponding to discharges in surface waters. Given its significant occurrence in the environment, DNP is regulated by many environmental-related laws, such as the Clean Air Act Amendments of 1990, which list DNP as a hazardous air pollutant (HAP) and the Clean Water Act, which sets the National Recommended Water Quality Criteria for Priority Toxic Pollutants (criterion of human health for consumption of water + organism) of 69 μg/L for DNP. DNP has also been included in the Comprehensive Environmental Response, Compensation, and Liability Act (CERCLA) priority list of hazardous substances as one of the top 100 substances most commonly found in Superfund sites and in the P list of acutely hazardous chemical products under the Resource Conservation and Recovery Act (RCRA). The main physical and chemical properties of DNP are listed in Table 8-1.

Table 8-1. Physical and Chemical Properties of 2,4-Dinitrophenol.

Characteristic	Value
Density	1.683 g/cm^3 (at 24°C) [1]
Dissociation constant (pK_a)	4.07 (at 25°C) [1]
Solubility in water	5,600 mg/l (18°C) [2]
Henry's law constant	1.57 x 10^{-8} atm-m^3/mole (18-20°C) [2]
Vapor pressure	0.00039 mm Hg (20°C) [2]
Log octanol/water partition coefficient	1.53 (1.51-1.54) [2]

[1] CRC Handbook of Chemistry and Physics (3rd Electronic Edition)

[2] Watts, R.J. (1998) Hazardous Wastes: Sources, Pathways, Receptors, John Wiley and Sons, Inc. New York.

DNP is a weak aromatic acid, which has been identified as an uncoupler of oxidative phosphorylation because of its ability of carrying protons across biological membranes. The pK_a value for this compound is around 4 and indicates that, at physiological pH, most DNP molecules are unprotonated, but the difference between intracellular and extracellular pH still allows for the small ratio of protonated molecules to carry and release their proton inside the cell. The low pK_a, however, makes it a weaker uncoupler than other substances with a more neutral pK_a (Lewis et al., 1994). The octanol/water partition coefficient for DNP has a value of approximately 35, suggesting that it has a tendency to become associated with more hydrophobic substances. This is in agreement with the fact that DNP has to diffuse through the hydrophobic phospholipid bilayer to effectively carry protons between the extracellular and intracellular environments in prokaryotes. Both the vapor pressure and the Henry's law constant indicate that DNP does not volatilize easily, suggesting that it tends to remain in a liquid or solid-associated phase rather than volatilize.

DNP has been extensively used in research due to its known effects as an uncoupler of oxidative phosphorylation. As an uncoupler, DNP has motivated researchers to attempt to reduce sludge production by chemically decreasing the yield of activated sludge systems. Rich and Yates (1955) conducted one of the first studies on the effects of 2,4-DNP on activated sludge. As cited by Mayhew and Stephenson (1997), the Rich and Yates study showed that DNP increased the substrate uptake rate by activated sludge when dosed at concentrations lower than 25 mg/L, but organic matter uptake inhibition was observed at higher DNP concentrations. Both effects disappeared when the chemical was washed out of the system. In a second study, Mayhew and Stephenson (1998) ran a series of experiments using porous pot reactors with activated sludge from a local plant and settled wastewater as influent. A screening experiment with known uncouplers was carried out and showed that 4 mg/l DNP caused an increase in oxygen uptake rate up to 200% of the control after 24 hours. After 7 weeks of continuous feeding with primary effluent spiked with 4 mg/L DNP, the difference in BOD removal efficiency was not significantly different from the control and the growth yield was statistically lower for the treated reactor (0.3) than for the control (0.42), which lowered the biomass production. Liu (2000) studied the effects of DNP on the growth yield of activated sludge cultures. A mathematical model was developed and shown to fit the observed data reasonably well, suggesting that the strength of the uncoupler as a protonophore is correlated to the uncoupler/biomass ratio rather than the uncoupler total concentration.

Okey and Stensel (1993) described the effects of uncouplers on a full-scale activated sludge system where unidentified uncouplers were suspected to be present in the influent. They reported that the uncouplers increased the oxygen requirements of the treatment system by a factor of 100%–200%. This corresponded with a concomitant decrease in biomass synthesis to a net yield around zero, a modest decrease in the rate of substrate uptake and assimilation, and partial or complete elimination of nitrification. This is the only report on full-scale treatment plants and the only one that indicates an effect on nitrification in a full-scale activated sludge system.

Two major hypotheses exist for the uncoupling of oxidative phosphorylation. The first is based on Mitchell's theory of uncouplers as lipophilic proton transporters that dissipate the proton gradient. Mitchell and Moyle (1967) studied the effect of organic protonophores on the membrane potential of mitochondria and concluded that lipophilic weak acids such as DNP and carbonyl cyanide p-(trifluoromethoxy) phenylhydrazone (FCCP) increase the proton conductance into the inner mitochondrial membrane, thereby dissipating the proton gradient. The respiratory chain tended to respond to this effect by increasing the rate of protons being pumped out, which led to an increase in electron transfer rate and respiration. The second theory is that uncouplers interact with specific proteins and binding sites, causing conformational changes that compromise the cell's ability to generate energy. Brummett and Ordal (1977) found that proline and glycine transport in *Bacillus subtilis* were inhibited by three chemically different uncouplers, one of which was pentachlorophenol (PCP), and hypothesized that the inhibition effect was caused by specific interactions between the uncoupler and the bacterial transport systems. They proposed that these interactions would stabilize a particular protein conformation, preventing the necessary conformational changes to occur that would otherwise facilitate transport of the amino acids. Nicholas and Ordal (1978) studied the effects of several chlorophenols in *Bacillus subtilis* to understand the influence of the pK_a on the transport inhibition potency. Proline and glycine transport were inhibited most strongly by stronger acids (lower pK_a), implying that the anion is the inhibitory species. The locations of interaction were postulated to be within the membrane. Glycine and glutamine transport were also found to be inhibited by uncouplers of oxidative phosphorylation in *Bacillus megaterium* wild-type cells and in a mutant described to be resistant to uncouplers of oxidative phosphorylation (Decker and Lang, 1977).

Other effects of uncouplers have been studied and include nutrient transport inhibition (Bakker and Randall, 1984) and interference with the translocation of proteins destined for insertion into the cytoplasmic membrane (Enequist et al., 1981; Geller, 1991). In general, uncouplers will most likely affect all bacterial processes that rely on the proton motive force as a source of energy (e.g., bacterial mobility), as these agents dissipate the proton motive force. Finally, it is conceivable that both hypothesized mechanisms of uncoupler action (proton gradient dissipation and direct interference with transport or energy-generating proteins) occur simultaneously.

The effects of uncouplers on nitrifying bacteria have also been studied. Hooper and Terry (1973) identified DNP as a specific inhibitor of the AMO-mediated ammonia oxidation process in *Nitrosomonas*. Using cell free extracts of *Nitrosomonas europaea*, Aleem (1966) observed that the energy-dependent reduction of pyridine nucleotides (NAD^+ and $NADP^+$), which are used by cells to run reverse electron transport, is almost completely inhibited by uncouplers of oxidative phosphorylation. Specifically, 18 mg/l DNP inhibited the process by 95%. Iizumi et al. (1998) used DNP and PCP to test the response of a bioluminescence assay developed to detect nitrification inhibitors. The assay consisted of *N. europaea* cells carrying the *luxAB* genes from

Vibrio harveyi. A rather high concentration of DNP (100 mg/l) caused a 50% decrease in the nitrite production rate of *N. europaea* cells, but a much lower concentration (around 12 mg/l) caused a 50% decrease in the light production rate.

Pure cultures of bacteria have been shown to acclimate to uncouplers. Work conducted on the adaptation of *E. coli* to DNP showed that the bacterial growth rate decreased immediately after addition of DNP, but that the growth rate recovered within 4 hours to the original rate for DNP concentrations up to 140 mg/l (Gage and Neidhardt, 1993). Using two-dimensional SDS polyacrylamide gel electrophoresis (SDS-PAGE), the authors showed that 53 proteins were synthesized when *E. coli* cells were exposed to DNP. Among the 53 induced proteins were 3 heat shock proteins, 6 carbon starvation proteins, 9 nitrogen starvation proteins and 11 phosphorus starvation proteins. Therefore, 24 of the 53 proteins were unknown proteins that had not been identified previously as being synthesized under stress, and they may be unique to DNP stress. After cells recovered, synthesis of this group of proteins ceased. The *E. coli* adaptation mechanism for uncouplers could be related to a "low-energy shock response," which was proposed by Lewis et al. (1994).

8.2 Results

8.2.1 Inhibitory Concentrations (ICxx)

The effects of varying DNP concentrations on the SOUR of the 2- and 10-day SRT biomasses and equivalent respiration inhibition are shown in Appendix G. The inhibitory concentrations corresponding to 15%, 25%, and 50% inhibition were determined and are indicated in Table 8-2 below. These concentrations were used to shock the reactors at the beginning of the source-effect experiment, as a single-pulse event.

Table 8-2. IC_{15}, IC_{25} and IC_{50} Values for DNP in 2- and SRT and 10-Day SRT Mixed Liquor.

	2-Day SRT biomass		10-Day SRT biomass	
	mg/L	mg DNP/g MLVSS	mg/L	mg DNP/g MLVSS
IC_{15}	30	20	21	7.4
IC_{25}	39	26	39	14
IC_{50}	50	34	107	39

8.2.2 Influent Characteristics

The influent characteristics are summarized in Table 8-3. The influent quality was relatively constant over the time of the experiments. The influent composition reveals that the raw wastewater used as the feed for the source-effect experiments with DNP was medium to strong in strength (Metcalf and Eddy, 2003).

Table 8-3. Characteristics of Wastewater Used During DNP Source-Effect Experiments.

Characteristic	2-Day SRT Average ± std dev	10-Day SRT Average ± std dev
Alkalinity (mg/L as $CaCO_3$)	500 ± 66	275 ± 35
COD, soluble (mg/L)	854 ± 50	423 ± 69
COD, total (mg/L)	1,003 ± 75	529 ± 82
pH	7.6 ± 0.3	7.3 ± 0.4
TSS (mg/L)	225 ± 70	79 ± 39
VSS (mg/L)	93 ± 22	64 ± 32

8.2.3 Effluent Characteristics

Data for the 10-day SRT reactors are reported for the first 37 cycles only. An error in operation occurred after this point. Most water quality parameters had recovered by 37 cycles, therefore the data presented is still suitable to evaluate the effects of DNP shock on activated sludge treatment performance and biomass characteristics.

8.2.3.1 Soluble COD

Effluent sCOD concentrations are shown in Figures 8-1 and 8-2 below, for the 2- and 10-day SRT reactors, respectively. Because DNP contributes to overall COD in solution according to a ratio of 1.22 g COD/g DNP, the total sCOD values (Figures 8-1 and 8-2) were corrected by subtracting the theoretical COD concentration of DNP based on measured DNP values (Figures 8-3 and 8-4). In the 2-day SRT reactors, the sCOD removal efficiency was not affected by the DNP shock (Figure 8-3). In fact, during the first cycle, the sCOD removal efficiency in the IC_{50} reactor was higher than in the control reactor. The observed effect was short-lived and only noticeable for the highest concentration of DNP (~50 mg/L). In the 10-day SRT reactors, the sCOD removal efficiency based on corrected values was not affected by the DNP shock for concentrations up to about 40 mg/L DNP (IC_{15} and IC_{25} reactors), but a statistically significant ($\alpha = 0.05$) decrease in organic matter removal was observed for the reactor subjected to 100 mg/l DNP (IC_{50}). This significant loss of sCOD removal efficiency as compared to the control reactor lasted for only 1.25 days (five cycles) and it was dissipated before the soluble DNP values decreased to undetectable limits. As shown in Figure 8-5, the DNP concentration in the effluent of the IC_{50} reactor at cycle 6 was still 28 mg/l. These results indicate that DNP inhibited substrate uptake at high concentrations, but at lower concentrations that inhibition effect was not observed.

The effluent soluble DNP concentrations decreased in a pattern almost identical to what was theoretically expected due to dilution (Figures 8-5 and 8-6). This indicates that an association of the toxin with the biomass, which would prolong the presence of DNP in the reactors, did not occur or was negligible, and that DNP degradation did not occur.

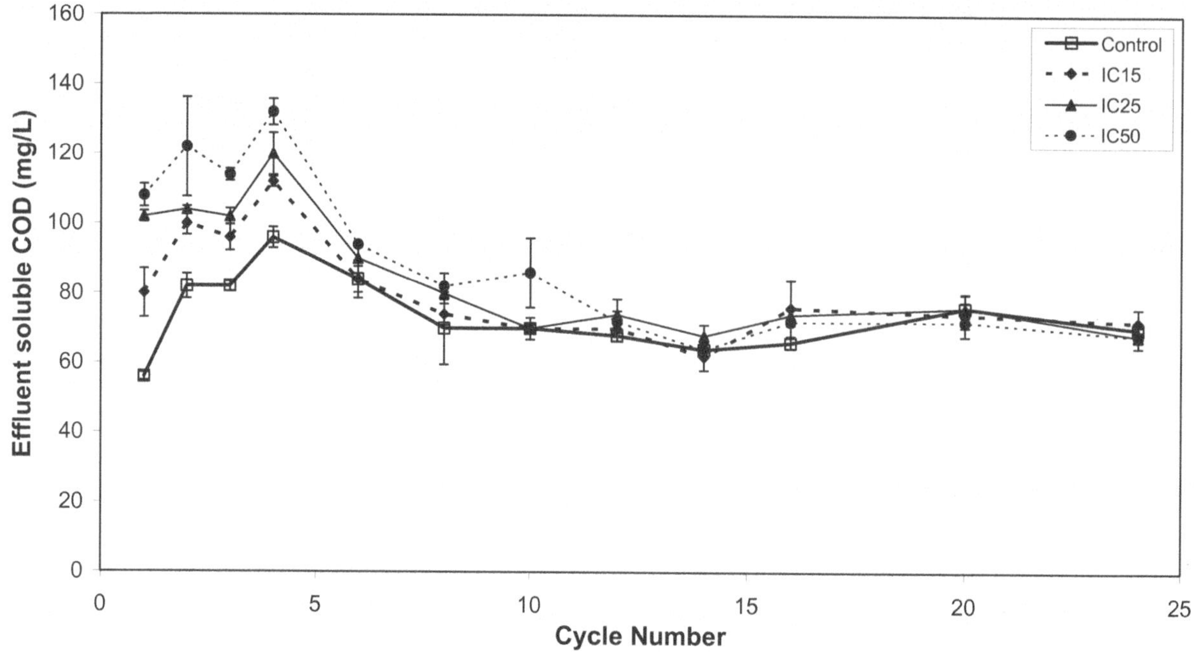

Figure 8-1. Effluent Soluble COD for the 2-Day SRT Reactors Exposed to DNP Shock.

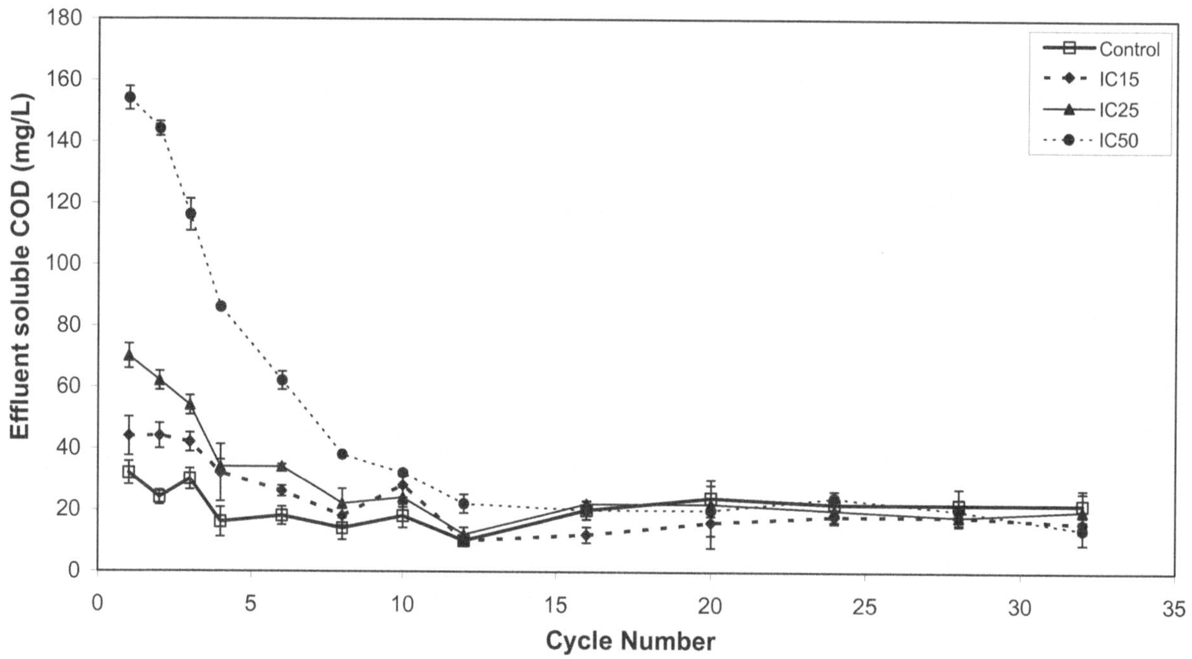

Figure 8-2. Effluent Soluble COD for the 10-Day SRT Reactors Exposed to DNP Shock.

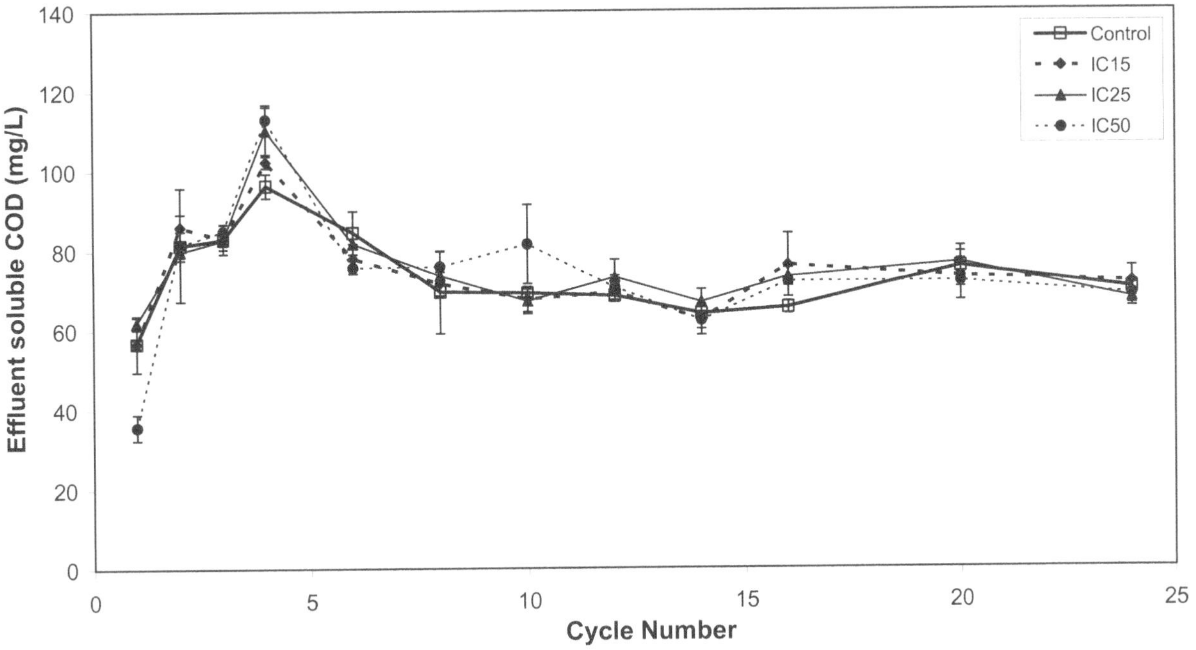

Figure 8-3. Effluent Soluble COD for the 2-Day SRT Reactors Exposed to DNP Shock; Values Corrected by Subtracting Measured Soluble DNP Contribution to sCOD.

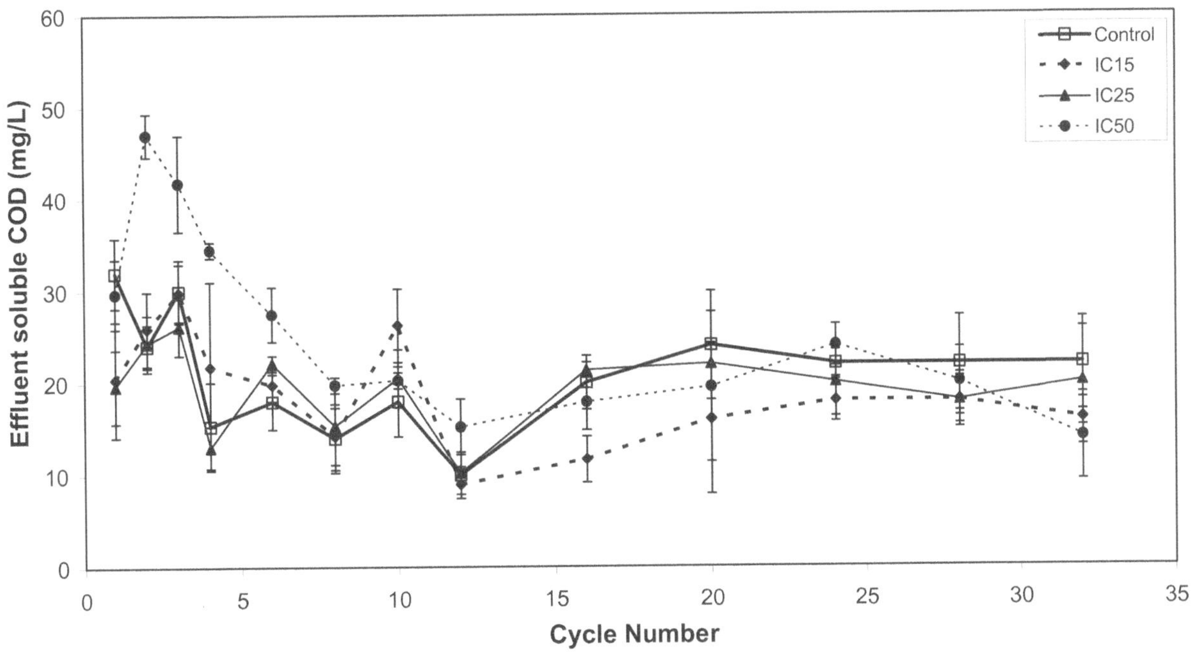

Figure 8-4. Effluent Soluble COD for the 10-Day SRT Reactors Exposed to DNP Shock; Values Corrected by Subtracting Measured Soluble DNP Contribution to sCOD.

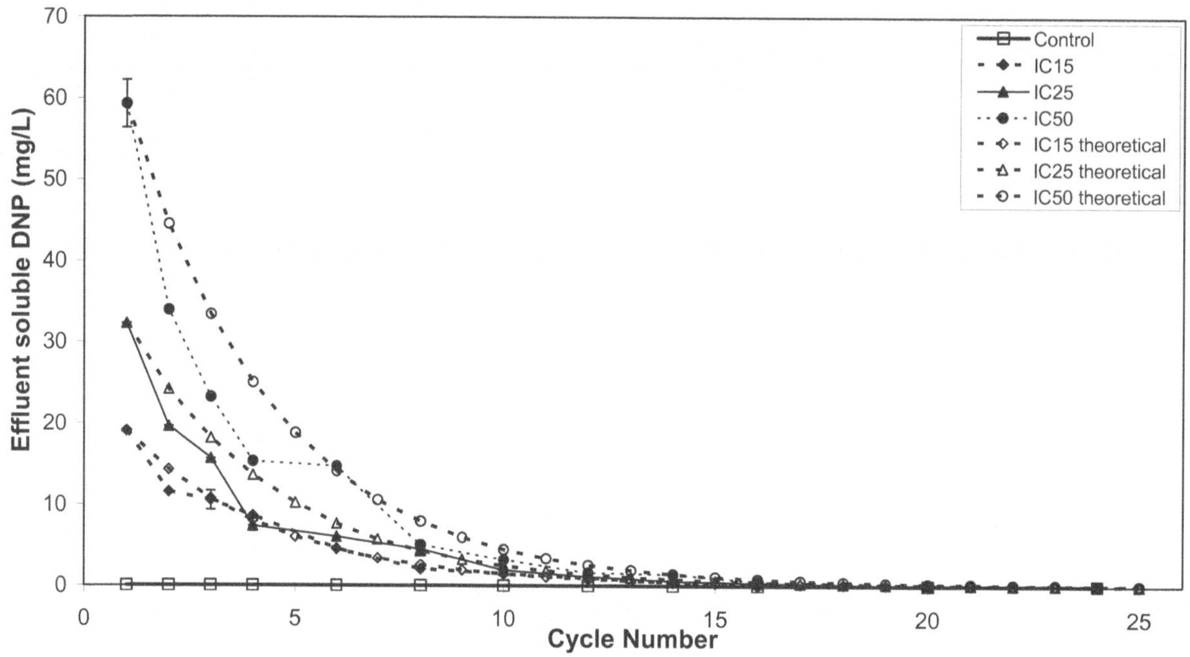

Figure 8-5. Effluent Soluble DNP for the 2-Day SRT Reactors Exposed to DNP Shock. The Dashed Lines Represent Theoretical Dilution Curves.

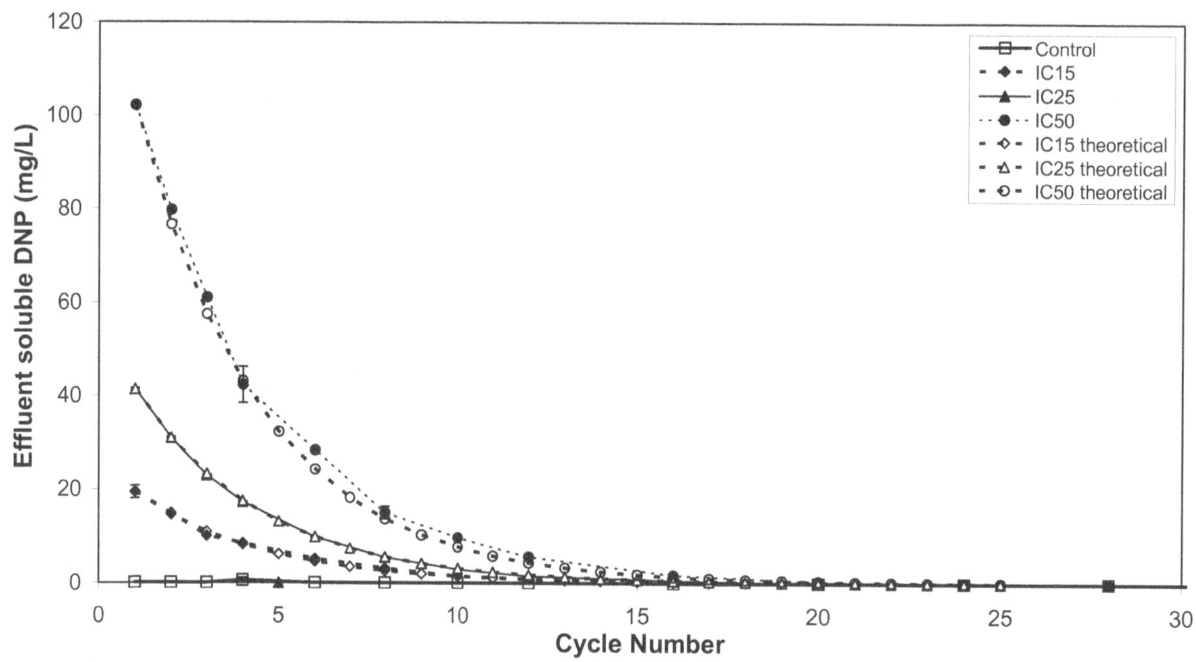

Figure 8-6. Effluent Soluble DNP for the 10-Day SRT Reactors Exposed to DNP Shock. The Dashed Lines Represent Theoretical Dilution Curves.

8.2.3.2 TSS and Potassium Concentrations

Effluent TSS concentrations for the 2-day SRT system showed no impact of DNP on the ability of the biomass to remain flocculant (Appendix G). For the 10-day SRT system, effluent TSS concentrations showed a deflocculation event which occurred only in the IC_{50} reactor (Figure 8-7). For this reactor, the TSS concentrations were significantly higher than in the control reactor during cycles two through eight ($\alpha = 0.05$). This event was short-lived and its magnitude was moderate, as the maximum TSS concentrations in the IC_{50} reactor were about 55 mg/l. Interestingly, the effluent TSS and non-DNP associated effluent COD in the IC_{50} reactor decreased to near control levels at the same time (Figure 8-3). The effluent TSS in the IC_{15} and IC_{25} reactors did not show any significant variation in comparison to the control reactor.

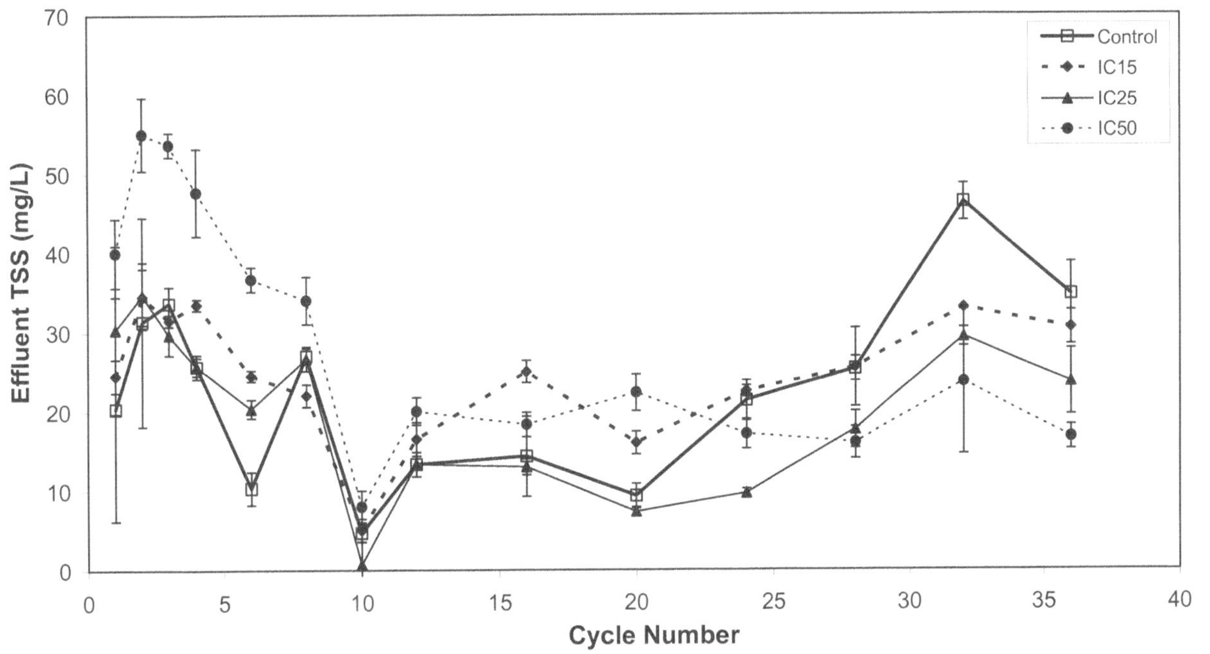

Figure 8-7. Effluent TSS for the 10-Day SRT Reactors Exposed to DNP Shock.

Effluent potassium concentrations in the 2- and 10-day SRT reactors showed no statistically significant difference from the levels in the control reactor for any of the DNP inhibitory concentrations. However, during the first four cycles, the potassium concentration in the effluent of the 10-day SRT IC_{50} reactor was slightly above that of the other reactors, including the control. The nitro substituent groups on the phenol ring could be sufficiently electrophilic to expect that cytoplasmic DNP would react with glutathione and cause a potassium efflux event. Further experiments need to be performed before concluding that DNP activates the GGKE response, but its chemical structure theoretically makes it amenable to activating this oxidative stress response.

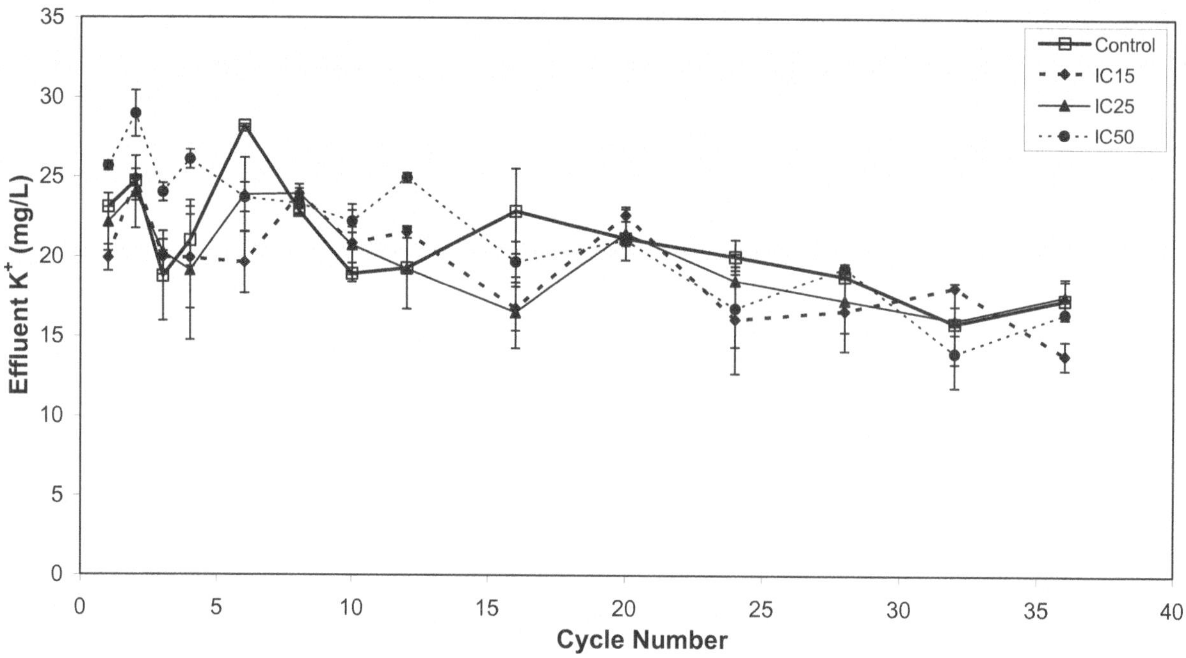

Figure 8-8. Effluent Potassium for the 10-Day SRT Reactors Exposed to DNP Shock.

8.2.3.3 Nitrification Inhibition

All the 2-day SRT reactors maintained nitrate levels below 1.5 mg/L during the entire period of the experiment (Appendix G). Additionally, undetectable nitrite levels were observed during the 6 day period. Together, these observations show that the system was not nitrifying and that DNP shock did not impact that characteristic of the biomass.

Figures 8-9 and 8-10 show concentration trends for effluent nitrate and nitrite in the 10-day SRT reactors. Effluent ammonia concentrations are not shown because all the reactors had undetectable ammonia levels during the entire experiment, which suggests that DNP did not have an inhibitory effect on the AOB. The nitrate and nitrite trends suggest that the NOB were negatively impacted by DNP in the IC_{50} reactor. Nitrite accumulated in the IC_{50} reactor during the first four cycles and disappeared after the DNP concentration declined to below 40 mg/L. The nitrate data do not correlate with the loss of nitrite oxidation because there was no significant difference in the nitrate concentrations observed in the control and IC_{50} reactors after the first cycle. Furthermore, the nitrate concentrations in the IC_{25} reactor were consistently below the control reactor and were found to be significantly lower than the control between cycles 10 and 28.

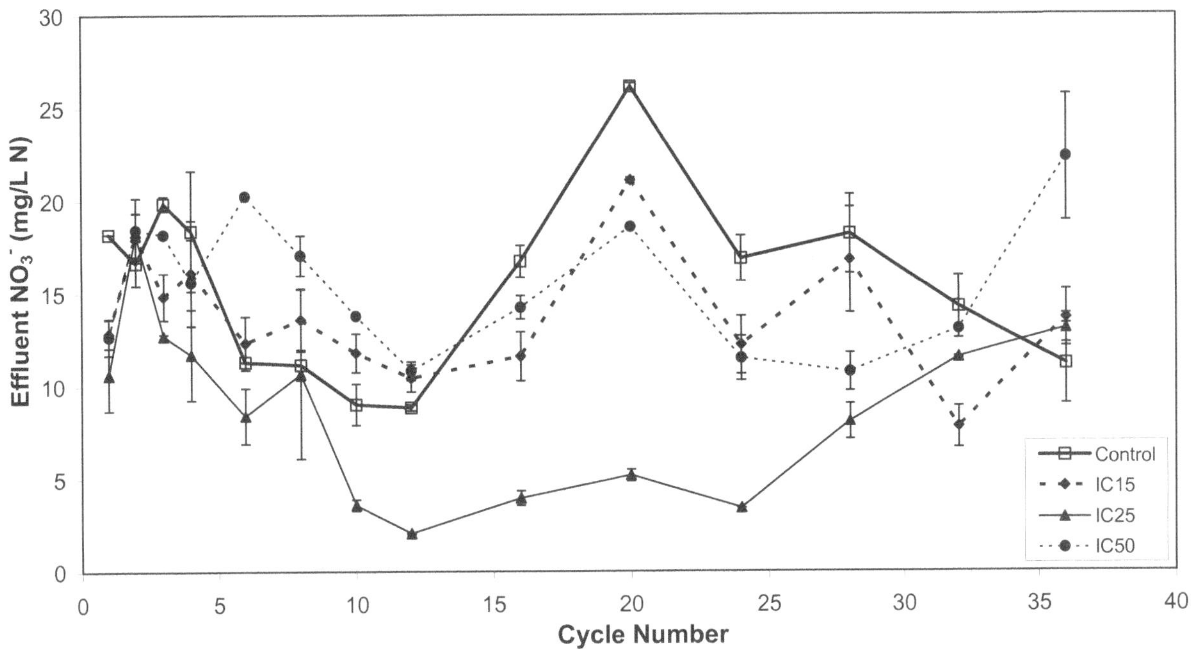

Figure 8-9. Effluent Nitrate-Nitrogen for the 10-Day SRT Reactors Exposed to DNP Shock.

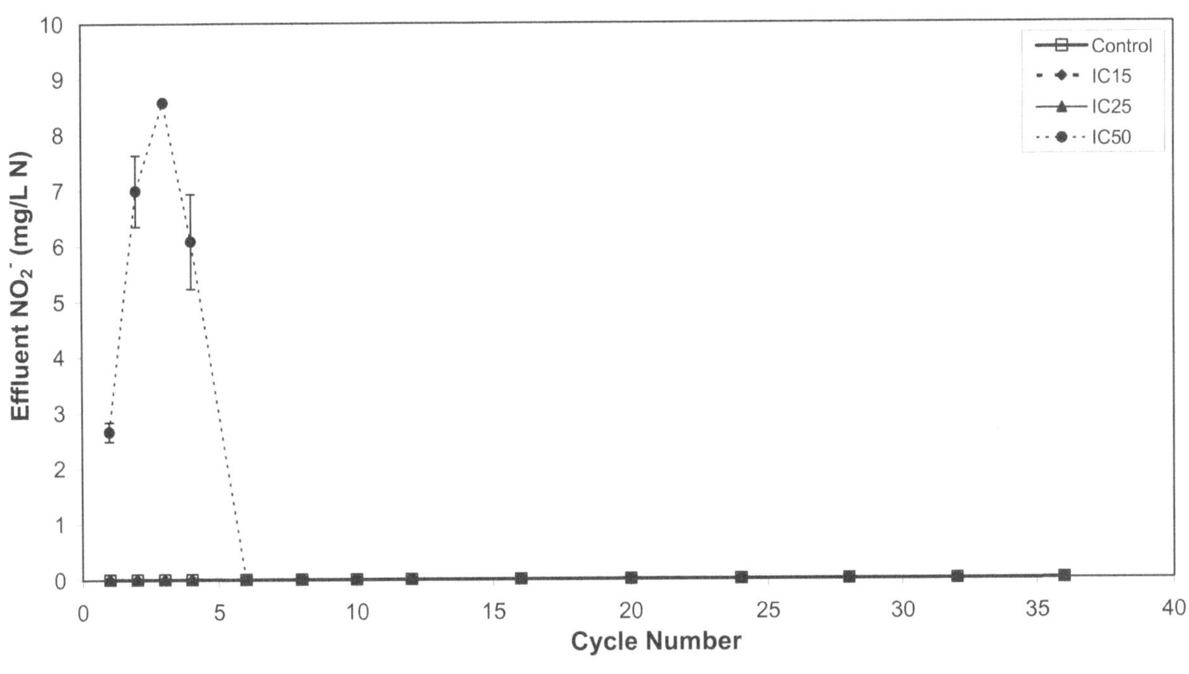

Figure 8-10. Effluent Nitrite-Nitrogen for the 10-Day SRT Reactors Exposed to DNP Shock.

Alkalinity was monitored throughout the source-effect experiments because it is consumed during ammonia oxidation and, therefore, should correlate with the degree of nitrification. As expected, no differences in alkalinity were noticed for the 2-day SRT-shocked reactors versus the control reactor (Appendix G). Figure 8-11 shows the alkalinity concentrations over time for the 10-day SRT reactors. The control reactor had similar alkalinities to the IC_{15} and IC_{50} experimental reactors until cycle 20, but the effluent alkalinity in the IC_{25} reactor was significantly higher than for the control from cycle two onward ($\alpha = 0.05$). This correlates well with the nitrate profile for this reactor, and suggests that less ammonia oxidation occurred (most alkalinity consumption is associated with ammonia consumption). However, given the lack of ammonia or nitrite build up in the IC_{25} reactor, these results do not correlate and the nitrogen mass balance does not close. A higher alkalinity was also recorded for the IC_{15} and IC_{50} reactors in comparison to the control, but only from cycle 20 onward. The difference in alkalinity levels between the shocked and control reactors suggests that AOB inhibition may have occurred due to DNP, particularly in the IC_{25} reactor, but the data is not sufficiently consistent to conclude this with certainty.

Effluent pH values were similar in the 2-day SRT shocked and control reactors (Appendix G). In the 10-day SRT reactor system, for the first two cycles, effluent pH values were significantly lower in the shocked reactors than in the control ($\alpha = 0.05$), as shown in Appendix G. This might be related to the fact that DNP is a weak acid and would tend to decrease the system pH, particularly in the first cycles when the concentration was still considerably high. The magnitude of the pH decline, however, was not very significant, and the pH values were maintained between 7.3 and 8.5 over the course of the experiment.

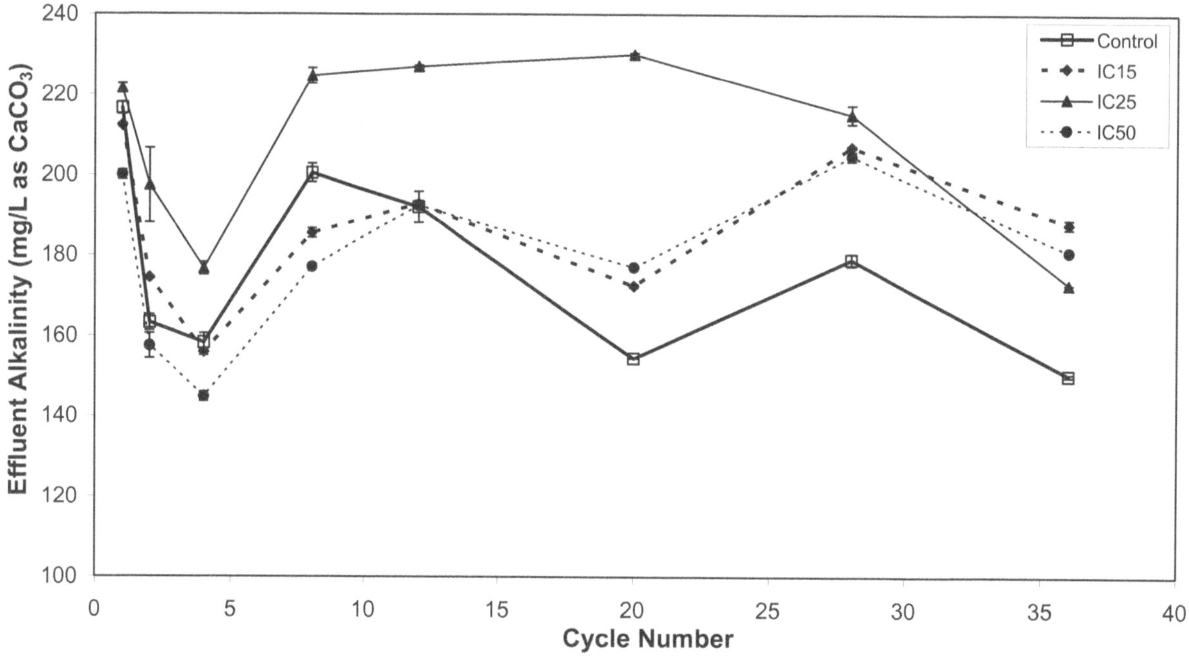

Figure 8-11. Effluent Alkalinity for the 10-Day SRT Reactors Exposed to DNP Shock.

8.2.4 Mixed Liquor Characteristics

8.2.4.1 Settling and Dewatering Characteristics

CST measurements in the 2-day SRT reactors suggest modest improvement to no impact on dewatering properties due to DNP (data in Appendix G). In the 10-day SRT-shocked reactors, however, CST measurements suggest a modest degree of impact on dewatering properties due to DNP (Figure 8-12). Both the IC_{25} and IC_{50} stressed reactors had increased capillary suction times relative to the control for cycle two and this difference continued in the IC_{50} reactor until cycle nine ($\alpha = 0.05$), suggesting that DNP may cause changes in floc structure, which affect the capacity of the biomass to retain water. However, these increased values were still reasonably low in terms of sludge dewaterability.

The SVI data for the 2-day SRT reactors remained in the good range, below 150 mL/g, and no significant differences were found between control and shocked reactors (Appendix G). The SVI data for the 10-day SRT reactors are shown in Figure 8-13. Although the SVI for the 10-day SRT reactors remained in the good range (below 150 mL/g), there was a visible trend indicating an increasing SVI with increasing DNP concentrations. Thus, settleability could be one of the system parameters affected by DNP. Loss of settleability could be related to the lower MLSS concentrations, as high biomass concentrations usually lead to better flocculation conditions.

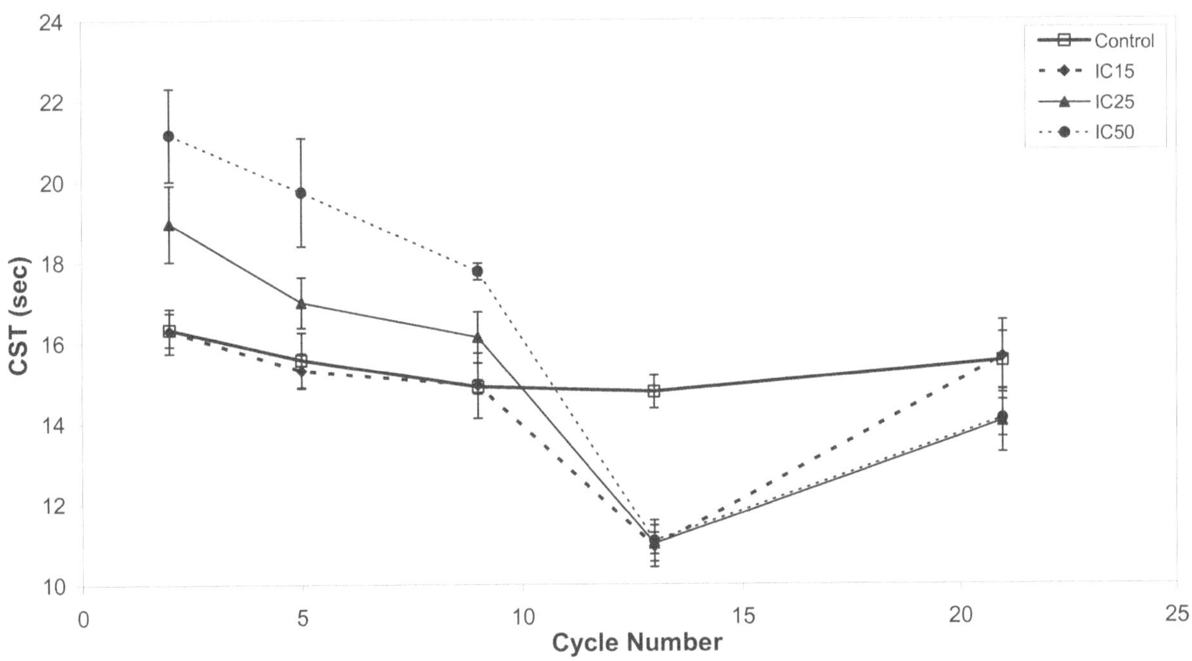

Figure 8-12. Mixed Liquor CST for the 10-Day SRT Reactors Exposed to DNP Shock.

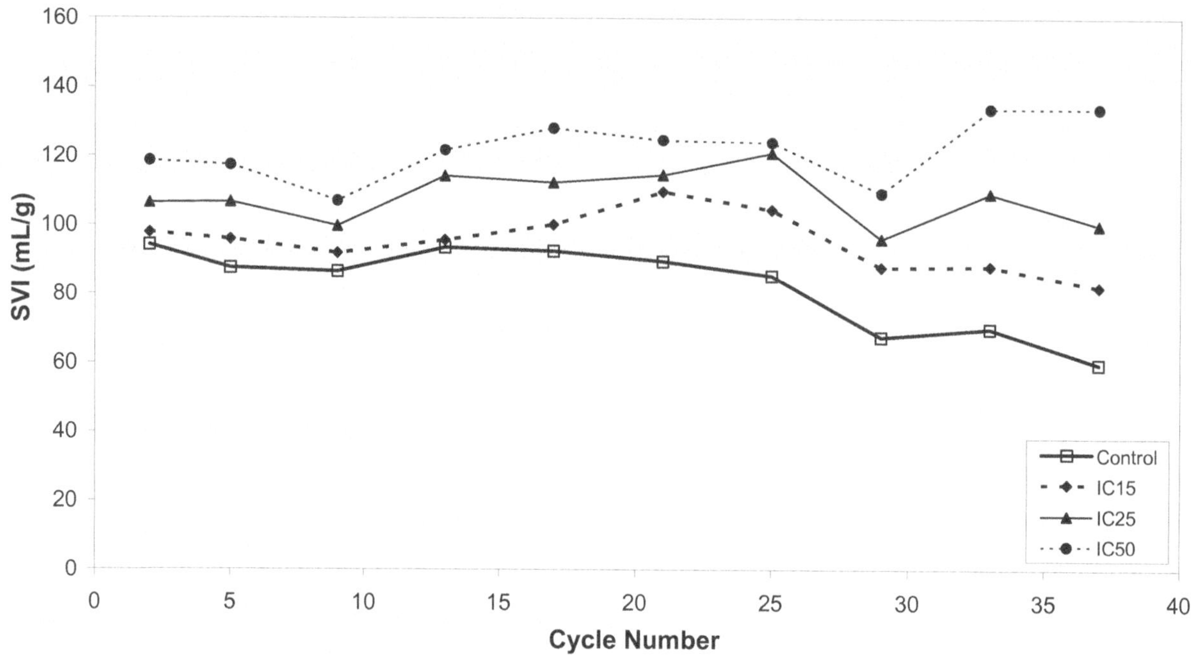

Figure 8-13. SVI for the 10-Day SRT Reactors Exposed to DNP Shock.

8.2.4.2 Mixed Liquor Solids Trends

MLSS and MLVSS data are shown in Table 8-4. In the 2-day SRT reactors, the suspended solids concentrations were rather variable in all reactors, but followed the same trends among control and shocked reactors. The MLSS decreased significantly during the entire experimental period from a concentration of about 1,500 mg/L during the first cycle to about 800 mg/L during the last cycle, a 46% decrease (data in Appendix G). This decrease was not reflected by decreasing influent COD or influent suspended solids, and was observed in all the reactors. It also could not be attributed to a higher SRT than the target two days at the pilot plant, since the average SRT in the pilot plant reactor during the month prior to the laboratory experiment was on target. The influent characteristics at the pilot plant and laboratory experiment were very similar, so discrepancies in reactor feed could not explain this behavior either. The fact that no differences appeared between the control and stressed reactors is not consistent with the uncoupling effects associated with DNP, which should cause energy spilling, thereby inhibiting cell growth.

The results concerning biomass growth in the 10-day SRT-shocked reactors indicate that a strong growth inhibition occurred since the very first cycle of the experiment. In comparison to the control, the MLSS and MLVSS concentrations (Figure 8-14) in the DNP treated reactors were statistically significantly lower since the beginning of the experiment and until cycle 29 ($\alpha = 0.05$). In addition, the difference between the IC_{15}, IC_{25}, and IC_{50} MLSS and MLVSS profiles was not significant, although the average impact in the IC_{50} reactor seemed to be slightly stronger. This behavior is consistent with the uncoupling effects associated with DNP. Figure 8-14 also shows that the mixed liquor suspended solids concentration decreased during the experiment for all reactors. This decrease could be attributed to a higher SRT than the target 10 days at the pilot plant (around 18 days) before the start of the laboratory experiment, while wasting in the lab was based on a 10-day SRT.

Table 8-4. Mixed Liquor Concentrations in 2- and 10-Day SRT Reactors Exposed to DNP.

	MLSS (mg/L)	MLVSS (mg/L)	MLVSS:MLSS Ratio
2-Day SRT reactors			
Control	1,084 ± 287	846 ± 202	0.79 ± 0.069
IC_{15}	1,130 ± 238	879 ± 162	0.78 ± 0.063
IC_{25}	1,077 ± 243	845 ± 178	0.79 ± 0.063
IC_{50}	1,120 ± 237	899 ± 174	0.81 ± 0.073
10-Day SRT reactors			
Control	2,538 ± 329	1,985 ± 268	0.78 ± 0.017
IC_{15}	2,182 ± 260	1,722 ± 190	0.79 ± 0.018
IC_{25}	2,189 ± 242	1,741 ± 181	0.80 ± 0.019
IC_{50}	2,135 ± 214	1,689 ± 155	0.79 ± 0.017

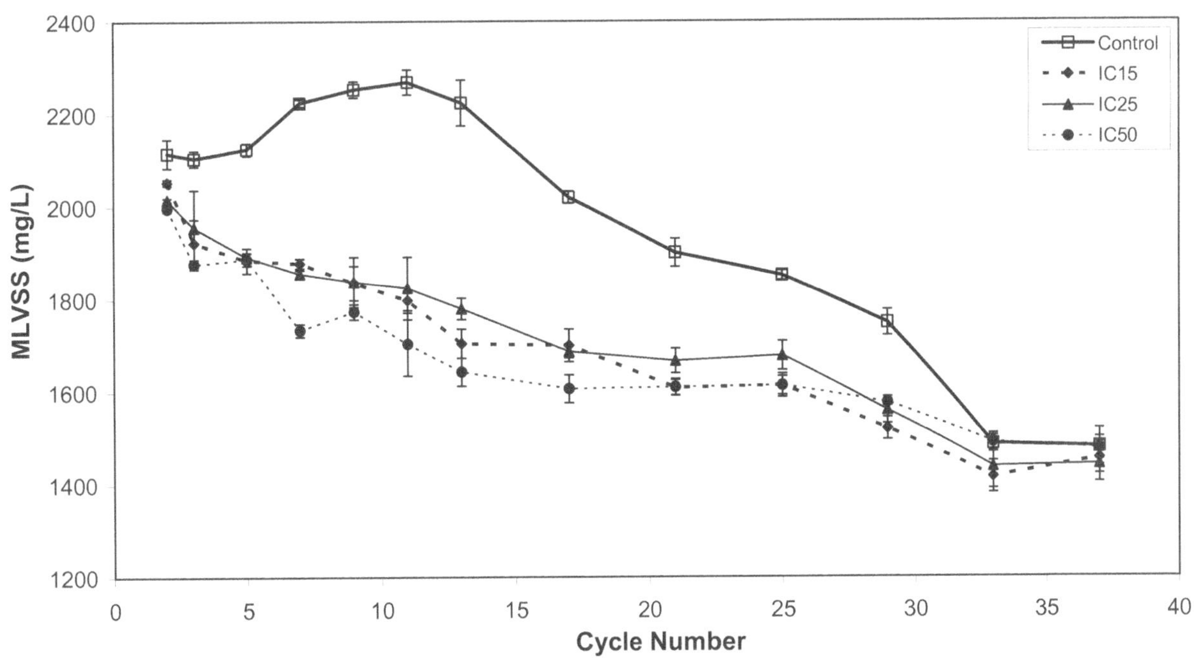

Figure 8-14. MLVSS for the 10-Day SRT Reactors Exposed to DNP Shock.

In the 10-day SRT reactor system, the DNP concentrations for mixed liquor samples collected during the aeration period of a cycle were very similar to the soluble DNP concentrations found in the effluent of that same cycle. This is in agreement with the profile of DNP release from the system, which was almost identical to the theoretical dilution curve, and corroborates the fact that a little or negligible fraction of the DNP adsorbed to the biomass in an irreversible way. Although the 2-day SRT DNP data is not as consistent as the data from the 10-day SRT reactors, the fact that the total DNP levels in the mixed liquor follow the theoretical dilution lines very closely (see Appendix G) is indicative of negligible interaction or sorption of DNP to biomass in activated sludge, just as was found for the 10-day SRT reactor system.

8.2.4.3 Kinetic Indicators of Inhibition

The effects of DNP shock differed between the 2-day SRT non-nitrifying mixed liquor and the 10-day SRT nitrifying biomass in terms of kinetic indicators of inhibition. The SOUR trends in the 2-day SRT reactors showed that moderate inhibition of oxygen uptake was caused by DNP shock (Figure 8-15). The initial concentrations clearly inhibited cellular respiration, but the degree of inhibition 6 hours after the spike was slightly different from the 15%, 25%, and 50% levels: the actual values for the IC_{15}, IC_{25}, and IC_{50} reactors were 24%, 32%, and 42%, respectively. Although low SOUR levels in the shocked reactors were statistically significant only during cycle 2, the respiration rates remained lower than the control throughout the entire experimental period. This behavior was unexpected because, as stated above, the main characteristic of an uncoupler of oxidative phosphorylation is a stimulation of respiration rates; however, increased SOURs relative to the control in the 2-day SRT stressed reactors were not observed.

In contrast, SOUR measurements for the 10-day SRT reactor system showed the characteristic effect of a classical uncoupler on biomass activity. As shown in Figure 8-16, the initial concentrations clearly inhibited cellular respiration, but at cycle nine the trend reversed and the respiration rates per unit of biomass started to increase above the control levels in all shocked reactors, which was indicative of uncoupling. At this point, the DNP concentrations in the IC_{15}, IC_{25} and IC_{50} reactors were approximately 2, 4, and 10 mg/L DNP, respectively. The effect continued to be observed throughout the remaining cycles, even though DNP was not detected in the reactors after cycle 20. It is unclear why the uncoupling effect continued to occur, even after DNP was no longer detected. One contributing factor may relate to a change in the population structure of the activated sludge, with bacteria more resistant to DNP being selected to the detriment of those less resistant.

The SNGR data correspond with patterns observed with effluent nitrogen species concentrations, but do not clarify the nature of the inhibition any further. For the IC_{50} reactor, the SNGR was lower than that measured in the control during the first few cycles (Figure 8-17), and is consistent with the nitrite spikes observed in this reactor during cycles 1 to 4 (Figure 8-10). The IC_{25} reactor showed a lower rate of nitrification relative to the control, and is consistent with lower effluent nitrate (Figure 8-9) and higher residual alkalinity (Figure 8-11) discussed previously. The control and IC_{15} reactors maintained fairly similar SNGRs throughout the experiment. Although the results from the SNGR validate the patterns observed earlier when effluent nitrogen species concentrations were evaluated, they do not shine any new light on the nature of the inhibition.

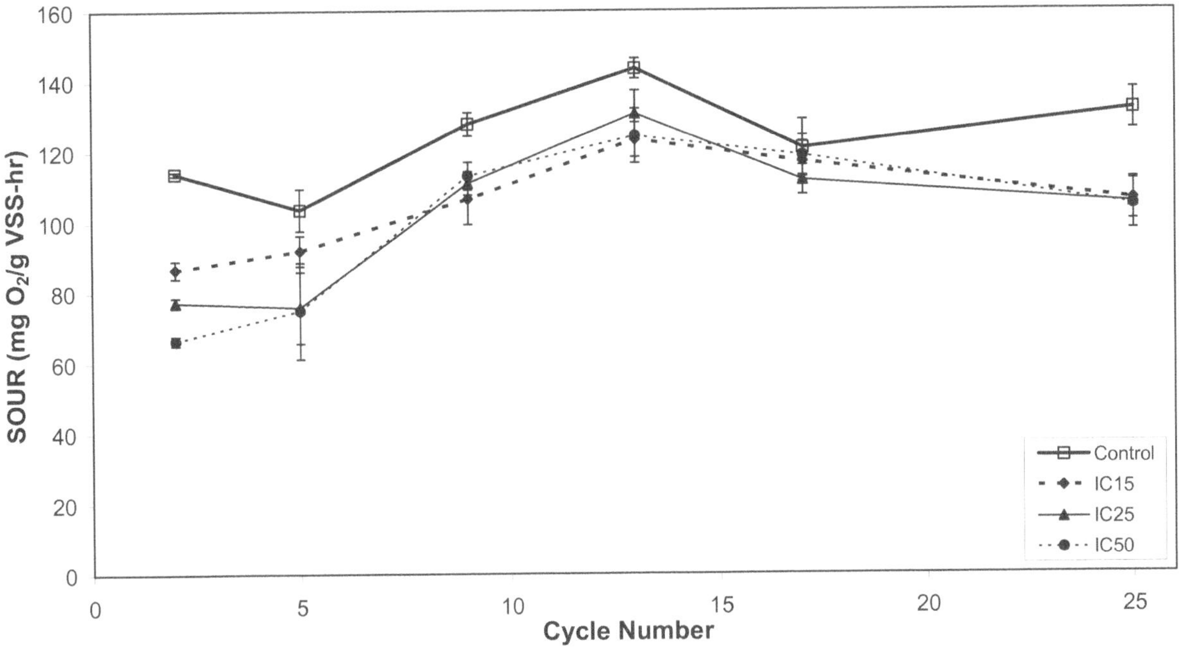

Figure 8-15. SOURs for the 2-Day SRT Reactors Exposed to DNP Shock.

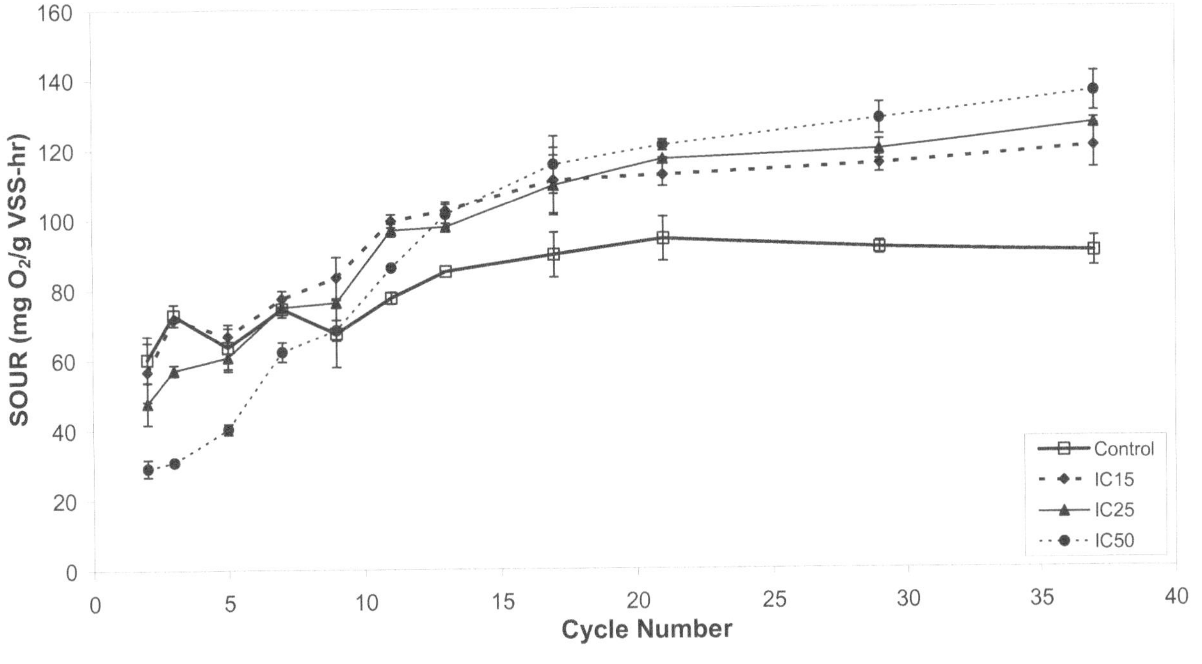

Figure 8-16. SOURs for the 10-Day SRT Reactors Exposed to DNP Shock.

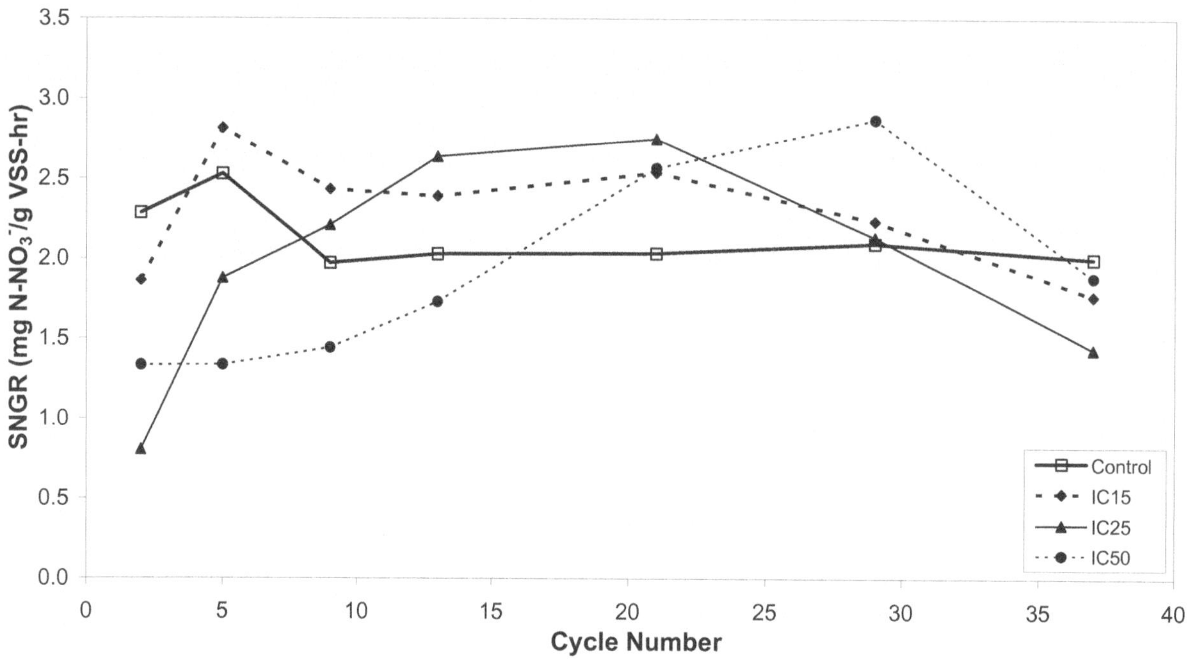

Figure 8-17. SNGRs for the 10-Day SRT Reactors Exposed to DNP Shock.

8.3 Discussion

Overall, DNP shock affected the 10-day SRT biomass much more strongly than the 2-day SRT non-nitrifying biomass. In the DNP-treated 10-day SRT reactors, particularly in the IC_{50} reactor, loss of treatment performance was observed through several parameters. Both soluble COD and effluent suspended solids were affected at the IC_{50} (100 mg/L DNP) shock level. Interestingly, these two effluent quality parameters recovered to control levels at approximately the same time, 1.5–2 days after the shock was inflicted. In addition, nitrification inhibition was noted for a short period in the same reactor and translated into a sudden increase in nitrite levels (which reached levels as high as 9 mg/L), indicating that NOBs were the primary nitrifying microorganism that was inhibited by DNP. Recovery of this inhibitory effect took place about 1.5 days after the shock. It seemed that there was a tendency for these effects to recover at approximately the same time, which may indicate that recovery is mostly dependent on the DNP concentration in solution, which reached values below 40 mg/L in the IC_{50} reactor around cycle six (1.5 days); in other words, as soon as the DNP levels in solution dropped, the substrate uptake capacities of both the heterotrophic and nitrifying populations and the flocculation ability of the biomass recovered immediately.

The most interesting effects noted from the 10-day SRT biomass source-effect experiment related to the uncoupling mechanism that caused inhibition of biomass growth and stimulation of respiration rates. These effects have been heavily documented in the literature as a result of uncouplers of oxidative phosphorylation and were expected to occur during this experiment. Initially, DNP shock caused an inhibition of respiratory functions at high DNP concentrations (first cycles after the shock) during a period that is consistent with the period when loss of COD removal, nitrification and flocculation ability of the biomass (effluent TSS) were observed. It is interesting to note that the stimulation of respiration functions only started after the DNP concentrations in the reactors decreased to levels below 10 mg/L and that the effect continued to be observed even at undetectable DNP concentrations in solution. Since no association of DNP with the biomass was observed, this may have been due to the existence of different populations in the shocked reactors versus the control. Although no direct evidence for this exists, the fact that the settleability of the different biomasses continued to differ during the entire experimental period indicates possible differences in floc structure, which may have resulted from the growth of different populations.

The DNP shock imposed on the 2-day SRT biomass did not affect treatment process efficiency based on effluent quality parameters. The only significant observed effect in the 2-day SRT system was a decrease in oxygen uptake rate in the stressed reactors in comparison to the control reactor. This observation suggests that metabolic activity was inhibited due to DNP, but this inhibition did not result in deterioration of effluent quality or worsening of flocculation characteristics of the biomass. The results do show, however, that uncoupling was not observed in the 2-day SRT biomass. The DNP concentrations used in the experiment were based on the levels that caused an inhibition of SOUR and were higher than those typically used in uncoupler studies. Based on the results from the 10-day SRT system, when uncoupling effects kicked in after the DNP concentration decreased, it appeared that uncouplers can actually function as inhibitors at high enough concentrations. It is also possible that the pH in the reactors — on average 8.3 — was too close to the intracellular pH, which would prevent the uncoupling mechanism from occurring.

8.4 Conclusions

- The impact of DNP shock was more pronounced on the nitrifying 10-day SRT biomass than on the non-nitrifying 2-day SRT biomass for all the parameters tested.
- In the 10-day SRT system, COD removal efficiency and effluent suspended solids were significantly affected by the DNP shock at the IC_{50} level (100 mg/L DNP). Concentrations as high as 40 mg/L (IC_{25} dose) did not adversely affect these parameters. In the 2-day SRT system, no effects on COD removal or effluent TSS were observed at DNP concentrations as high as 50 mg/L. Therefore, rather high DNP concentrations are needed to impact these parameters in activated sludge systems.
- DNP degradation, either chemical or biological, and sorption to biomass were negligible.
- Inhibition of NOB by DNP occurred in the 10-day SRT IC_{50} reactor. The inhibition effect lasted only while DNP concentrations were above 40 mg/l DNP.
- Potassium efflux measurements did not produce results that consistently correlated with effluent turbidity measures except in the case of the IC50 dose for the 10-day SRT system. The higher DNP concentration and higher biomass concentration may have been sufficient to allow detection of a statistically significant increase in effluent potassium concentration, which correlated with an increase in effluent turbidity, suggesting that DNP functioned as a moderate electrophile and activator of GGKE.
- Both total and volatile mixed liquor suspended solids were strongly affected by the DNP shock, which resulted in a significant growth inhibition and consequent decline in biomass concentration in all the 10-day SRT stressed reactors (wastage was maintained constant). In contrast, this effect was not observed in the 2-day SRT biomass.

- Respiration inhibition by DNP was observed through decreases in SOUR during the first cycles after the shock, and was followed by a stimulation in SOUR in all 10-day SRT shocked reactors after cycle nine when DNP concentrations fell below 10 mg/L. This response was observed for the remaining of the experiment, even though DNP concentrations in the shocked reactors were below detection after cycle 20. SOURs of the 2-day SRT biomass were inhibited during the entire period of the experiment; the expected increase in respiration rates due to uncoupling of oxidative phosphorylation was not observed at the lower SRT.

- Both sludge dewaterability and settleability were modestly affected by DNP shock, and may indicate a change in the floc structure caused by the chemical, but these effects were only seen for the 10-day SRT system.

CHAPTER 9.0

pH SHOCK EXPERIMENTS

9.1 Literature Review

Domestic wastewater contains a range of salts that provide buffering capacity, which keeps the pH near or slightly above neutral values (Metcalf and Eddy, 2003). In contrast, industrial wastewaters can have variable or extreme pH values, depending upon the wastewater source, and often require buffering or pH adjustments (Eckenfelder, 2000). Industrial wastewaters that are treated alone or blended with domestic wastewater can undergo dramatic fluctuations in influent pH, thereby challenging the buffering capacity of the wastewater and causing pH changes in downstream bioreactors. Such changes in pH can hinder system performance by retarding the activity of both autotrophic and heterotrophic bacteria (Chong et al., 1997). The optimum pH range for most activated sludge treatment systems is usually between 6.5 and 8.0 (Grady et al., 1999). Mixed liquor organisms can die or experience retarded function if the pH drops below 4 or increases above 10 (Chong et al., 1997). Acidic pH levels can promote the growth of filamentous fungi and decrease the rate of organic stabilization in aerobic digestion, while alkaline pH levels can lead to microbial death (Grady et al., 1999).

Changes in pH can have both direct and indirect effects on microorganisms. Extreme pH values can directly damage external cell structures outside the cytoplasmic membrane, such as flagella, chemoreceptors, and cell walls (Dilworth and Glenn, 1999). There are few options for the cell to protect these structures from external chemical toxins, so the cell must produce structures that are resistant to adverse conditions, or eliminate the function altogether. This may mean, for example, that cells can become nonmotile and unable to relocate to more favorable environments in the presence of chemical stress (Dilworth and Glenn, 1999). At acidic or alkaline pH, bacteria are challenged not just by excesses of H^+ or OH^-, but also by excesses or shortages of metal ions with solubilities that are affected by the pH shift. Under acidic conditions, it is necessary to consider the impact of the increased availability of aluminum, copper, zinc, cadmium, and manganese. At pH levels below five, cells may be unable to prevent the uptake of these metal ions, which can activate stress mechanisms. To survive under low pH conditions, cells need to acquire detoxification mechanisms like those of acidophiles. Under alkaline conditions, many of these same heavy metal ions are less bioavailable. Since some are micronutrients (e.g., Na, Ca, Mg, Fe, Mn, Zn, Ni, Mo, Cu), limitations can result in limited cell growth and possible changes in biomass physiology.

Intracellular components are affected by changes in pH in a variety of ways. It is important for cells to hold intracellular pH fairly constant. They are able to achieve a relatively stable pH, even with significant changes to extracellular pH, through a range of mechanisms. The first defense against unfavorable pH is the minimization of membrane permeability and maintenance of intracellular pH homeostasis, which is the ability to sustain intracellular pH within a narrow range of values despite variations in the pH of the environment (Booth, 1999; Dilworth and Glenn, 1999). It is generally accepted that this is achieved by controlling the activity of ion transport systems that facilitate proton entry and loss (Booth, 1985). A second defense option involves energy-requiring pumping systems. In bacteria, protons are translocated outwards across the cytoplasmic membrane by means of primary proton pumps linked to electron transport and ATP hydrolysis. It has been suggested that these pumps play a role in controlling internal pH in coordination with K^+ (a major cellular cation) transport (Nakamura et al., 1984). Another defense against unfavorable pH is internal buffering. The ability of cells to use internal buffering capacity to recover from intracellular pH perturbation depends on the magnitude of the change (Booth, 1999). Cytoplasmic buffering is most likely due to the creation and accumulation of metabolites that have buffering features (Dilworth and Glenn, 1999). For most cells, the buffering capacity is derived through titration of the phosphate groups and carboxylates of physiologically abundant metabolite species (Booth, 1999). This may be a good way to delay adverse pH effects, but can be difficult to maintain as a long-term survival strategy (Dilworth and Glenn, 1999).

Nitrification can be greatly affected by changes in pH. Although all bacteria grow poorly outside the normal physiological pH range of 6.0 to 8.0, nitrifying bacteria are particularly sensitive to pH (Grady et al., 1999). The pH range for growth of pure cultures of ammonia oxidizers is 5.8 to 8.5, and the pH range for growth of nitrite oxidizers is 6.5 to 8.5 (Prinčič et al., 1998). However, a key factor to consider when defining the preferred pH range for growth of these organisms is the total concentration of inorganic nitrogen species (ammonia/ammonium and nitrite/nitrous acid), which speciate as a function of pH. For example, at high but industrially realistic ammonia/um concentrations, sufficient free ammonia will exist in solution at pH 8 to inhibit AOBs (Anthonisen et al., 1976). Therefore, "ideal" pH ranges need to be interpreted in a broader context.

The efficient operation of biological treatment systems depends largely on the interaction between bacteria and the community of various microfauna present. This microfaunal community is dominated by protozoa (ciliates, flagellates, shelled amoeba), fungi, rotifers, and nematodes (Baldwin and Campbell, 2001). Each of these different types of microorganisms exhibits different tolerance to pH shifts and responds differently to changes in pH. It has been found that most protozoan species function best at pH levels between 6 and 8, but can tolerate pH levels as acidic as 4.5 (Baldwin and Campbell, 2001). Under acidic conditions, protozoan cells may incur increased permeability, which raises energy costs associated with osmoregulation. As a result, a larger mass of food (bacteria) may be needed to maintain protozoan growth at low pH in wastewater treatment systems.

Higgins and Novak (1997a) established that cation concentrations in the influent of laboratory-scale biological treatment system reactors had a significant effect on the settling and dewatering properties of activated sludge. From this information, it was concluded that there exists a cation balance for a given system, which optimizes settling and dewatering properties. It was suggested that this was especially important for industrial systems, where a cation imbalance is more likely to occur. It was determined that a ratio of monovalent to divalent cations (M:D) greater than about 4, expressed on a milliequivalent basis, resulted in deteriorated settling and dewatering characteristics (Higgins and Novak, 1997b). It is important to understand and consider this effect when designing source-effect experiments to test the impact of alkaline pH values on activated sludge system performance because bases contain monovalent and divalent cations.

9.2 Results

9.2.1 pH Levels

The pH maintained in the control reactor for previous experiments averaged between 7 and 8. The pH levels chosen for this experiment illustrate the effect of pH over a wide range. An acidic pH of 5 and alkaline pH values of 9 and 11 were chosen, while the control reactor pH values averaged 7.58 $\pm$ 0.16 and 7.67 $\pm$ 0.15 for the 2- and 10-day systems, respectively. A solution containing equal concentrations (on a milliequivalent basis) of NaOH and $Ca(OH)_2$ was used for the base pH adjustment experiments to avoid causing deflocculation through an imbalanced M:D. For the acid shocked reactor, 1N sulfuric acid was used instead of HCl to prevent chloride interference during COD and ion chromatography procedures.

9.2.2 Influent Characteristics

Average influent characteristics over the course of the pH source-effect experiment are given in Table 9-1. The influent quality was relatively constant over the period of the 10-day SRT experiment, but not during the 2-day SRT experiment, as indicated by the high standard deviations for almost all the quality parameters tested. Influent supplementation was used during this experiment. Consequently, the influent composition reveals that the wastewater used as the feed for the source-effect experiments was medium in strength (Metcalf and Eddy, 2003).

Table 9-1. Characteristics of Wastewater Used During pH Source-Effect Experiments.

Characteristic	2-Day SRT Average $\pm$ std dev	10-Day SRT Average $\pm$ std dev
Alkalinity (mg/L as $CaCO_3$)	734 $\pm$ 367	286 $\pm$ 21
Total Ammonia (mg/L as N)	14.3 $\pm$ 3.4	23.7 $\pm$ 16.1
COD, soluble (mg/L)	639 $\pm$ 275	537 $\pm$ 85
COD, total (mg/L)	889 $\pm$ 142	698 $\pm$ 120
pH	7.6 $\pm$ 0.2	7.5 $\pm$ 0.3
TSS (mg/L)	284 $\pm$ 176	116 $\pm$ 37
VSS (mg/L)	99 $\pm$ 90	91 $\pm$ 38

9.2.3 Effluent Characteristics

9.2.3.1 Soluble COD

Effluent sCOD data shows that the alkaline pH shock greatly impacted treatment efficiency for both the 2- and 10-day SRT reactors, as shown in Figures 9-1 and 9-2, respectively. There was no significant change in the effluent sCOD of pH 5 and 9 stressed reactors for both biomass sources compared to the controls ($\alpha = 0.05$). In contrast, the effluent sCOD remained elevated in the pH 11 stressed reactor relative to the control throughout the 2-day SRT experiment, and until cycle 52 (1.3X SRT) for the 10-day SRT system.

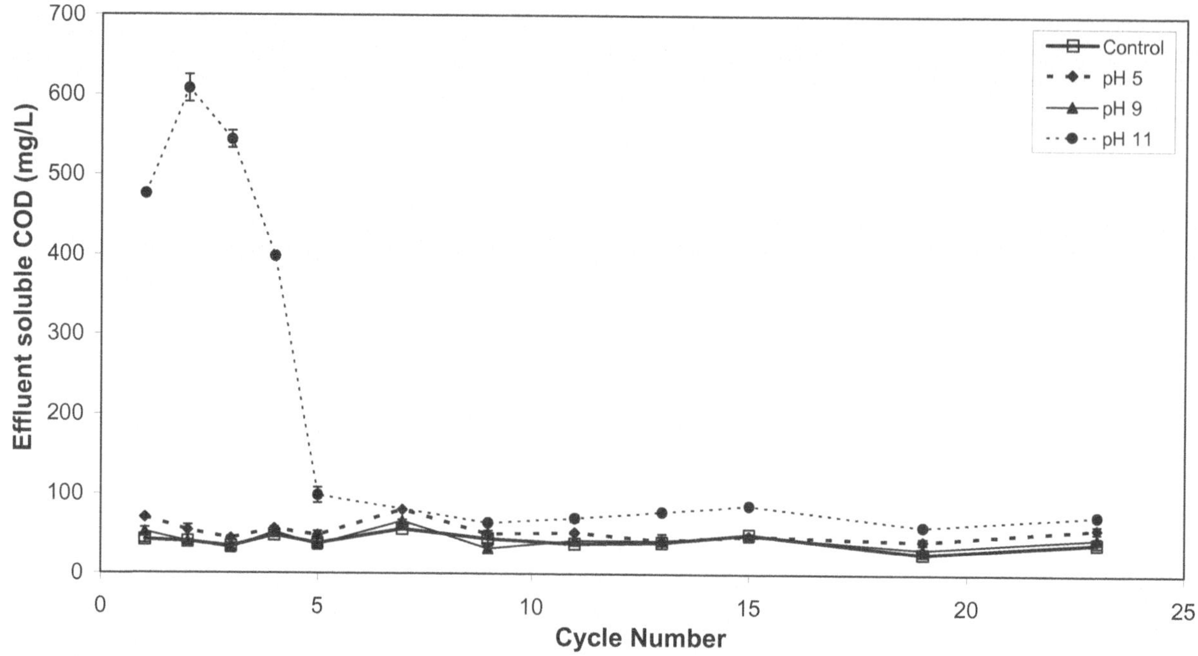

Figure 9-1. Effluent Soluble COD for the 2-Day SRT Reactors Exposed to Various pH Values.

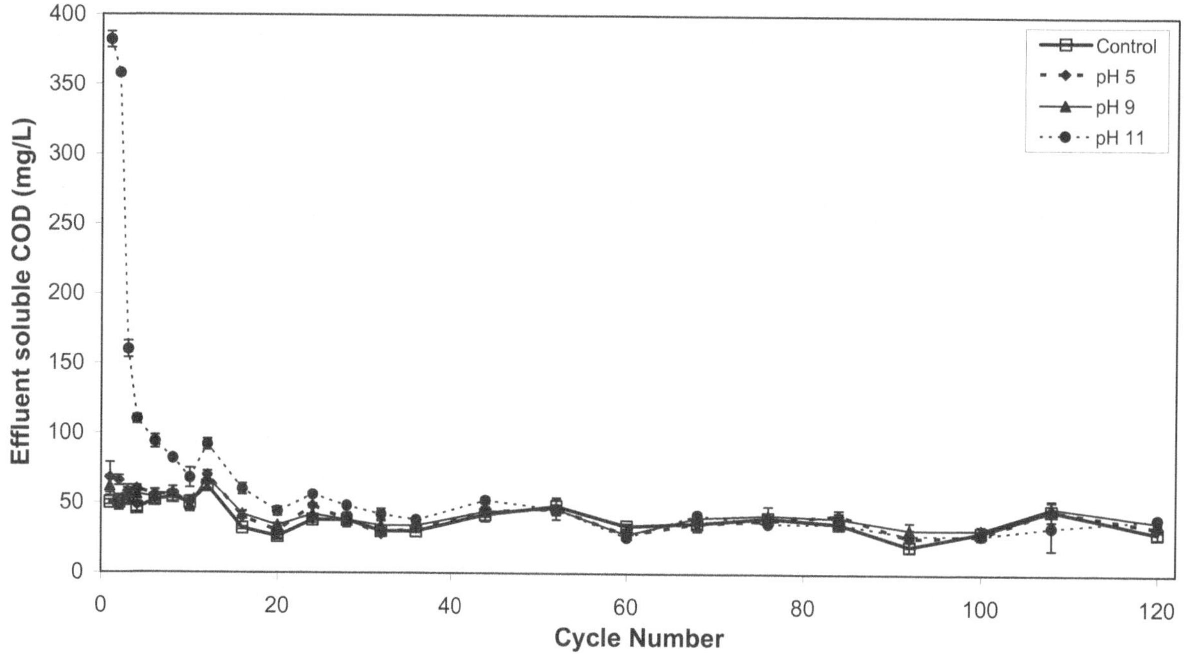

Figure 9-2. Effluent Soluble COD for the 10-Day SRT Reactors Exposed to Various pH Values.

9.2.3.2 TSS and Potassium Concentrations

Effluent TSS was significantly impacted initially by the alkaline shock for the 2-day SRT system, and the acidic and alkaline shocks for the 10-day SRT system. In the case of the pH 11 reactors, the increase in effluent TSS was due to high turbidity that appeared to be caused by deflocculation. Deflocculation was observed beginning with the first cycle after the shock in both the 2- and 10-day SRT pH 11 shocked reactors, resulting in effluent TSS concentrations that were about 2.5 times higher than the control in both cases (Figures 9-3 and 9-4). Statistical recovery from the pH 11 shock occurred in the 2-day SRT reactor by cycle five; interestingly, effluent TSS levels continued to decrease relative to the control until the last cycle. Since high pH values can cause precipitation of carbonate salts, it is speculated that coagulation of suspended solids by carbonate precipitates may have contributed to the lower effluent TSS in the pH 11 reactor towards the end of the experiment. The pH 5-stressed reactor showed a pattern that was the opposite of that seen with the pH 11 shocked biomass in the 2-day SRT system. Effluent TSS was significantly less than the control for the first nine cycles, then was statistically higher than the control for the remainder of the experiment ($\alpha = 0.05$). Despite the statistical significance of this trend, it is unclear if it was due to the pH 5 shock or just random mixed liquor performance. The lack of a similar trend during a previous 2-day SRT pH shock experiment (data not shown) supports the notion that the variation seen in Figure 9-3 was not due to acidic pH shock.

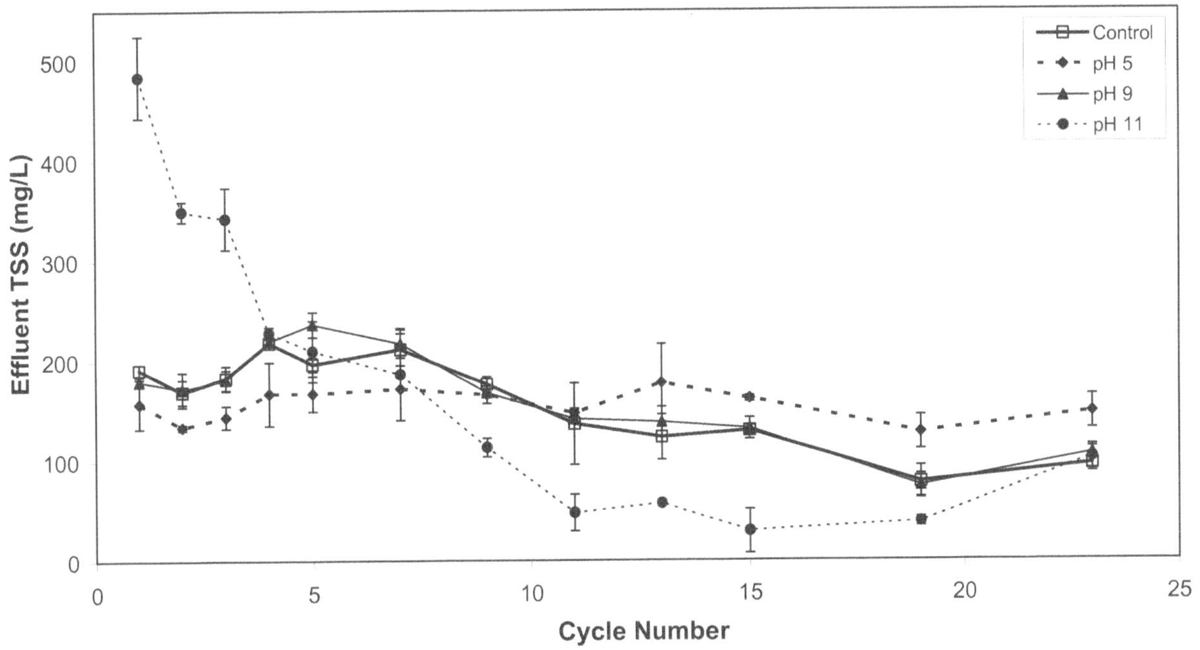

Figure 9-3. Effluent TSS for the 2-Day SRT Reactors Exposed to Various pH Values.

Similar to the 2-day SRT system, very high alkaline conditions corresponded with increasing effluent turbidity during the first few cycles for the 10-day SRT biomass (Figure 9-4); however, by cycle 12 (day three), there was no statistical distinction between the effluent TSS concentrations for the pH 11 stressed reactor and the control reactor ($\alpha = 0.05$). From that point on, the 10-day SRT stressed reactors followed the general trend of the control reactor, but all remained at slightly (but often statistically significant) lower levels. The pH 5-stressed reactor in the 10-day SRT system also showed a significant response in terms of high effluent turbidity, but recovered by cycle three and performed similar to or better than the control for the remainder of the experiment.

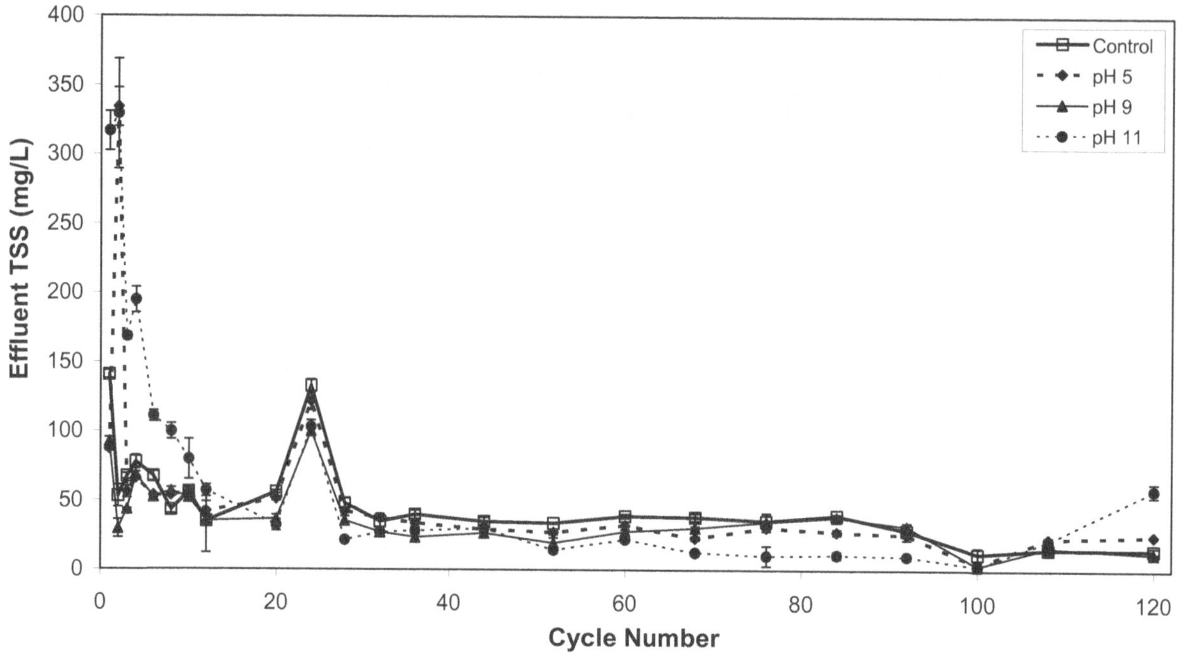

Figure 9-4. Effluent TSS for the 10-Day SRT Reactors Exposed to Various pH Values.

Effluent potassium concentration results for the 2-day SRT system showed increased potassium levels for the pH 11 reactor relative to the control for the first five cycles ($\alpha = 0.05$), while no significant differences were found between the control and the pH 5 and 9 shocked reactors (Appendix H). Increased potassium levels in the pH 11 reactor may have been due to cell lysis or result from the activation of stress response mechanisms in the bacterial cells to counteract the effects of the high extracellular pH. Potassium data for the 10-day SRT were inconclusive due to problems with the analytical method.

9.2.3.3 Nitrification Inhibition

The 2-day pilot plant reactor was operated to prevent nitrification and was not nitrifying the day before the experiment began; however, the biomass was partially nitrifying at the beginning of the laboratory experiment. The presence of nitrification in the 2-day SRT system was surprising because the reactors were maintained at a reduced temperature to suppress the growth of nitrifiers. Although the pilot plant's actual SRT averaged two days, the 2-day SRT laboratory reactors were operated at an SRT of less than two days (Figure 9-5) due primarily to lower effluent TSS concentrations in the 2-day SRT pilot plant reactor (average effluent TSS around 58 mg/L). Despite these unfavorable growth conditions for nitrifiers, the control reactor nitrified completely and no ammonia remained in the effluent throughout most of the experiment (Figure 9-6). Except for the first cycle, the pH 9 shocked reactor had similar ammonia concentrations as the control reactor, which was expected since the pH was very close to that measured in the control. The pH 5 acid shock initially inhibited AOB but recovered by cycle six when the levels began to follow the recovering control reactor. The pH 11 shock greatly inhibited ammonia oxidation. Ammonia concentrations in the effluent of the pH 11 shocked reactor remained elevated above those in the control, except between cycles six and 12. The elevated ammonia concentrations observed during the first six cycles of the pH 11 shocked reactor were higher than those observed in the non-nitrifying pilot plant reactor and are consistent with the notion that heterotrophic bacteria, which use ammonia as a nitrogen source, were inhibited along with the AOB. This is corroborated by the elevated effluent sCOD concentrations in Figure 9-1. Additionally, the loss of TSS in the effluent may have contributed to a washout effect for nitrifiers early during the experiment when the true SRT was particularly low (Figure 9-5). Nitrite concentrations were consistently detectable in all reactors including the control, but remained below 1.8 mg/L as N, which further indicates that incomplete nitrification occurred in all systems (see Appendix H). Nitrate concentrations were generally below 2 mg/L for the control reactor, but higher in all the shocked reactors throughout the experiment (see Appendix H). The pH 11 and 9 reactors generated the highest nitrate concentrations, while the pH 5 showed higher nitrate relative to the control at the beginning of the experiment. This indicates that the shocked reactors started nitrifying to a greater degree than the control. The reason for this stimulation is unclear. In summary, partial nitrification was occurring throughout the 2-day SRT pH shock experiment and inhibition of ammonia oxidation was primarily seen in the pH 11 and pH 5 shocked reactors. Inhibition of heterotrophic bacteria could have also contributed to the increase in effluent ammonia concentrations, especially in the pH 11 shocked reactor. Other complicating factors due to partial nitrification in a low SRT biomass may have also contributed to the observed effects.

In the nitrifying 10-day SRT system, nitrification was significantly affected by a pH 11 shock event. Effluent ammonia concentrations spiked very quickly in the pH 11 reactor (Figure 9-7) and stayed at levels above 20 mg/L from cycles 2 - 36. After cycle 36 (day 9) effluent nitrite levels quickly increased in the pH 11 shocked reactor (Figure 9-8) and continued to stay high until cycle 60 (day 15). The initial high ammonia levels in the pH 11 reactor indicate that the AOB were inhibited. The main nitrite peak occurred after rapid ammonia oxidation was initiated and was probably due to a rapid influx of nitrite, which the NOB could not dissimilate fast enough. From cycle 36 onward, the pH 11 stressed reactor nitrate levels increased slowly, which correlated with a decrease in ammonium. The slight nitrite buildup did not prevent nitrite oxidation, and nitrate was formed. Overall, the rate of recovery from deflocculation or effluent sCOD disruption was more rapid than that from nitrification inhibition in the pH 11 shocked system. Effluent nitrate concentrations in the pH 5 and 9 reactors generally remained below those in the control reactor, suggesting that a modest level of inhibition may have occurred that was not indicated by the effluent ammonia data.

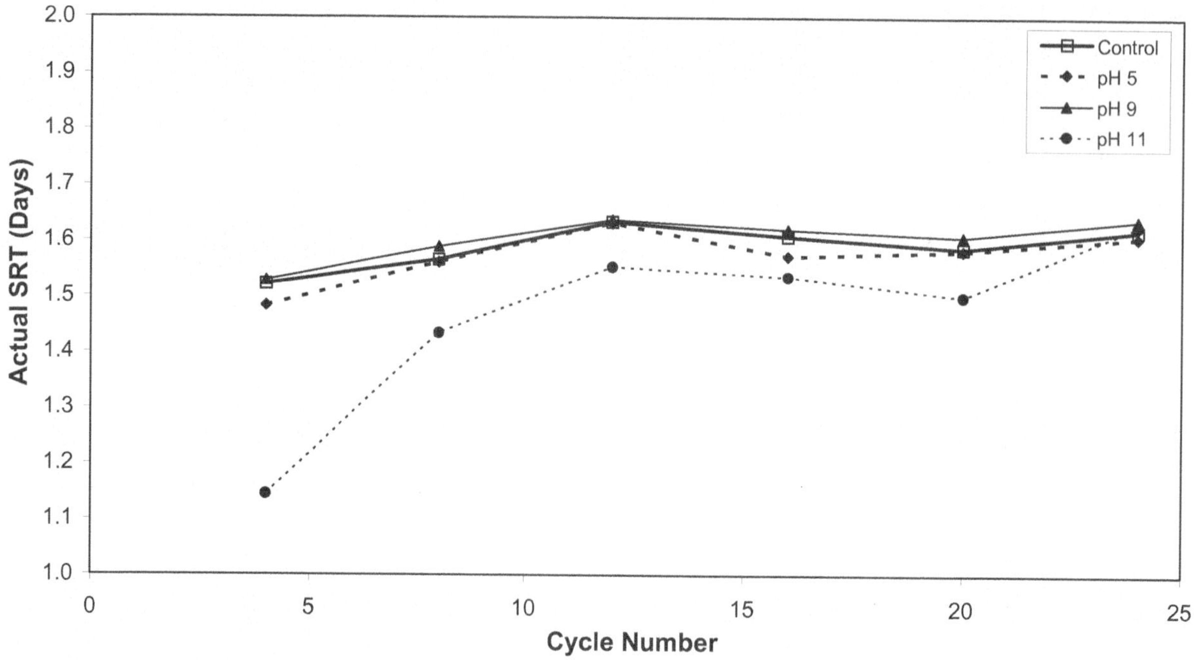

Figure 9-5. Actual SRT for the 2-Day SRT Reactors Exposed to Various pH Values.

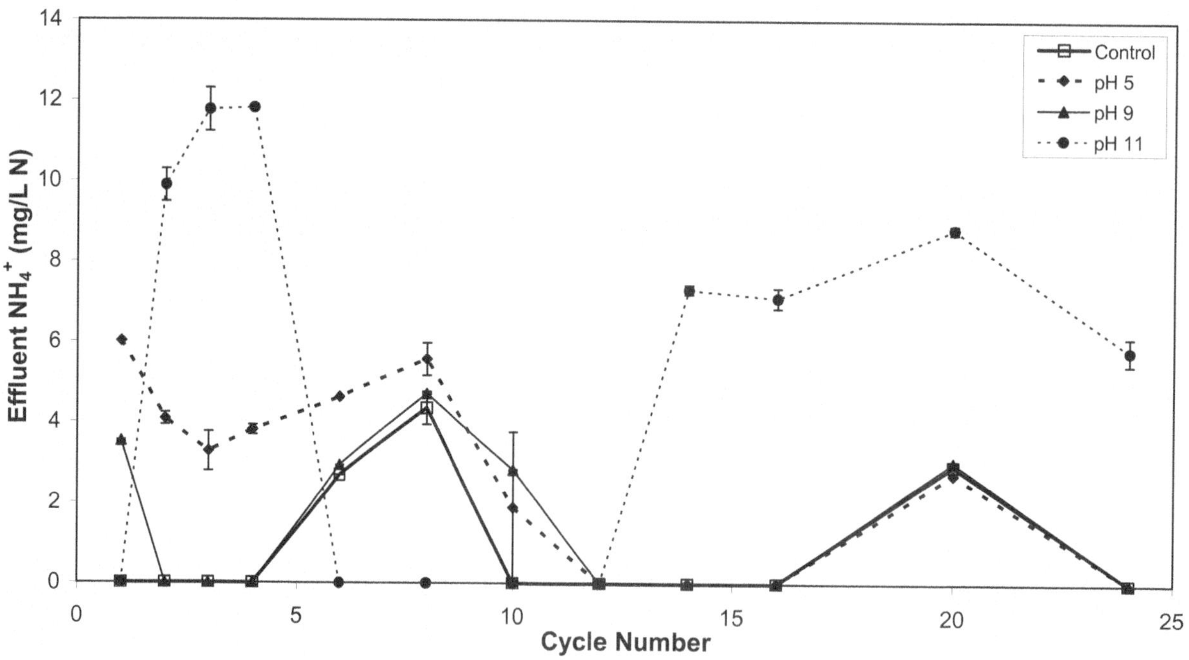

Figure 9-6. Effluent Ammonia-Nitrogen for the 2-Day SRT Reactors Exposed to Various pH Values.

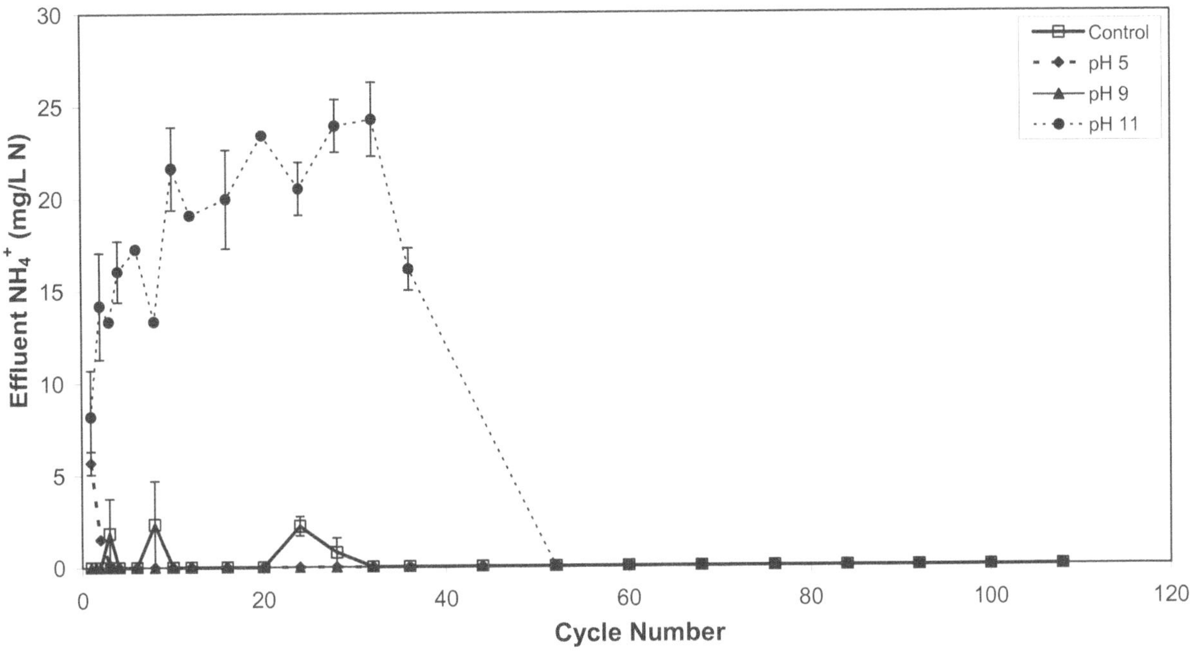

Figure 9-7. Effluent Ammonia-Nitrogen for the 10-Day SRT Reactors Exposed to Various pH Values.

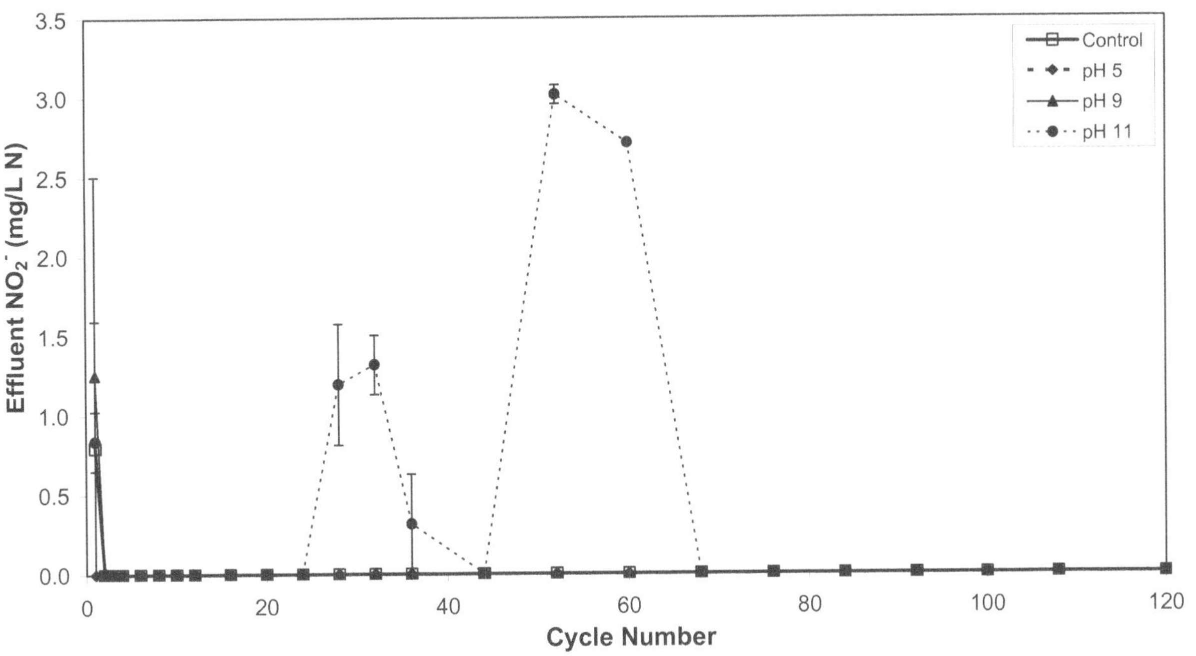

Figure 9-8. Effluent Nitrite-Nitrogen for the 10-Day SRT Reactors Exposed to Various pH Values.

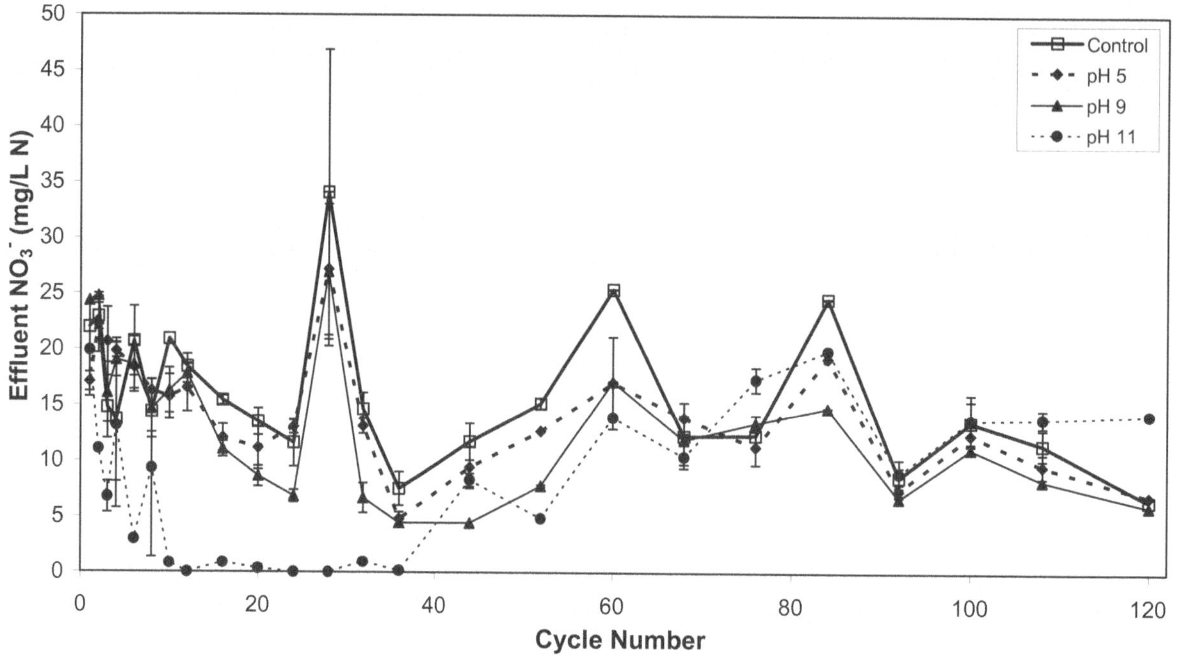

Figure 9-9. Effluent Nitrate-Nitrogen for the 10-Day SRT Reactors Exposed to Various pH Values.

Since effluent alkalinity concentrations were affected by both the degree of nitrification occurring and the pH in the reactor due to the initial shock, it was difficult to use alkalinity as an independent indicator of nitrification inhibition for these source-effect experiments. Therefore, an intensive analysis of alkalinity data relative to nitrification inhibition was not completed. It is interesting to note, though, that the alkalinity in the pH 11 shocked 10-day SRT reactor returned to control levels (Figure 9-12) around the same time that ammonia oxidation was initiated, according to Figure 9-7.

In both the 2- and 10-day SRT systems, Figures 9-10 and 9-12 show that acidic shocks resulted in decreased effluent alkalinity. The slow increase in alkalinity to control levels in the pH 5 shocked systems correlated with a slow increase in pH from acidic conditions to neutral conditions (Figures 9-11 and 9-13), and suggests that some of this alkalinity loss was due to CO_2 stripping. Slight increases in alkalinity in pH 9 shocked reactors relative to the control were brief and did not correlate with significant ammonia oxidation inhibition effects. Finally, the effluent pH values in both the 2- and 10-day SRT systems returned to control levels in less than one SRT for both systems (Figures 9-11 and 9-13).

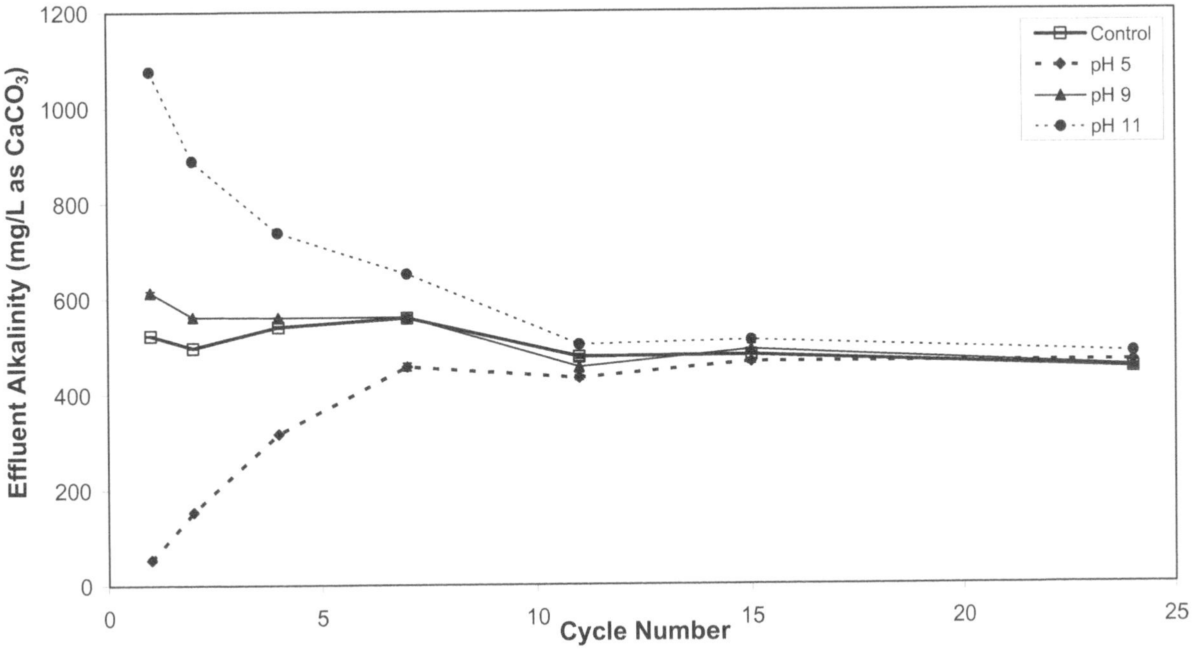

Figure 9-10. Effluent Alkalinity for the 2-Day SRT Reactors Exposed to Various pH Values.

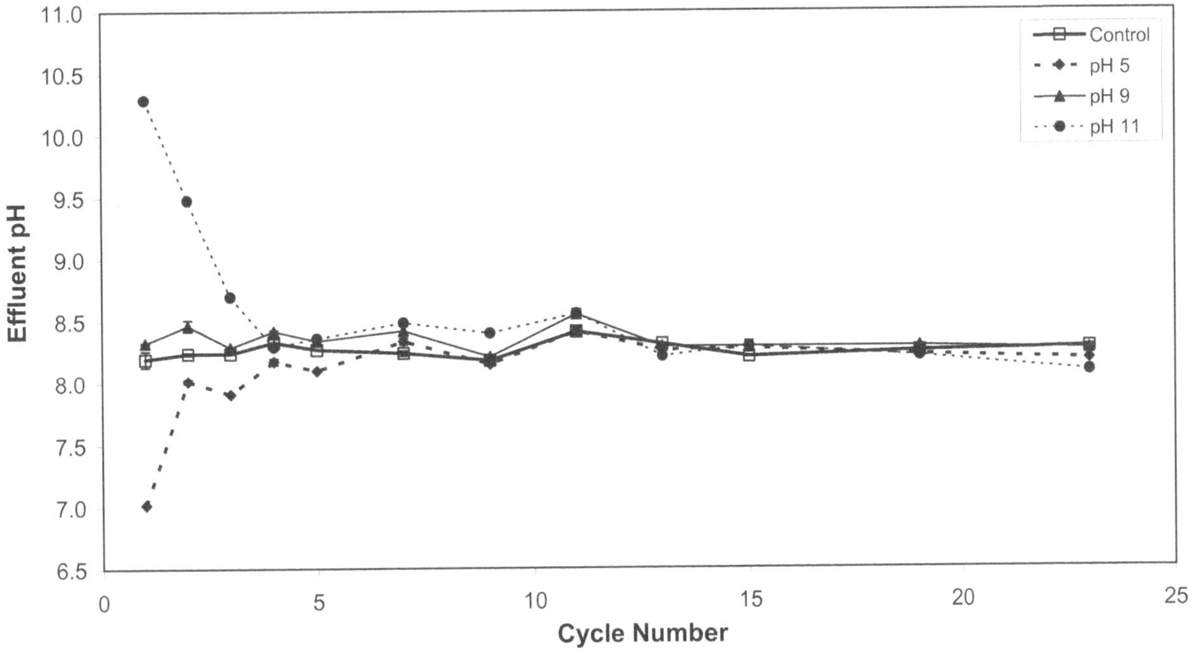

Figure 9-11. Effluent pH for the 2-Day SRT Reactors Exposed to Various pH Values.

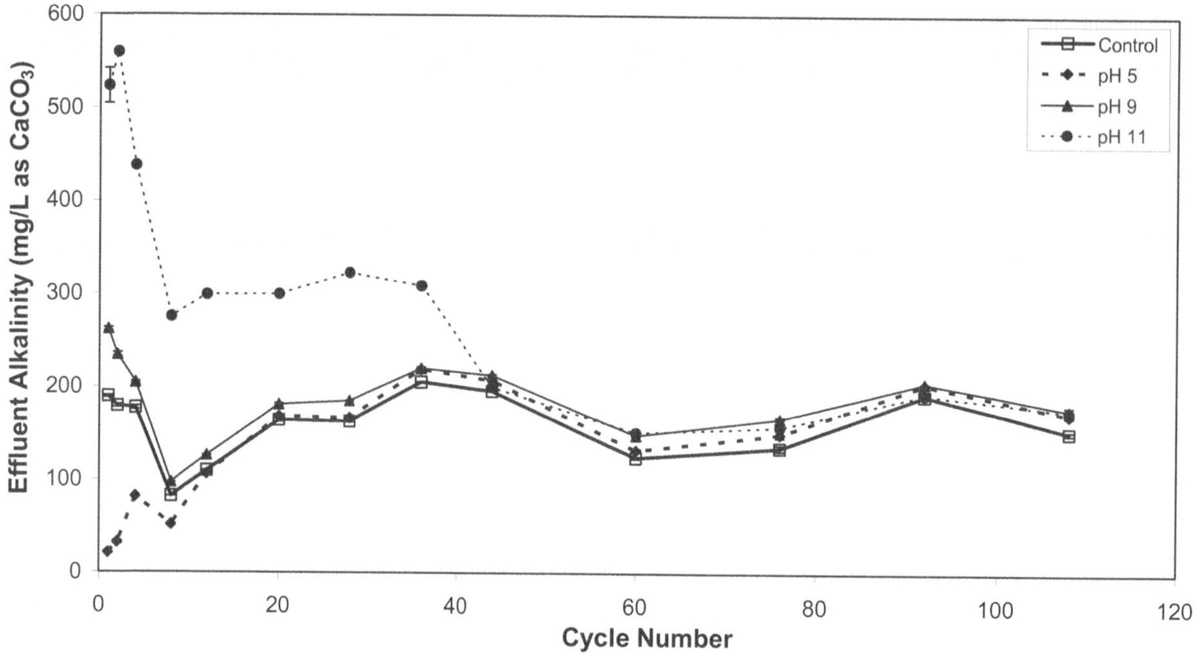

Figure 9-12. Effluent Alkalinity for the 10-Day SRT Reactors Exposed to Various pH Values.

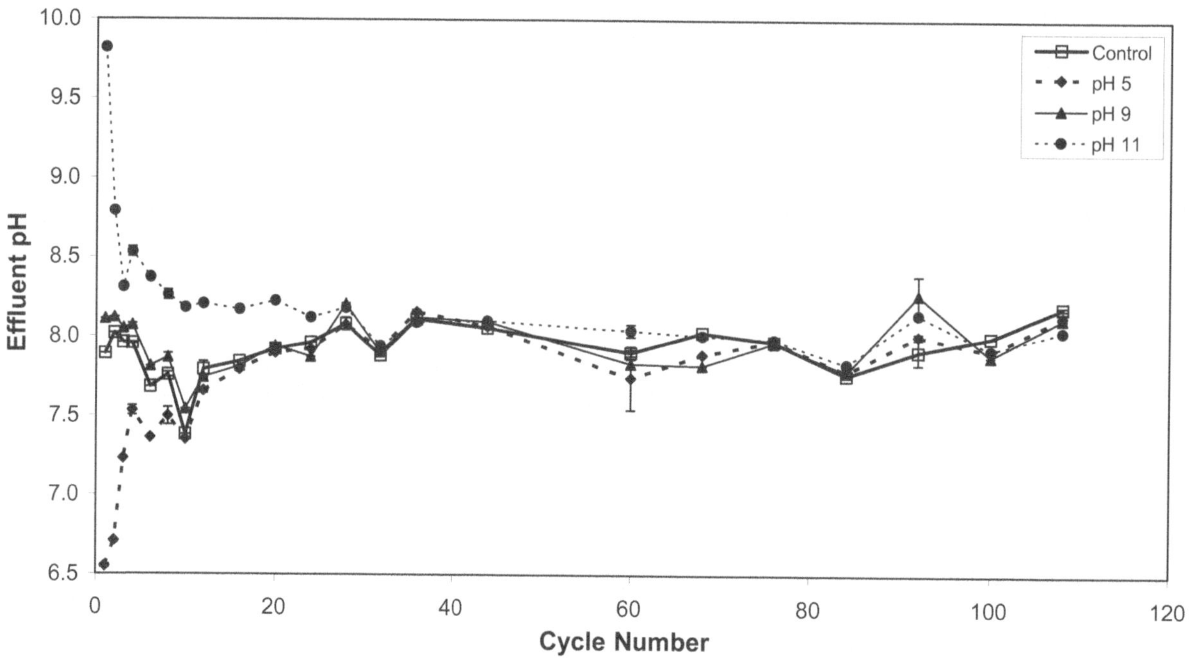

Figure 9-13. Effluent pH for the 10-Day SRT Reactors Exposed to Various pH Values.

9.2.4 Mixed Liquor Characteristics

9.2.4.1 Settling and Dewatering Characteristics

SVI trends in the 2-day SRT reactors showed that the pH 11-exposed biomass had a lower SVI value than that of the control and other stressed reactors throughout the experiment

(Figure 9-14). Improved solids compaction may have occurred due to precipitation of carbonate or phosphate salts, which could have increased the density of the biomass. Although the pH declined to levels measured in the control biomass by cycle five (Figure 9-11), the SVI did not return to higher values typical of those measured for the control biomass by the end of the experiment. This suggests that dense inorganic precipitates remained entrapped within flocs and influenced the compaction of the biomass for the duration of the experiment. This hypothesis is supported by the MLVSS/MLSS ratio reported in Table 9-2 (next section), which was significantly lower in the pH 11 shocked biomass relative to the control (t-test, p=0.013). In addition, the supernatant above the settled biomass in the SVI cylinder was turbid for the pH 11 stressed reactor, which was consistent with the high turbidity observed in the effluent TSS measurement (Figure 9-3). In contrast, the MLVSS:MLSS ratio for the pH 5 shocked biomass was significantly higher than the control (t-test, $p = 4.7 \times 10^{-7}$). This could reflect a tendency toward dissolving inorganic precipitated particles early on when the mixed liquor pH remained lower than the control, and a slow return to conditions present in the control. For the 2-day SRT pH 5 shocked reactor, the biomass had a higher SVI than the control for the first 11 cycles (Figure 9-14). There was no observed difference in the SVI for the pH 9 shocked reactor relative to the control. In summary, one effect of basic pH shock loads for marginally-nitrifying sludges may be an increase in the density of flocs due to precipitation of inorganic minerals. However, as has been shown for the pH 11 shocked biomasses, a lower SVI does not necessarily translate into better clarifier performance because the SVI test does not detect a turbid effluent due to deflocculation. Additionally, acidic shock may have the reverse effect on floc density but it did not consistently translate into poorer SVI values.

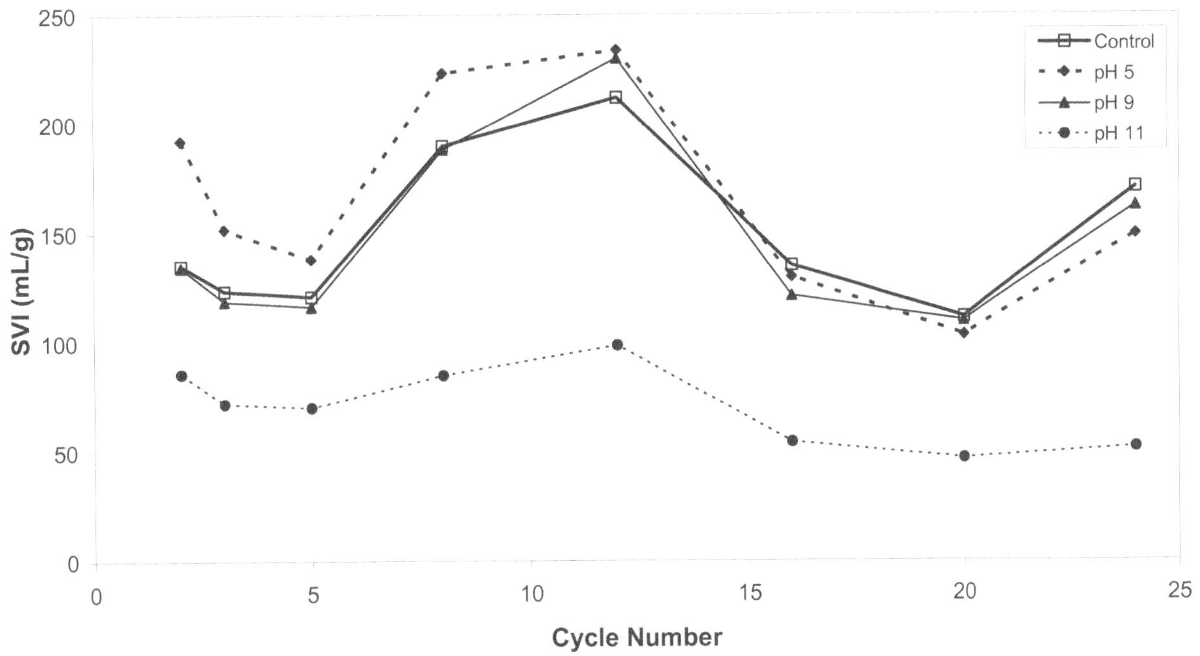

Figure 9-14. SVI for the 2-Day SRT Reactors Exposed to Various pH Values.

The SVI in the 10-day SRT shocked reactors and control were in the excellent range (<125 mL/g) for the first 34 cycles, as shown in Figure 9-15, indicating that acid and base shock did not have an immediate impact on SVI and, therefore, settleability. Despite the excellent SVI values, microscopic evaluation of the pH 11 shocked mixed liquor revealed an increase in the amount of free cells in suspension and an accompanied murkiness in the settled supernatant (consistent with high effluent TSS in Figure 9-4). After cycle 34, there was a large increase in the SVI of the pH 11 shocked reactor, which continued until the end of the experiment. This large SVI increase was most likely not due to the initial pH 11 shock, because the effluent pH had already recovered by that time. Additionally, the SVI increase did not correlate with a dramatic change in floc size, floc morphology or protozoan abundance. However, it did correspond with the dynamic increase in nitrite levels in the system (Figure 9-8). The reason for this correlation is not clear. Overall, acidic and basic shock did not appear to have a direct impact on settleability in terms of sludge blanket compaction for the 10-day SRT biomass.

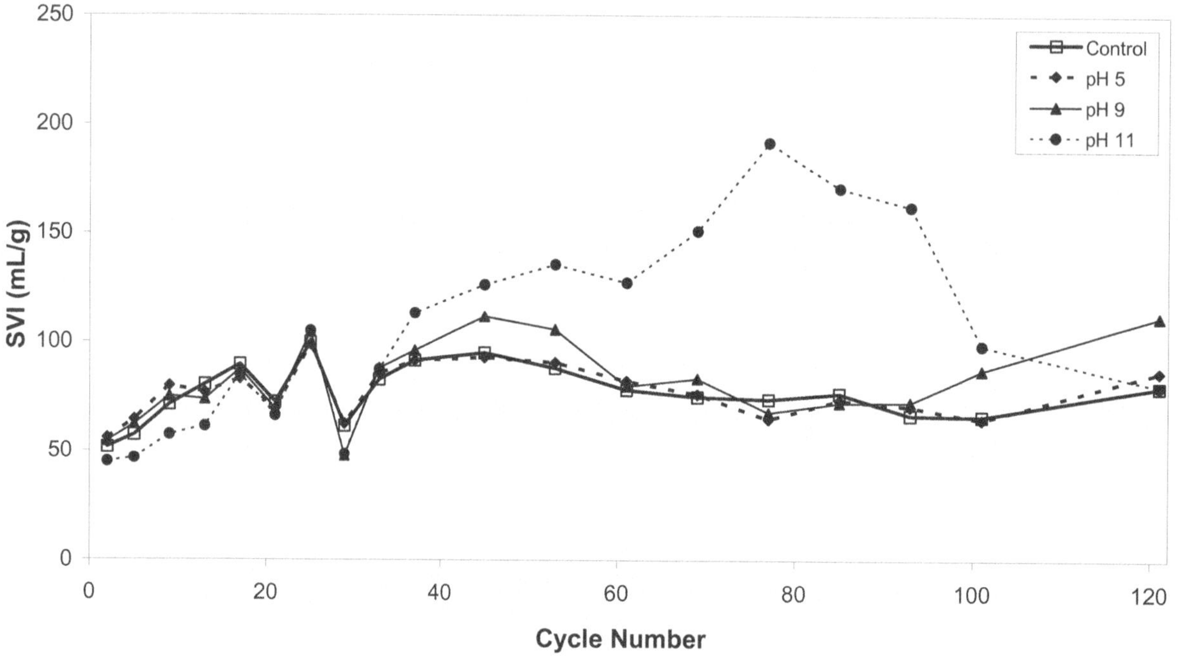

Figure 9-15. SVI for the 10-Day SRT Reactors Exposed to Various pH Values.

The impact of pH shock on biomass dewaterability is shown in Figure 9-16 for the 2-day SRT biomass. A significant increase in CST was observed in the pH 11 shocked reactor during the first 12 cycles ($\alpha = 0.05$); the values were as much as seven times those in the control reactor during cycle 2. This behavior could be due to protein denaturation and release from the floc, which increases the viscosity of the mixed liquor and, therefore, deteriorates its dewatering characteristics. No significant impact on CST was noted for the other two pH levels for the 2-day SRT biomass. The 10-day SRT biomass showed a slight increase in the pH 11 shocked biomass CST after the second cycle (average 40 seconds), but subsided by the end of the experiment to a CST of 15 sec, which was consistent with that measured in the control. Overall, the CST values for the 10-day SRT biomass were low and indicate no significant impact of acidic or basic shock on dewaterability in a nitrifying, 10-day SRT biomass (data not shown).

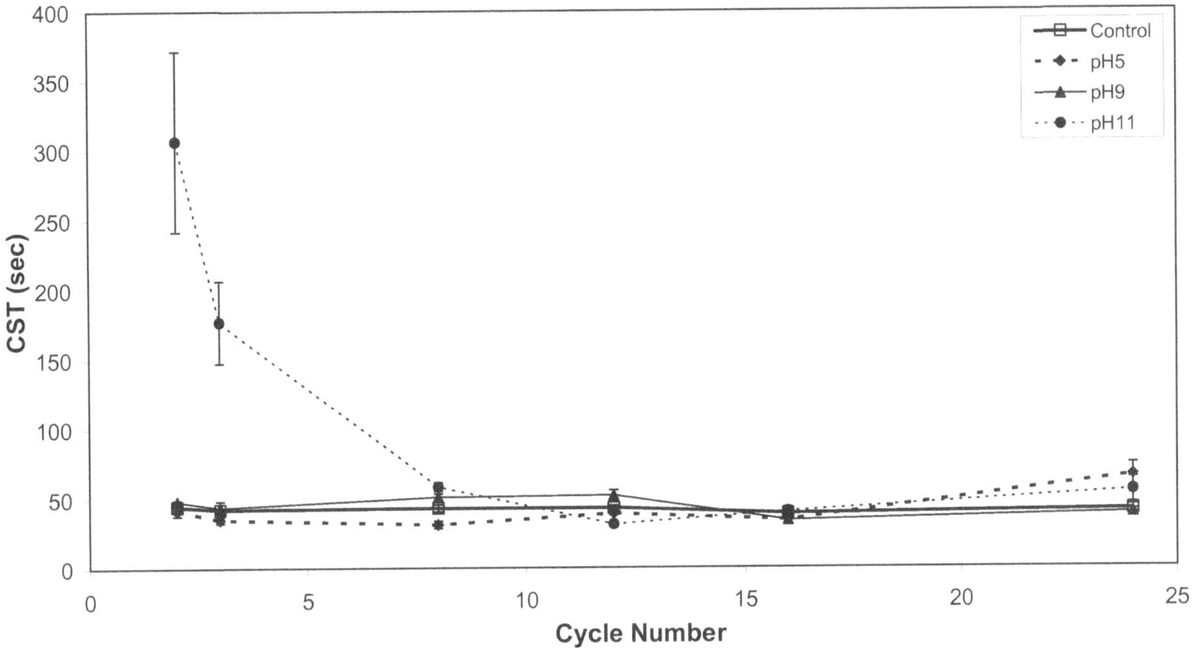

Figure 9-16. Mixed Liquor CST for the 2-Day SRT Reactors Exposed to Various pH Values.

9.2.4.2 Mixed Liquor Solids Trends

MLSS and MLVSS data are shown in Table 9-2 for the 2- and 10-day SRT reactors. The concentrations are consistent with the COD load to the systems with nutrient supplementation. Wastage volumes were held constant throughout the experiments and, therefore, the impact of changes in effluent TSS could be directly evaluated. The increase in effluent suspended solids in the 10-day SRT system shocked with pH 11 caused a moderate (around 10%) decrease in MLSS for the first 7 days of the 10-day SRT experiment, but then recovered to the level observed in the control and, in fact, surpassed that concentration toward the later part of the experiment (data in Appendix H). In the 2-day SRT system, however, the pH 11 shocked reactor MLVSS decreased approximately 30% during the first five cycles and never recovered to a concentration comparable to that in the control (Figure 9-17). The ratio of MLVSS:MLSS decreased in the pH 11 shocked reactors in both the 2- and 10-day SRT systems, presumably because of the precipitation of inorganic salts that are less soluble at higher pH levels. In addition to the effects in the pH 11 reactors, the 2-day SRT pH 5 reactor also exhibited decreased MLVSS relative to the control for the first eight cycles ($\alpha = 0.05$), indicating that growth inhibition occurred in this reactor during the initial period following the shock.

Table 9-2. Mixed Liquor Concentrations in 2- and 10-Day SRT Reactors Exposed to Various pH Levels.

	MLSS (mg/L)	MLVSS (mg/L)	MLVSS:MLSS Ratio
2-Day SRT reactors			
Control	1,370 ± 520	940 ± 334	0.69 ± 0.01
pH 5	1,118 ± 334	877 ± 281	0.78 ± 0.02
pH 9	1,397 ± 511	946 ± 314	0.68 ± 0.01
pH 11	1,056 ± 480	614 ± 202	0.61 ± 0.02
10-Day SRT reactors			
Control	1,695 ± 287	1,443 ± 261	0.85 ± 0.008
pH 5	1,847 ± 396	1,574 ± 351	0.85 ± 0.008
pH 9	1,879 ± 392	1,580 ± 353	0.84 ± 0.009
pH 11	1,843 ± 514	1,514 ± 482	0.81 ± 0.006

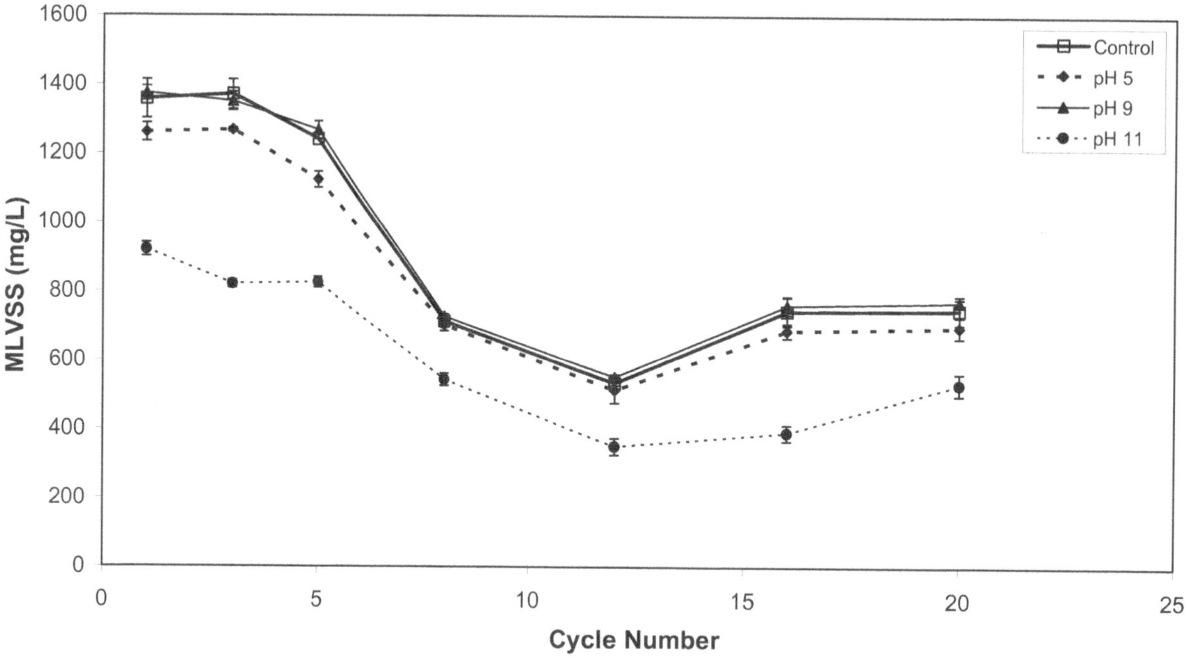

Figure 9-17. MLVSS for the 2-Day SRT Reactors Exposed to Various pH Values.

9.2.4.3 Kinetic Indicators of Inhibition

The 2-day SRT system (Figure 9-18) experienced a substantial impact on SOUR in the pH 11 shocked reactor. Specifically, the SOUR for the pH 11 shocked biomass increased until it was similar to the control level around cycle 8 when effluent ammonia concentrations were zero (Figure 9-6). SOUR values were similar in the pH 5 and 9 shocked biomass and the control biomass. The magnitude of the SOUR measured for the pH 11 shocked biomass 5 hours after the initial shock (end of cycle one reaction phase) suggested that significant cell lysis may have occurred, but the biomass recovered somewhat by the end of the first day after the shock (cycle four). This result demonstrates the hardiness of the biomass and its ability to recover from severe shock events.

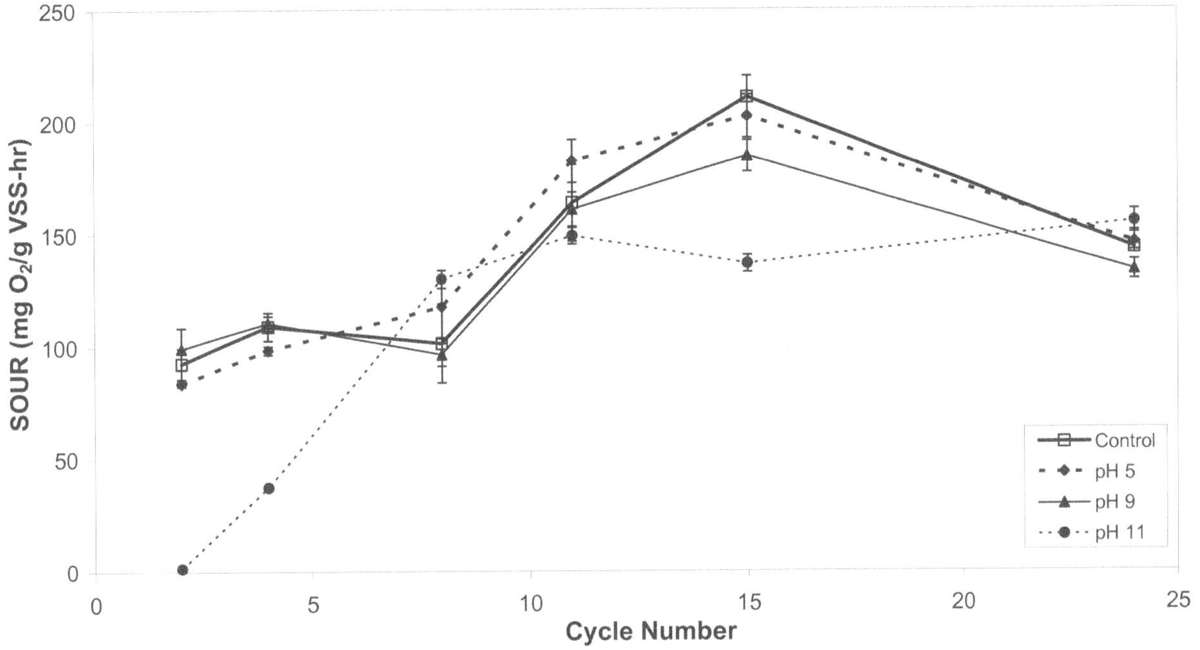

Figure 9-18. SOURs for the 2-Day SRT Reactors Exposed to Various pH Values.

SOUR results for the first cycle of the 10-day SRT reactors show that respiration rates decreased in both acidic and basic pH shocked reactors, and the degree of respiration decrease correlated with an increase in pH (Figure 9-19). From the fifth (1.25 day) to the 25th cycle (24.25 days), the stressed reactor SOURs were comparable to those measured in the control reactor. Between the 38th (9.5 days) and 53rd cycle (13.25 days), the SOUR in the control reactor decreased, suggesting that respiration inhibition occurred. This coincided with the detection of a small degree of nitrification inhibition (Figure 9-7), but otherwise did not correlate with other measured parameters. There was a severe decrease in SOUR for the pH 11 stressed biomass beginning after cycle 57 and continuing past cycle 80. Although delayed somewhat, this coincides with the point when AOB recovered but nitrite accumulated in this reactor (Figure 9-8). Complete recovery was not achieved in the time frame of the experiment.

The SNGR test was performed on biomass from the 10-day SRT system. As shown in Figure 9-20, the control reactor did not maintain a constant SNGR, but the pH 5 and 9 shocked reactors followed the same general trend as the control for the first 60 cycles, which suggests that fluctuations in these two stressed reactors were not due to pH shock. The pH 11 shocked reactor response suggests that low rates of nitrification were occurring through cycle 40, after which nitrification kinetics improved. This transition point corresponds with the point when nitrite accumulation started, followed by nitrate accumulation and decreased nitrite concentrations in the pH 11 shocked reactor.

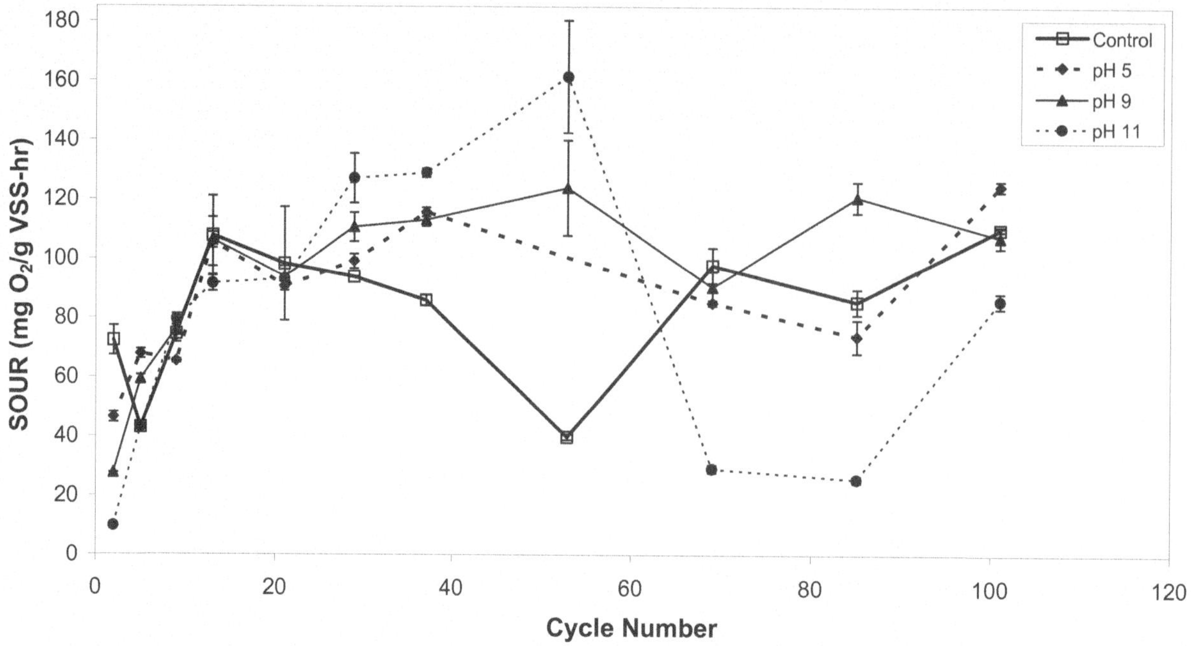

Figure 9-19. SOURs for the 10-Day SRT Reactors Exposed to Various pH Values.

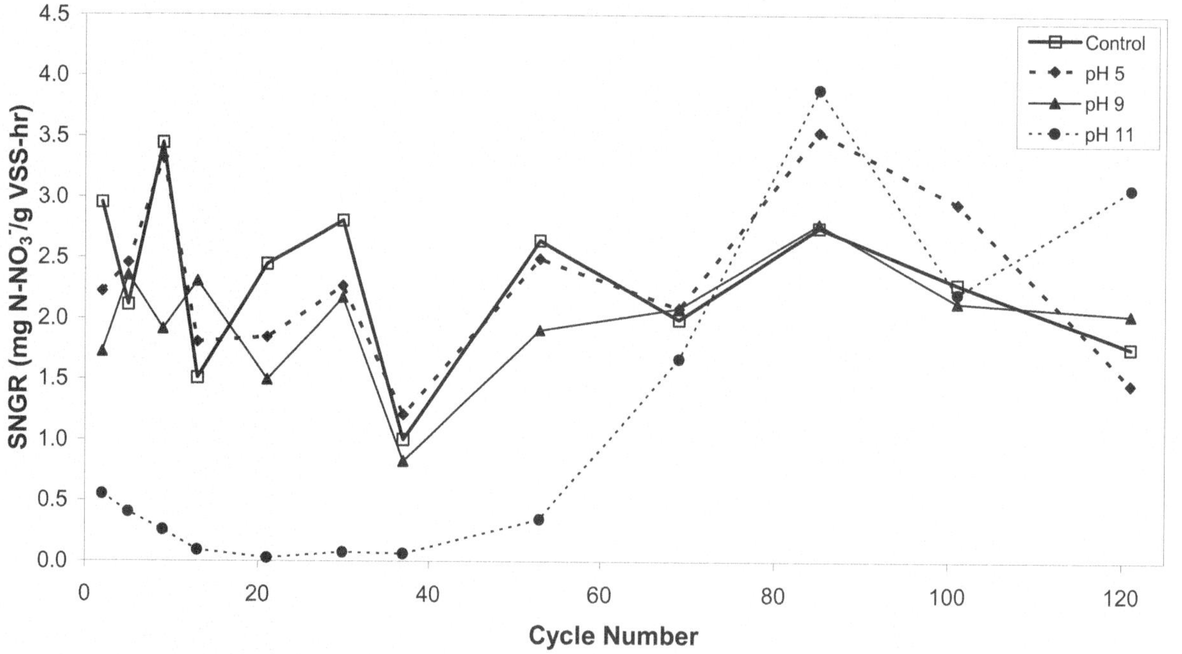

Figure 9-20. SNGRs for the 10-Day SRT Reactors Exposed to Various pH Values.

9.3 Discussion

The results from the pH source-effect experiment suggest that extreme alkaline pH is toxic to activated sludge systems and affects the treatment process at various levels. The predominant process effects that were observed included deflocculation, nitrification inhibition,

COD removal inhibition (heterotrophic inhibition), and short-term but significant decreased dewatering potential. A modest acidic shock (pH 5) had less of an impact on biomass performance and only seemed to adversely affect ammonia oxidation and effluent TSS in the 10-day SRT system.

Optimal pH in activated sludge systems is defined by three factors: 1) activation-deactivation of nitrifying bacteria; 2) nutritional effects connected with alkalinity; and 3) inhibition through free ammonia and free nitrous acid (Villaverde et al., 1997). Nutritional effects can occur if carbonate (carbon source) becomes limiting at low pH because alkalinity is primarily in the form of CO_2, which will be stripped by aeration. At high pH, alkalinity is primarily in the form of carbonate, and most carbonates are insoluble and can also limit the availability of the carbon source for nitrification. This experiment showed significant nitrification inhibition during extreme alkaline conditions (pH 11), which resulted in very high ammonia levels for the first 40 cycles (10 days). This suggests that nitrification inhibition due directly to the pH shock was linked most strongly to interference with metabolism of the AOB rather than the NOB, since the pH of the pH 11 shocked mixed liquor had returned to that of the control by the time inhibition of AOBs had subsided. This inhibition took more than 3X SRT to recover. Prinčič et al. (1998) found retarded nitrification in low pH environments (pH 6.0), but the nitrification results from this pH shock test showed only minor retardation at the very start of both the 2- and 10-day SRT acidification experiments, which recovered quickly to control levels. Nitrification in the pH 9-stressed reactors was affected very little, as the pH values of these systems were very close to the control reactors and recovered to control levels very quickly.

The pH 11 shocks were found to cause significant deflocculation in both the 2- and 10-day SRT reactors. In the case of the pH 11 shocked systems, the deflocculation was manifested as a sharp increase in effluent suspended solids and high effluent turbidity. Deflocculation was observed beginning with the first cycle after shock in both the 2- and 10-day SRT reactors, resulting in effluent TSS concentrations that were about 2.5 times higher for both. Little deflocculation was seen in either the pH 5 or 9 shocked systems. It is likely that the cause of alkaline-derived turbidity relates to dissolution of floc-binding proteins in the exocellular polymers contained in the flocs.

Effluent sCOD removal was inhibited severely by the initial shock of pH 11 in both the 2- and 10-day SRT systems. This coincided with a jump in effluent suspended solids and a near 30% decrease in MLVSS concentration in the 2-day SRT pH 11 shocked reactor. For the 2-day SRT system, the large decrease in biomass undoubtedly contributed to the decline in sCOD removal capability. However, it is also apparent from the 10-day SRT system that the alkaline shock caused toxicity to the biomass that resulted in reduced COD removal capacity as well. The toxicity is probably related to denaturation of critical proteins in cell membranes that contribute to catabolic functions. COD removal was not affected in either the pH 5 or 9 stressed systems.

9.4 Conclusions

- The brief acid (pH 5) shock imposed during this study did not cause significant or sustained process effects.
- pH shock caused significant negative effects on biosolids settleability in the pH 11 stressed reactors as seen though effluent TSS data. Dewatering was also significantly impacted early on in the 2-day SRT biomass at high pH. Cell membrane structures can be

solubilized at high pH values, so it is possible that the high turbidity observed with the pH 11 shock was due to significant cell lysis. Additionally, exocellular polymers are solubilized by high pH so the turbidity could also occur because of a loss of floc structure and release of smaller flocs — a true deflocculation event. This response can also impact dewaterability by binding water. It is unknown if other mechanisms associated with the loss of metabolic capacity (sCOD removal, ammonia oxidation) contributed to the observed increase in effluent TSS.

- Ammonia oxidation was significantly affected by pH 11 shock in the nitrifying 10-day SRT biomass. Ammonia levels increased almost immediately and a corresponding decrease in nitrate was observed. Significant nitrite accumulation was observed after cycle 40 (day 10), which suggests that nitrification occurred sequentially as a result of the alkaline shock condition. Nitrification inhibition was reversed for the 10-day SRT reactors within 3 x SRT, as determined by effluent nitrate concentrations. Nitrification inhibition patterns in the 2-day SRT biomass were complicated by the fact that the biomass was partially nitrifying and the heterotrophs, which also use ammonia as a nitrogen source, were inhibited during the early phase of the experiment as well (as noted by the poor sCOD removal efficiency).
- Soluble COD removal efficiency decreased when shocked at pH 11 for both 2- and 10-day SRT biomass. The cause for this decrease is believed to be partly due to loss of biomass via deflocculation and also denaturation of critical catabolic biomolecules needed to sustain effective metabolism.
- Despite severe impacts on respiration after the initial pH 11 shock for the 2-day SRT biomass (Figure 9-18), recover was initiated within one day and ultimately recovered to a level observed in the unstressed control within two days. This recovery was too fast to be due entirely to new biomass growth, which was not observed via the MLSS concentrations, and demonstrates the ability of biomass to recover from severe alkaline pH inhibition events.
- High pH shock caused precipitation of minerals into the mixed liquor, which in turn increased the density of flocs and improved settling velocity and compaction. However, it may have also caused a temporary nutrient limitation in the alkaline shocked reactors, leading to apparent inhibition of metabolism.

CHAPTER 10.0

CONCLUSIONS AND RECOMMENDATIONS

10.1 Conclusions

A summary of results from the 14 source-effect experiments completed for this study are provided in Tables 10-1 and 10-2, which show that a range of process effect patterns were observed. Based on overall impact, the order in which chemicals caused adverse effects on process performance parameters for the non-nitrifying 2-day SRT biomass, the order in which the chemicals caused adverse effects was (least detrimental to most detrimental): pH 5 < octanol < ammonia < DNP < cyanide < CDNB < cadmium < pH 11. For the nitrifying, 10-day SRT biomass was: pH 5 < ammonium < octanol < DNP < cyanide < CDNB < cadmium and pH 11. In determining the relative order of chemical impact, a chemical that caused the substantial deterioration of one aspect of process performance was considered to be more detrimental than a chemical that caused multiple processes to deteriorate moderately or slightly. This approach was used because failure of one element in a process is sufficient to wreak havoc on an activated sludge plant, and long recovery periods may be required.

Impacts on process effects such as SOUR and COD removal were expected, in large part because of the way the experiment was designed. The chemical shock doses were based on preliminary short-term respirometry results and, therefore, adverse impacts on SOUR were expected. Of interest was the observation that long-term SOURs yielded different results in many instances from what was observed with the short-term SOURs. This observation is important because it reveals that respirometry may, in many cases, underestimate the impact of a chemical shock if the biomass being tested has only been exposed to the chemical for several minutes rather than several hours. Poor COD removal is an expected consequence of inhibited respiration, as they are coupled. Indeed, in all but one case where near severe (three arrows) to severe (four arrows) SOUR inhibition was observed, significant deterioration in COD-removal efficiency was also observed (Tables 10-1 and 10-2).

The impact of chemical shock on nitrification was most dramatic among the process effects in that it was impacted moderately to severely by all chemical stressors tested. The step in the nitrification process that was inhibited varied among contaminants. In particular, cadmium, octanol, metal complexed cyanide, and basic pH shocks had a greater affect on AOBs whereas ammonia, CDNB, and DNP appeared to have a greater affect on NOBs. Since nitrification is such a critical metabolic process to so many treatment facilities, especially when biological nitrogen removal is practiced, the broad range of chemicals that lead to adverse impacts on nitrification is considered to be a critically important finding of this work. Several hypotheses have been generated to explain the broad range of chemicals that inhibited nitrification during this study, and are discussed in Section 10.2.

Severe increases in effluent turbidity that were presumably due to deflocculation were found in response to shock events in three cases for the 2-day SRT biomass and in four cases for the 10-day SRT biomass. This phenomenon was also considered to be important, because it can take several sludge age equivalents for a plant to recover from a deflocculation event. Depending upon the nature of the deflocculating contaminant, such a process response can also increase the effluent toxicity if the contaminant itself sorbs to the small turbid particles that pass out the effluent. Chemically induced deflocculation has been studied for chemicals that have electrophilic properties, which can activate K^+ efflux from biomass via the GGKE stress response (Bott and Love, 2002; Bott and Love, 2004). In all but one case (high ammonia shock with 2-day SRT biomass), detection of long-term (at least 6 hours) K^+ efflux corresponded with deflocculation. However, deflocculation occurred in some cases where long-term K^+ efflux was not observed. Therefore, electrophile-induced K^+ efflux does tend to correlate with deflocculation events but does not explain all cases where deflocculation occurs. Nevertheless, deflocculation is very detrimental to a biological treatment process, independent of the mechanisms involved. Initiation of deflocculation could expose more biomass surface to the toxin, thereby making the biomass even more susceptible to chemical exposure (Henriques et al., in press). Several hypotheses are put forth regarding how floc structure defines biomass response to toxic shock in Section 10.2.

Settling compactness, as described by SVI, and dewaterability, as defined by CST, did not appear to be impacted by the toxic shock events studied. Although both assays are surrogate indicators that do not exactly replicate the situation in the field, the responses suggest that, in general, toxins do not interfere with the thickening process of secondary clarifiers or unconditioned dewatering practices. These results suggest that, on average, some commonly employed monitoring practices (especially SVI) may not yield much information relative to other assays when it comes to detecting process upset.

In summary, the results of this study suggest that there are multiple ways in which chemicals can impact activated sludge plants and suggest that corrective action practices need to be tailored based on the nature of the chemical causing the upset. Identifying the appropriate corrective action is often difficult for operators because they do not know the composition of the chemical or chemicals that caused problems until days later, if ever. The study results strongly suggest that further work needs to be done in this area, as defined in Section 10-3.

Table 10-1. Summary of Process Effects for Seven Source Chemicals for the 2-Day SRT Biomass Using a Pseudo-Quantitative Scale[1].

Measurement	Observed Problematic Process Effect Relative to Control	Hypothesized Causal Mechanism	Ammonia 300 mg/L-N	Cadmium	CDNB	Cyanide	DNP	Octanol	pH 11
Effluent TSS/VSS	Increase in Effluent TSS/VSS	Flocs deteriorate and cause smaller, less dense particles (deflocculation)	0	↓↓↓↓	↓↓↓	0	0	↓	↓↓↓↓
Effluent COD	Decrease in COD removal	Inhibition of metabolic pathways	0	↓↓↓↓	↓↓↓↓	↓↓	0	↓	↓↓↓↓
SOUR	Decrease in SOUR	Inhibition of catabolic pathways	↓↓	↓↓↓↓	↓↓↓↓	↓↓↓↓	↓↓↓	↓	↓↓↓↓
Soluble Potassium	Increase in soluble K^+ concentration	Glutathione-gated K^+ efflux	↓↓	↓↓↓↓	↓↓	0	0	0	↓↓↓
SVI	Increase in SVI	Poor biosolids compression, or settleability	0	++	++	0	0	0	++
CST	Increase in CST	Retention of bound water, leading to poor dewaterability	0	↓	X	0	0	↓	↓↓↓

[1] The qualitative scale reflects the intensity of the effect for the IC_{50}-shocked reactors and the indicated NH_3 and pH shock level, in comparison to a negative control. The intensity scale ranges from ↓↓↓↓ (most intense process deterioration effect), 0 (no effect), and ++++ (most intense process improvement effect).
X means inconclusive results.

Table 10-2. Summary of Process Effects for Seven Source Chemicals for the 10-Day SRT Biomass Using a Pseudo-Quantitative Scale[1].

Measurement	Observed Problematic Process Effect Relative to Control	Hypothesized Causal Mechanism	Ammonium ~ 400 mg/L	Cadmium	CDNB	Cyanide	DNP	Octanol	pH 11
Effluent TSS/VSS	Increase in effluent TSS/VSS	Flocs deteriorate and cause smaller, less dense particles (deflocculation)	0	↓↓↓↓	↓↓↓	0	↓↓	+	↓↓↓↓
Effluent COD	Decrease in COD removal	Inhibition of metabolic pathways	0	↓↓↓↓	↓↓↓	↓	↓↓	+	↓↓↓↓
SOUR	Decrease in SOUR	Inhibition of catabolic pathways	0	↓↓↓↓	↓↓↓	↓↓	↓↓/++	↓↓	↓↓↓
Soluble Potassium	Increase in soluble K^+ concentration	Glutathione-gated K^+ efflux	0	↓↓	↓↓↓	0	0	0	X
Inorganic N effluent concentrations and NGR	Nitrification inhibition	See Table 10-4	↓	↓↓	↓↓↓	↓↓↓	↓↓	↓↓	↓↓↓↓
SVI	Increase in SVI	Poor biosolids compression, or settleability	0	++	+	↓	0	0	+
CST	Increase in CST	Retention of bound water, leading to poor dewaterability	0	0	0	0	0	0	0

[1] The qualitative scale reflects the intensity of the effect for the IC_{50}-shocked reactors and the indicated NH_3 and pH shock level, in comparison to a negative control. The intensity scale ranges from ↓↓↓↓ (most intense process deterioration effect), 0 (no effect), and ++++ (most intense process improvement effect).
X means inconclusive results

10.2 Hypotheses Generated from Results

An evaluation of the results presented in Tables 10-1 and 10-2 reveals that deflocculation was common among four of the seven stressors. It follows that floc structure may play an important role in protecting cells from stressful chemical shock events. While deflocculation may reflect a primitive "fight or flight" response by bacteria, it actually has the potential to make the situation worse by exposing more surface area to the offending toxin. From this study, respirometric inhibitory assay results indicated that the perceived severity of toxicity varied among contaminants for biomasses containing different floc sizes because different SRTs were used (2-day SRT biomass flocs were much smaller than 10-day SRT biomass). It is unclear whether physiology or floc structure predominated in defining the different outcomes with the 2-day and 10-day SRT biomasses, but the relative contribution of both needs to be considered. Specifically, it is proposed that floc structure can vary, depending upon the reactor configuration, wastewater composition, and operational practices, and such variations play a role in controlling the sensitivity of complex biological activated sludge systems to chemical stress. A concise and prioritized list of hypotheses that target this topic is provided in Table 10-3.

The results also show that nitrification inhibition was a moderate to significant process effect for all of the contaminants applied to the 10-day SRT nitrifying biomass. Although cyanide, heavy metals, and high ammonia (free ammonia) have been previously shown to cause nitrification inhibition, the electrophilic toxin CDNB and the hydrophobic toxin octanol have not. Preliminary hypotheses that elucidate the causal mechanisms behind effects from electrophilic and hydrophobic toxins are summarized in Table 10-4. There is a dearth of information in the research literature about the protective or remedial stress mechanisms used by nitrifying bacteria; however, this has the potential to change rapidly as the genome for *N. europaea* has been sequenced and published (Chain et al., 2003). Given the critical and growing importance of nitrifying bacteria in biological wastewater treatment, we propose that further elucidating key stress mechanisms that control the response of nitrifying bacteria to selected toxins is an important area of study.

Table 10-3. Hypotheses Relating to How Floc Structure Influences Floc Function in Response to Chemical Stress.

Research Questions	Hypotheses
Does EPS protect bacterial cells embedded in a floc from chemical toxins?	EPS play a role in protecting bacterial cells from toxins that interact with the polymer matrix; however, floc structure is not an important factor controlling the toxicity response for chemicals that do not interact with EPS.
What is deflocculation, and what materials are released from flocs when effluent TSS increases due to chemical toxic shocks?	Deflocculation is caused by different mechanisms, which depend on the agent that causes the floc disruption. The composition and characteristics (or "fingerprint") of the particles and materials released from the activated sludge flocs during deflocculation events are characteristic of the toxin that elicits deflocculation and may help predict the nature of the toxin.
What is the role of cell density-dependent mechanisms on the response and recovery of flocs from shocks of chemical toxins?	Cell density-regulated mechanisms, such as quorum sensing, occur in activated sludge and impact the toxicity response of the community to stress, as well as the recovery response that occurs after exposure to stress conditions.

Table 10-4. Hypotheses Relating to How Nitrifying Bacteria Respond to Chemical Stress.

Research Question	Hypotheses
Do GSH and the GGKE system play a major role in protecting nitrifying bacteria from electrophilic (oxidative) chemical toxins?	Autotrophic nitrifying bacteria are more sensitive to shock loads of electrophilic chemical toxins because they lack the glutathione gated potassium efflux mechanism for cytoplasmic acidification and protection.
Can nitrifying bacteria acidify their cytoplasm in response to oxidative toxins?	Acidification activates secondary protective stress responses and occurs in stressed nitrifying bacteria independent of the GGKE mechanism to confer protection.
Do chemical stressors interfere directly or indirectly with enzymatic and energy generating reactions in cells?	Disruption of the proton motive force, energy production, and enzymes required for nitrification (nitrite oxidoreductase, ammonia monooxygenase, or hydroxylamine oxidoreductase) are directly inhibited by chemical stressors, thereby causing nitrification disruption.
How do hydrophobic organic chemicals inhibit nitrifying bacteria?	Hydrophobic organic chemicals are toxic to nitrifying organisms due to the insertion of these chemicals into the cell membranes, which causes a disruption of the proton/membrane potential of the cells and protein-protein interactions of key enzymes (such as membrane-bound ammonia monooxygenase), affecting energy production.
How do chemical stressors impact cell viability?	Nitrification inhibition occurs in part because of a decrease in nitrifying biomass through cell lysis and death.

10.3 Recommendations for Future Work

There are several benefits to conducting studies that further elucidate the source-cause-effect relationships between chemical toxins, bioreactor biomass and process upset effects. Understanding how chemicals elicit stress responses can result in a better understanding of appropriate, counteracting corrective actions that reduce the impact of the toxin. Furthermore, it can lead to the development of new sensing technologies that will improve the wastewater treatment industry's ability to rapidly detect and respond to these toxins. Therefore, it is recommended that additional studies be conducted to correlate mechanisms of biological response to toxic upset with the source-effect information presented in this report. The hypotheses presented in Tables 10-3 and 10-4 are being evaluated through on-going research and will be reported on in a future report. Other recommendations for future work include:

- Additional studies are needed that include a broader range of chemicals within chemical classes (e.g., copper, zinc, chromium in addition to cadmium as a heavy metal) to determine variations in response within a class.
- Many toxic upset events occur when multiple toxins are present. It is necessary to determine the source-effect relationships that occur when activated sludge is exposed to mixtures of toxins, and how mixtures influence causal mechanism responses.
- The experiments reported here used completely aerobic activated sludge systems. However, many systems have anaerobic or anoxic zones incorporated into the front of the plant for nutrient removal, or biofilm-based treatment systems (stand-alone, or integrated fixed-film systems). A systematic study on the impact of chemical toxin shock on these systems is needed to expand the source-effect matrix to apply to a broader range of reactor configurations.

- The information collected from this and other studies needs to be integrated into a decision support system (DSS) to assist operators who have to deal with emergency toxic shock situations. DSS development relies upon improved sensing technologies that provide real time data that informs the DSS as to the nature of the toxicity.
- To facilitate information flow as part of a corrective action/DSS, robust, inexpensive, and field-compatible upset early warning devices need to be developed and implemented.

APPENDIX A

PILOT PLANT PERFORMANCE

A.1 Influent Wastewater Quality

The influent wastewater was composed of sewage collected from a sewer located adjacent to the pilot plant facility, plus supplemental organic carbon and nutrient sources, as described in Section 2.1.2. The influent wastewater was stored in two holding tanks; one fed the 2-day SRT non-nitrifying pilot plant reactor and the other fed the 10-day SRT nitrifying pilot plant reactor. The tanks were initially connected through a piece of tubing at the bottom, which was small enough to minimize the amount of liquid exchange between these tanks after they were filled. During this time, several of the 2- and 10-day SRT shock experiments were performed; however, the two2-day experiments conducted before day 580 were not used because problems were encountered in preventing nitrification, and the experiments were repeated after day 580. Therefore, pilot plant data for the 10-day SRT system is shown beginning one month after the pilot plant was started (day 30). Starting around day 500, the tanks were isolated from one another because a separate supplementation program was initiated for the two reactors as part of the effort to prevent nitrification in the 2-day SRT system. As a result, there were slight variations in the composition of the influent wastewater between the two reactor systems. The results of routine influent water quality analyses for the time period from day 580 to 708 in the 2-day SRT reactor, and from day 30 to 708 for the 10-day SRT reactor are summarized in Figures A-1 through A-6. Key dates for operational changes in the pilot plant are given in Table A-1. Samples were not collected from day 627 through 657 because of the holiday break. Sampling and source-effect experiments were discontinued during the time period when excessive filamentous bulking occurred from day 200 to 219.

Table A-1. Major Operational Changes in the Pilot Plant System.

Date	Day #	Operational change
September 26th 2001	182	Influent supplementation begun to achieve Total COD of 600 mg/L in both systems and maintain TKN/COD and M:D ratios approximately constant.
August 19th 2002	509	Influent supplementation in the 2-day SRT reactor changed to achieve Total COD of 1000 mg/L and maintain TKN/COD ratio. Influent tanks isolated from each other.
November 4th 2002	586	Supplementation of nitrogen decreased in the 2-day SRT reactor and increased in the 10-day SRT reactor.
March 1st 2003	703	Screens installed in influent tanks to prevent high solids concentration in the influent of the pilot plant reactors.

A.1.1 Chemical Oxygen Demand

The influent COD was inherently too low in the Virginia Tech sewer to maintain representative mixed liquor concentrations. Based on PSC recommendations, an influent wastewater supplementation program was initiated on day 182. The goal was to increase the total

COD from an average of around 400 mg/L to 600 mg/L. The supplementation program was initiated in a staged manner over a one week period. The average soluble and total COD for the 10-day SRT system from day 200 on was 292 $\pm$ 173 and 684 $\pm$ 335 mg/L, respectively. Additional supplementation to the 2-day SRT pilot unit was initiated on day 509 based on an additional PSC recommendation. The goal was to increase the total COD from an average of around 600 mg/l to 1,000 mg/L. The average influent soluble and total COD for the 2-day SRT reactor since day 613 was 493 $\pm$ 228 and 1,155 $\pm$ 576 mg/L, respectively. Influent COD data are shown in Figures A-1 and A-2 for the 2- and 10-day systems, respectively.

A.1.2 Suspended Solids

In general, the concentration of influent suspended solids was quite stable. There were infrequent, periodic events when high suspended solids of unknown origin were found in the sewage, especially toward the end of the study. The average influent TSS and VSS fed to the 2-day SRT reactor since day 580 was 665 $\pm$ 704 and 457 $\pm$ 468 mg/L, respectively. The average influent TSS and VSS fed to the 10-day SRT reactor since day 200 was 492 $\pm$ 565 and 373 $\pm$ 390 mg/L, respectively. Influent suspended solids data are shown in Figures A-3 and A-4 for the 2- and 10-day systems, respectively.

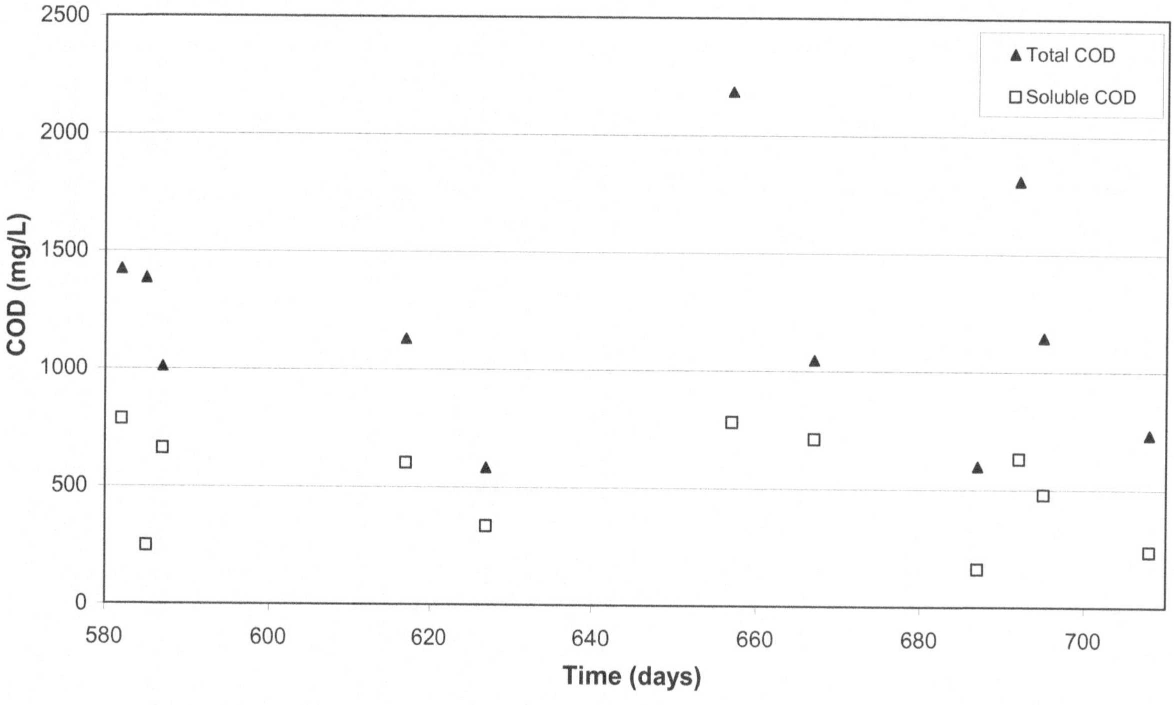

Figure A-1. Influent Total and Soluble COD for the 2-Day SRT Pilot Plant Reactor.

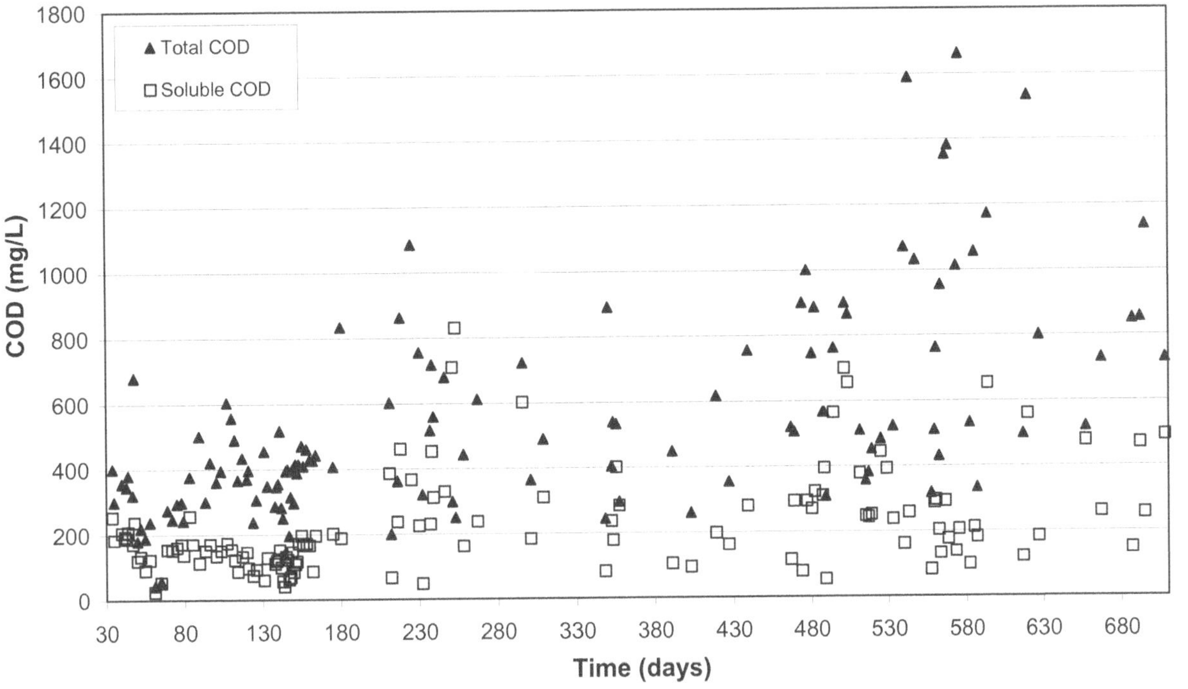

Figure A-2. Influent Total and Soluble COD for the 10-Day SRT Pilot Plant Reactor.

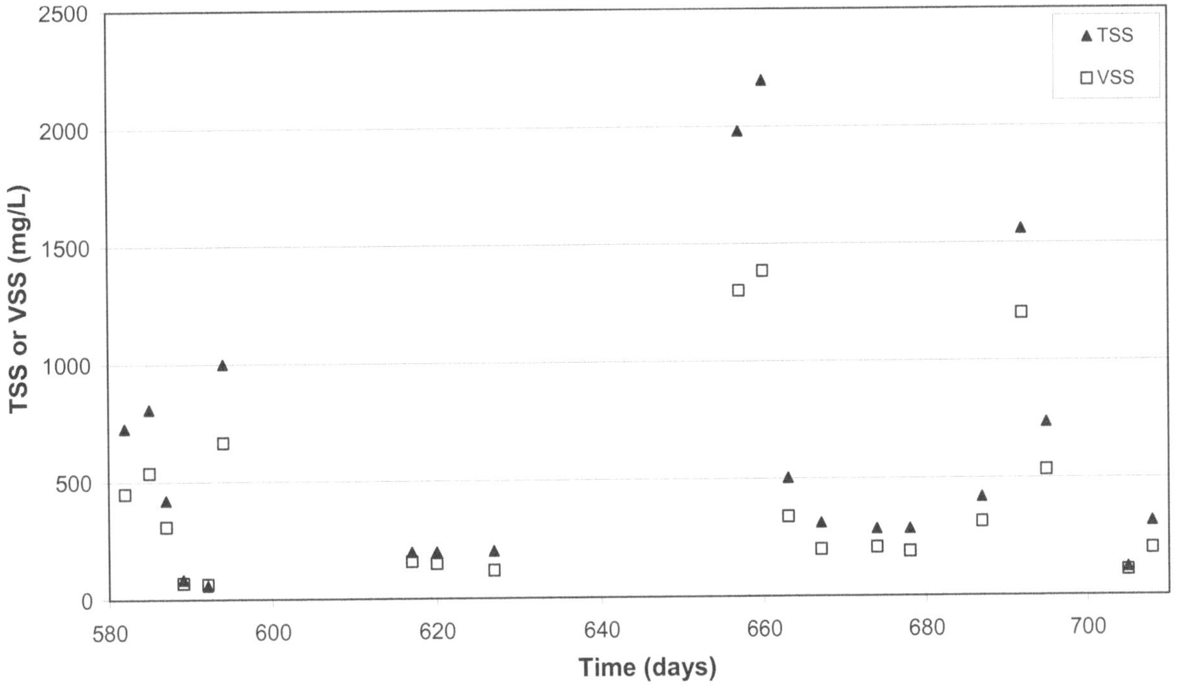

Figure A-3. Influent TSS and VSS for the 2-Day SRT Pilot Plant Reactor.

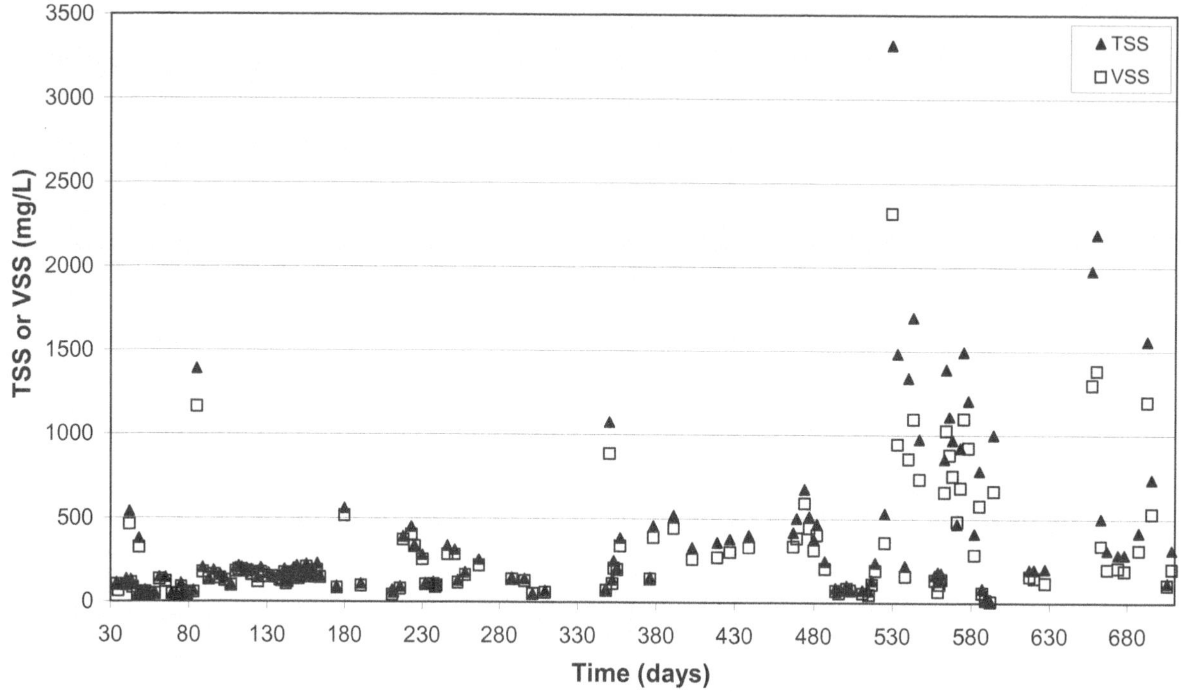

Figure A-4. Influent TSS and VSS for the 10-Day SRT Pilot Plant Reactor.

A.1.3 Ammonia

The influent ammonia concentration was consistent within both the 2- and 10-day SRT influent tanks for the duration of the studies reported here; however, their relative concentrations were different. The supplementation scheme that was initiated on day 497 included ammonia and urea to increase the TKN so that the influent COD:TKN ratio remained relatively constant (about 10 in the 2-day SRT system and 12 in the 10-day SRT system). On day 586, an additional supplement of ammonia was added to the 10-day SRT reactor while the ammonia supplement for the 2-day SRT reactor was decreased. The average influent ammonia for the 2-day SRT reactor since day 580 was 20.8 $\pm$ 10.2, and for the 10-day SRT reactor since day 200 was 31.1 $\pm$ 10.8 mg/L as N. The influent COD:TKN ratios became 16 for the 2-day SRT system and 11 for the 10-day SRT system. Influent ammonia data are shown in Figures A-5 and A-6 for the 2- and 10-day systems, respectively.

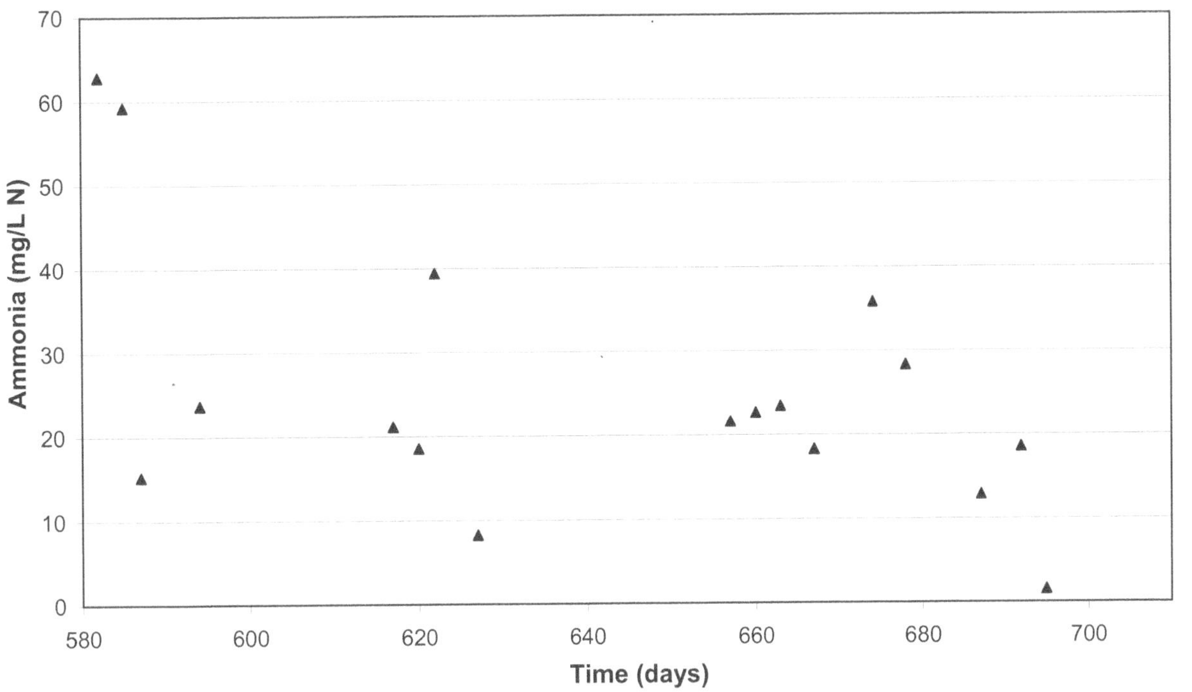

Figure A-5. Influent Ammonium-Nitrogen Concentration for the 2-Day SRT Pilot Plant Reactor.

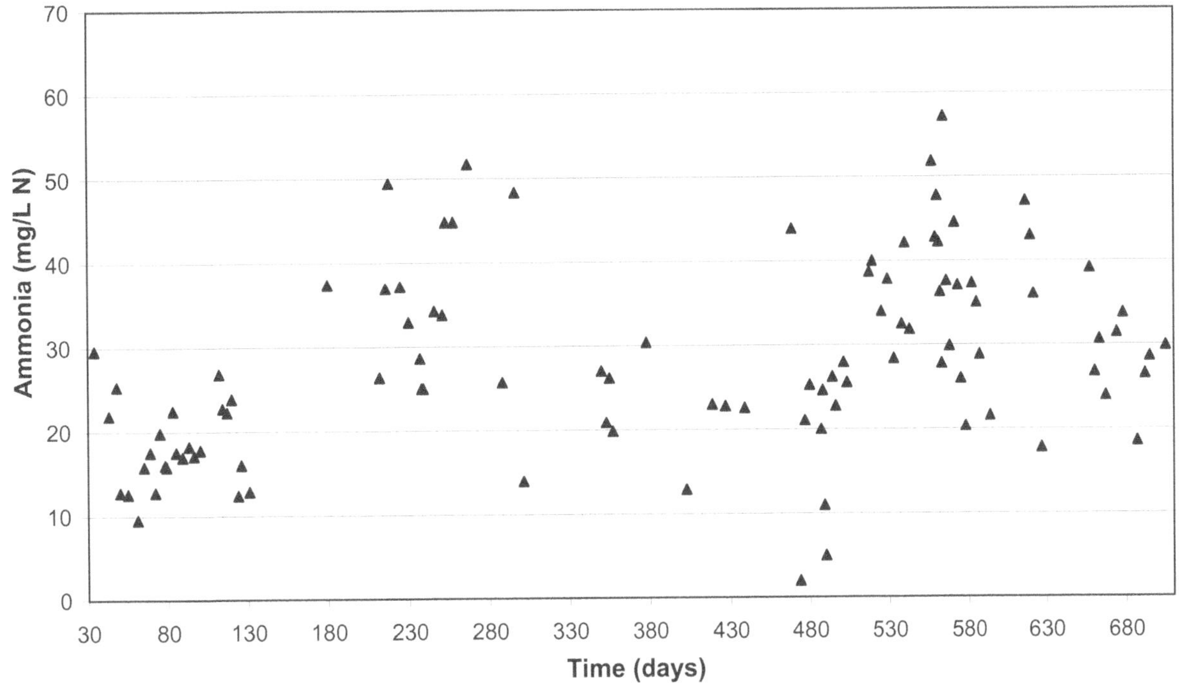

Figure A-6. Influent Ammonium-Nitrogen Concentration for the 10-Day SRT Pilot Plant Reactor.

A.2 2-DAY SRT Bioreactor Performance

A.2.1 Mixed Liquor

The MLSS and MLVSS profiles for the 2-day SRT pilot plant reactor are summarized in Figure A-7. The PSC noted that the MLSS in the 2-day SRT reactor was too low and recommended supplementing to bring the concentration up. Supplementation began on day 509, and the MLSS and MLVSS averaged 1,706 ± 587 and 1,241 ± 372 mg/L, respectively, for the time period shown in Figure A-7. Although a dispersed floc formed, good settling continued to occur and the system appeared to be stable under this operating condition.

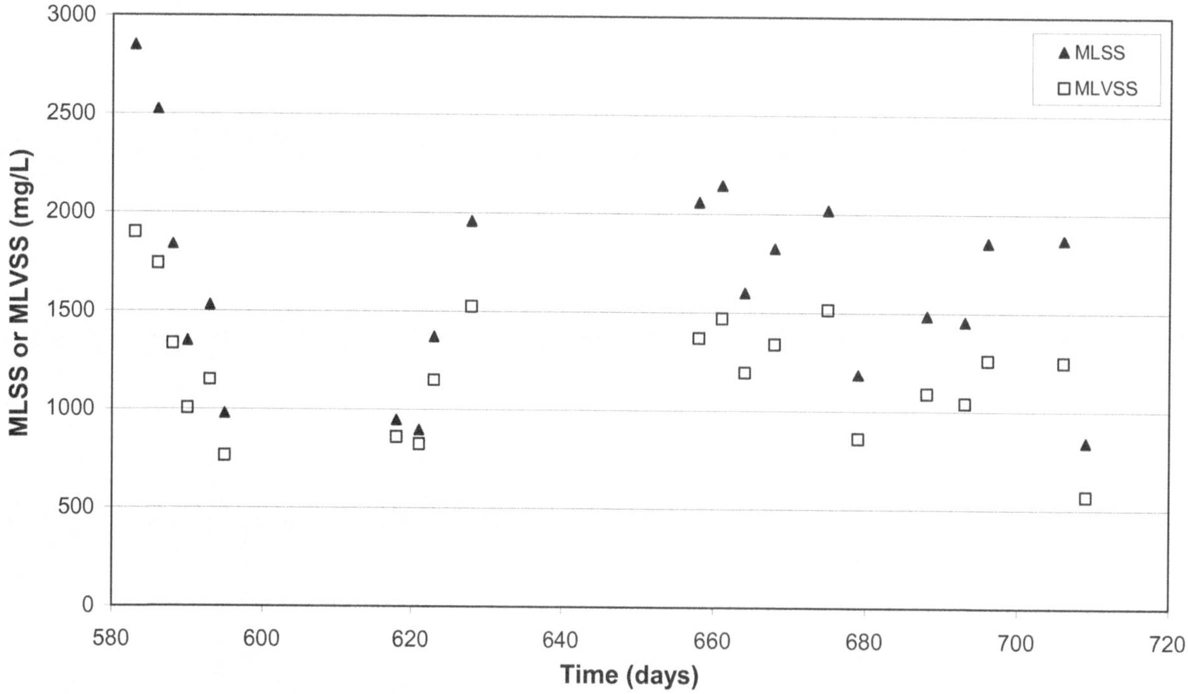

Figure A-7. Mixed Liquor Concentrations and Reactor History for the 2-Day SRT Pilot Plant Reactor.

A.2.2 Effluent Water Quality Parameters

Effluent performances for the 2-day SRT reactor are summarized in Figures A-8 through A-10. The 2-day SRT reactor performed well from day 580 until the end of the study, and maintained an average effluent soluble COD of 73 ± 23 mg/L, and TSS and VSS concentrations of 59 ± 33 and 53 ± 29 mg/L, respectively. Although the 2-day SRT reactor was designed to be a non-nitrifying system, there were periods when complete ammonia removal was observed. The effluent ammonia-nitrogen concentration averaged 5.3 ± 10 mg/L after day 580.

A.2.3 Solids Retention Time

A summary of the calculated SRT for the 2-day SRT (target) reactor is shown in Figure A-11. Since day 580, the calculated SRT averaged 2.0 ± 0.1 days. This highly stable and accurate SRT is a substantial improvement relative to earlier days in the project when the actual SRT was below target. The improvement was achieved by adjusting weekly wastage rates after using measured wastage volumes, MLSS concentrations and effluent TSS concentrations from the previous week to calculate a theoretical SRT and new wastage rate.

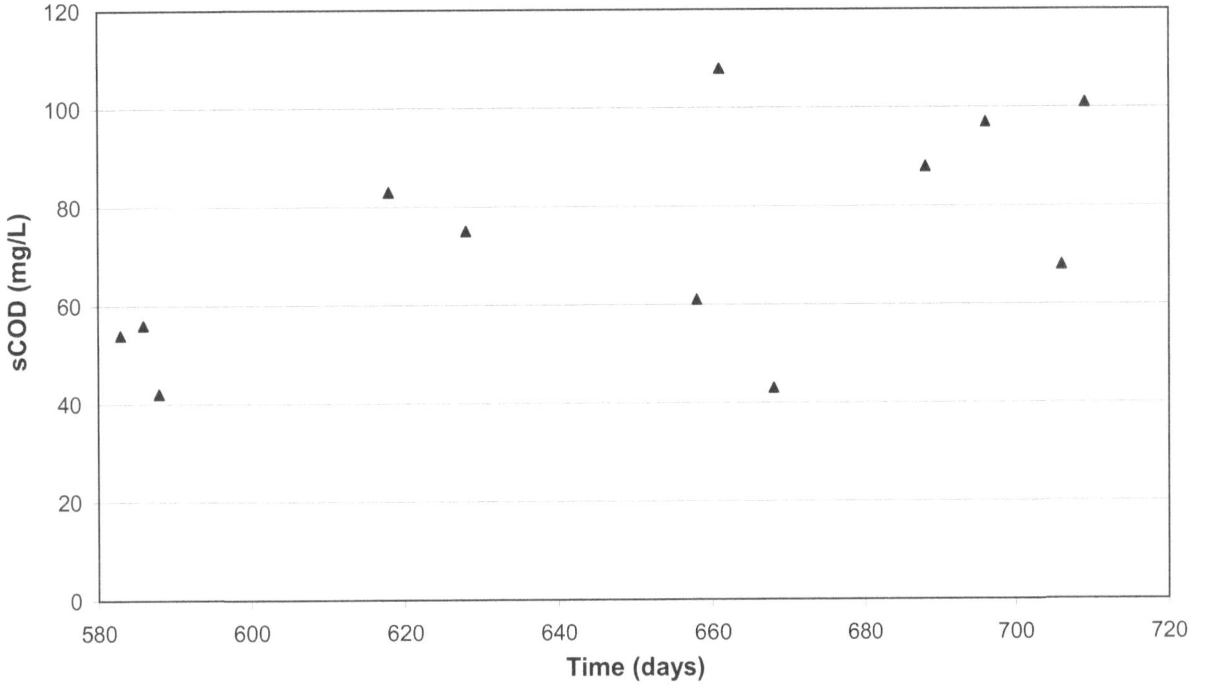

Figure A-8. Effluent Soluble COD for the 2-Day SRT Pilot Plant Reactor.

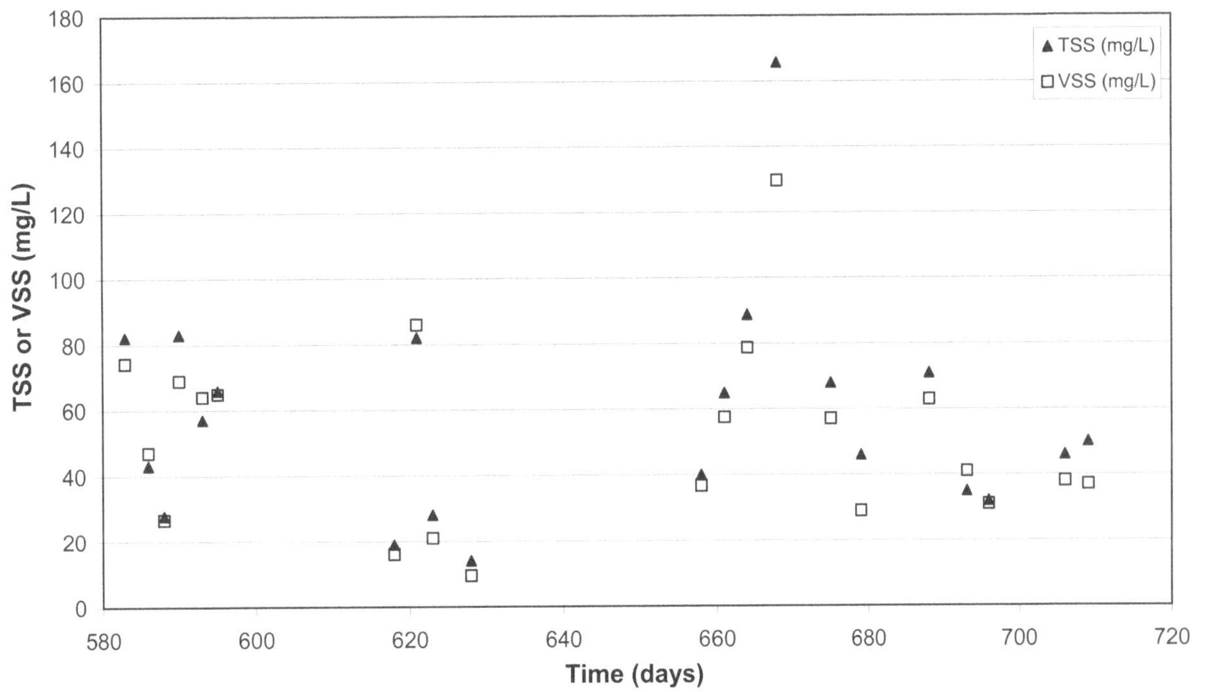

Figure A-9. Effluent TSS and VSS for the 2-Day SRT Pilot Plant Reactor.

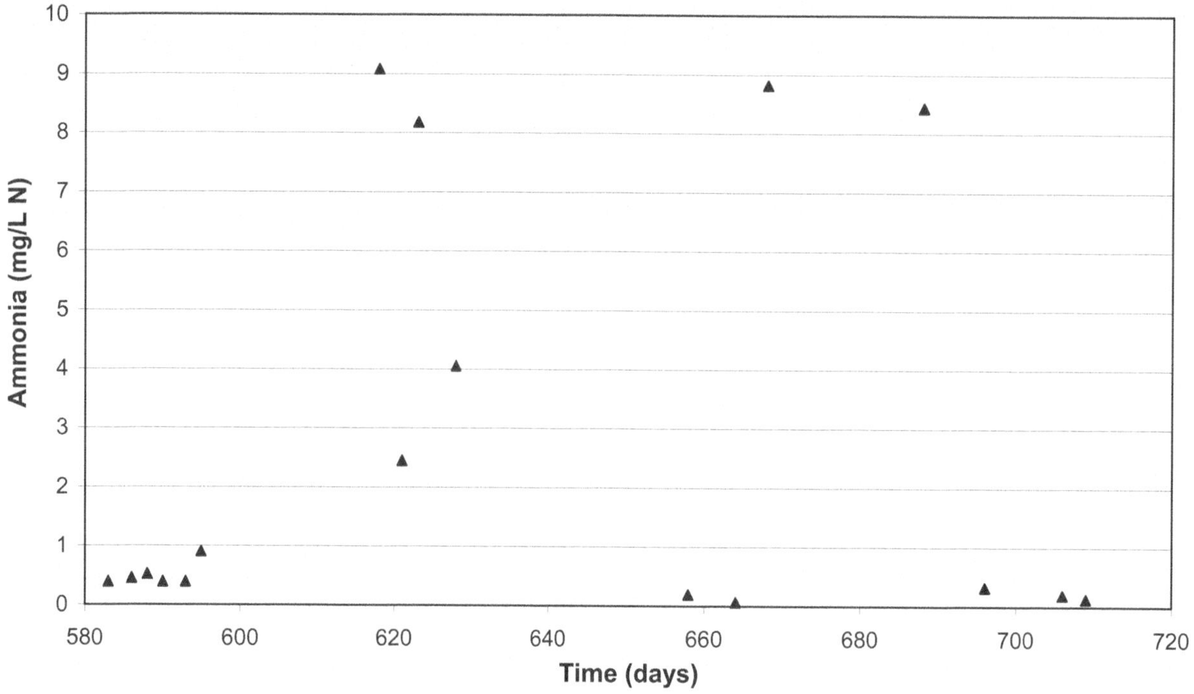

Figure A-10. Effluent Ammonium Nitrogen for the 2-Day SRT Pilot Plant Reactor.

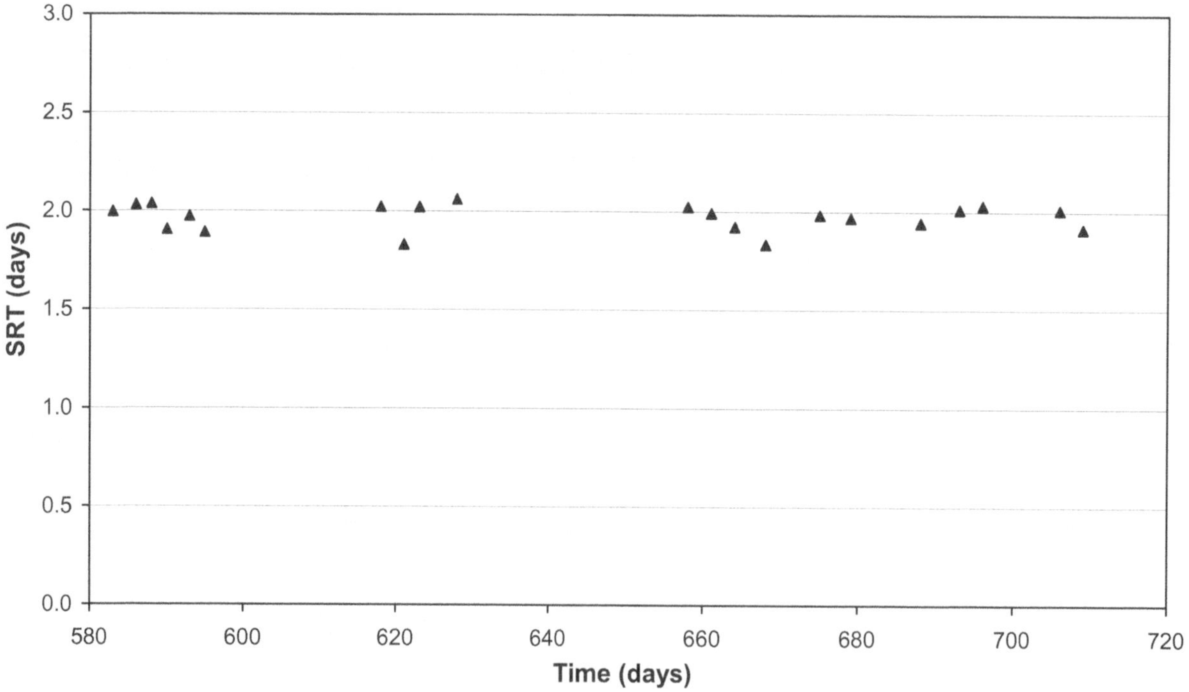

Figure A-11. Calculated Solids Residence Time (SRT) for the 2-Day SRT Pilot Plant Reactor.

A.3 10-DAY SRT Bioreactor Performance

A.3.1 Mixed Liquor

The MLSS and MLVSS profiles for the 10-day SRT pilot plant reactor are summarized in Figure A-12. Starting from day 200, the MLSS and MLVSS concentrations averaged 2,692 ± 1,299 and 2,024 ± 738 mg/L, respectively. The highest MLSS measurements probably reflect an accumulation of particles in the influent tank, which in turn accumulated in the bioreactor.

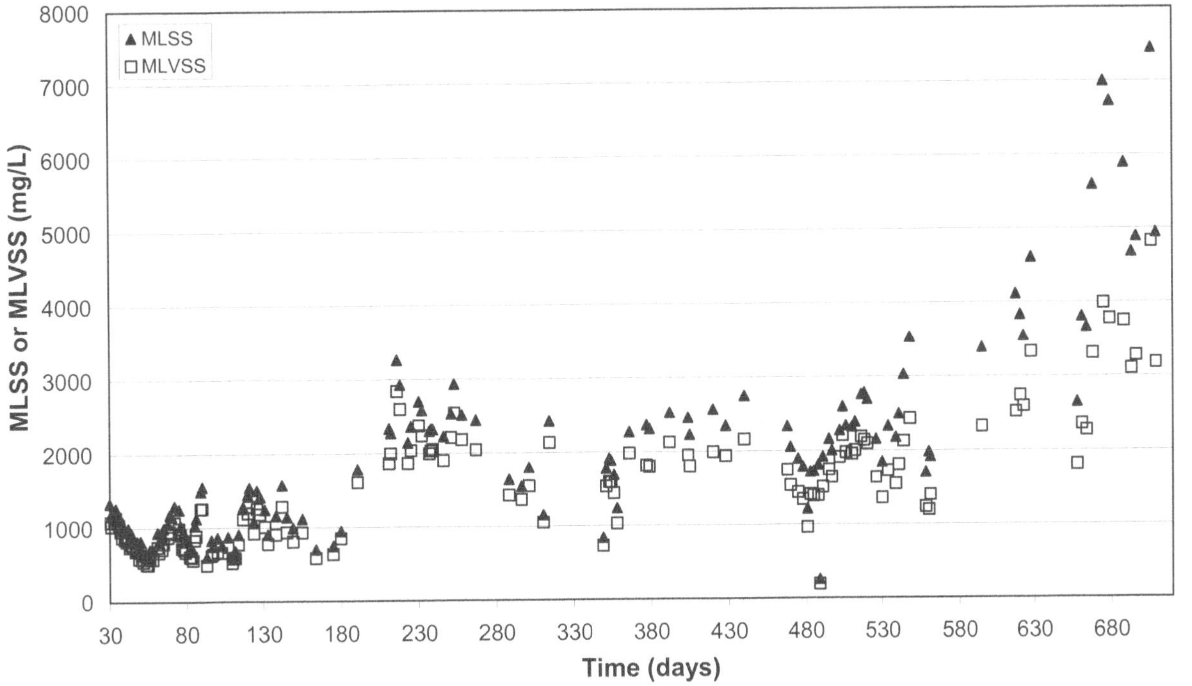

Figure A-12. Mixed Liquor Concentrations and Reactor History for the 10-Day SRT Pilot Plant Reactor.

A.3.2 Effluent Water Quality Parameters

Effluent performance for the 10-day SRT reactor are summarized in Figures A-13 through A-15. The 10-day SRT reactor performed well since the pilot plants were started. Since day 200, an average effluent soluble COD of 38 ± 30 mg/L and effluent TSS concentration of 18.3 ± 18.9 mg/L. The 10-day reactor was designed to be a nitrifying system, and performed very well during most of the year with an average effluent ammonia concentration of 0.7 ± 1.4 mg/L as N from day 200 onward. This indicates that excellent nitrification was achieved.

A.3.3 Solids Retention Time

A summary of the calculated SRT for the 10-day SRT (target) reactor is shown in Figure A-16. The SRT averaged 10.5 ± 2.4 days after operating day 200. Except for a three-month period in late spring when heavy rain was experienced in the region, there was very little variation in the actual SRT for the system.

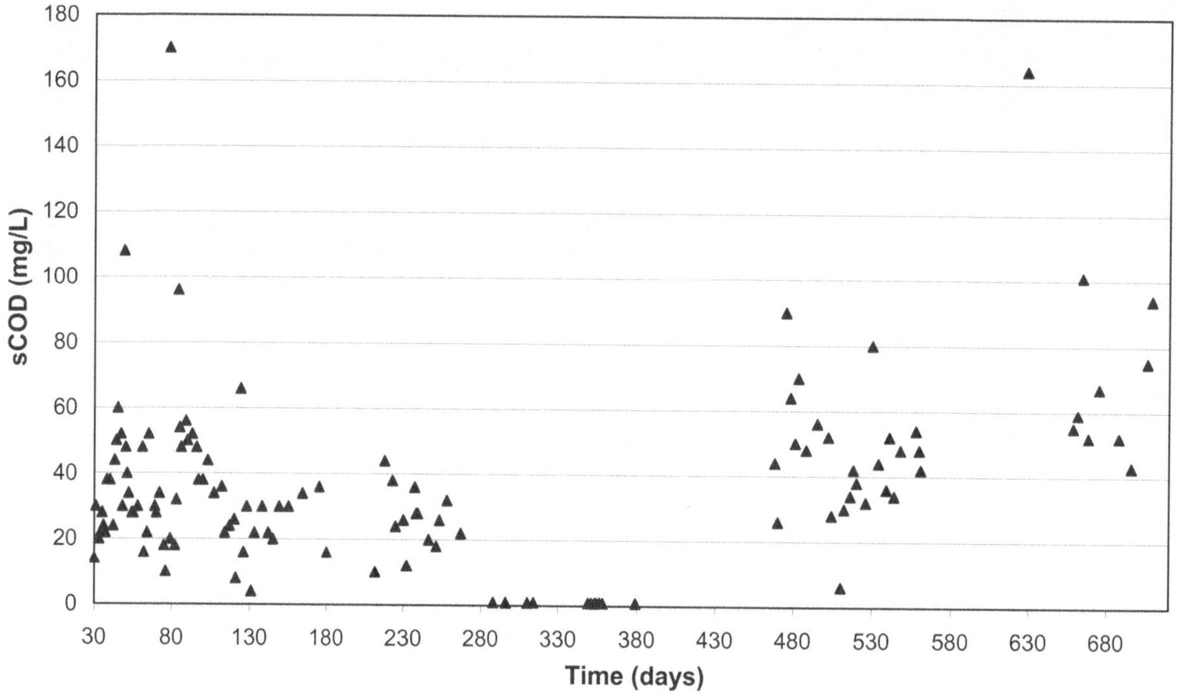

Figure A-13. Effluent Soluble COD for the 10-Day SRT Pilot Plant Reactor.

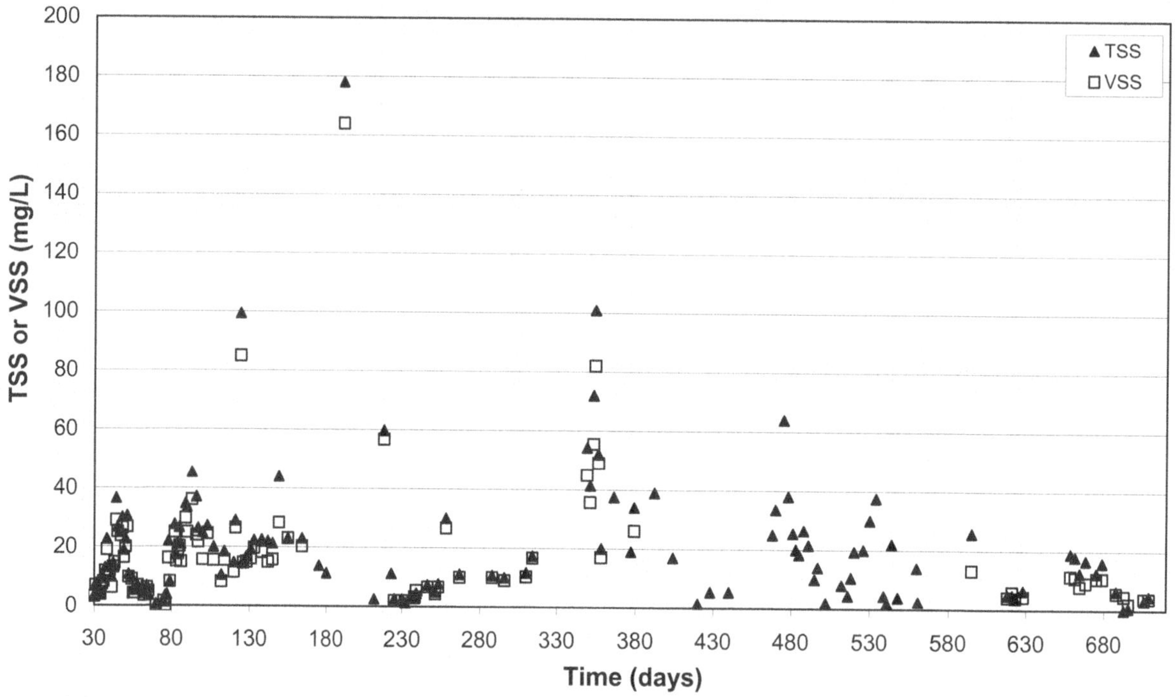

Figure A-14. Effluent TSS and VSS for the 10-Day SRT Pilot Plant Reactor.

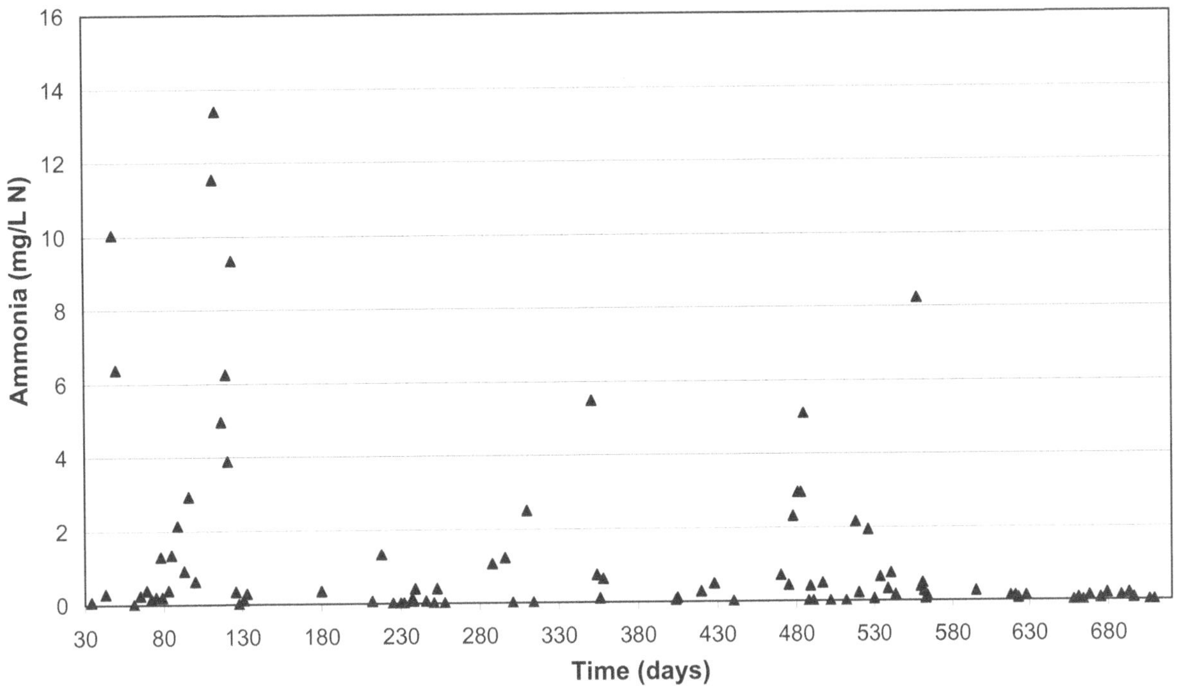

Figure A-15. Effluent Ammonium-Nitrogen for the 10-Day SRT Pilot Plant Reactor.

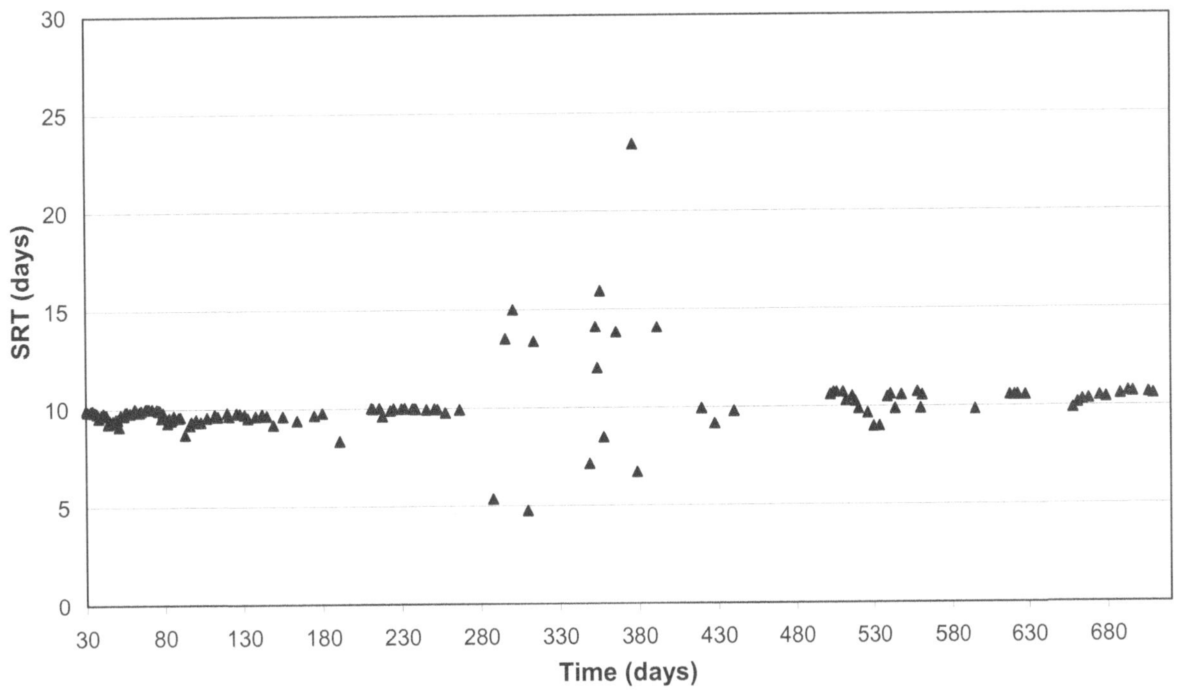

Figure A-16. Calculated Solids Residence Time (SRT) for the 10-Day SRT Pilot Plant Reactor.

A.4 Source-Effect Experiments

Table A-2 summarizes the schedule for collecting both mixed liquor and wastewater for the laboratory source-effect experiments for which results are reported here. All 2-day SRT experiments were initially run between days 147 and 436 but were rerun to ensure non-nitrifying conditions. Only data from the non-nitrifying period are included in this report.

Table A-2. Schedule for Collecting Mixed Liquor and Wastewater for Source-Effect Experiments. Basis for Days is Pilot Plant Schedule, with Day 0 Representing the Start of the Pilot Plant.

Source-Effect Experiment	2-day SRT: Duration of Source-Effect Experiment (inclusive days)	10-day SRT: Duration of Source-Effect Experiment (inclusive days)
Ammonia	588 to 594	219 to 249
pH	604 to 610	359 to 389
Cadmium	619 to 625	138 to 168
CDNB	675 to 681	303 to 331
Octanol	666 to 672	184 to 199
DNP	683 to 689	441 to 471
Cyanide	713 to 719	693 to 706

APPENDIX B

ADDITIONAL DATA FOR CADMIUM SHOCK EXPERIMENTS

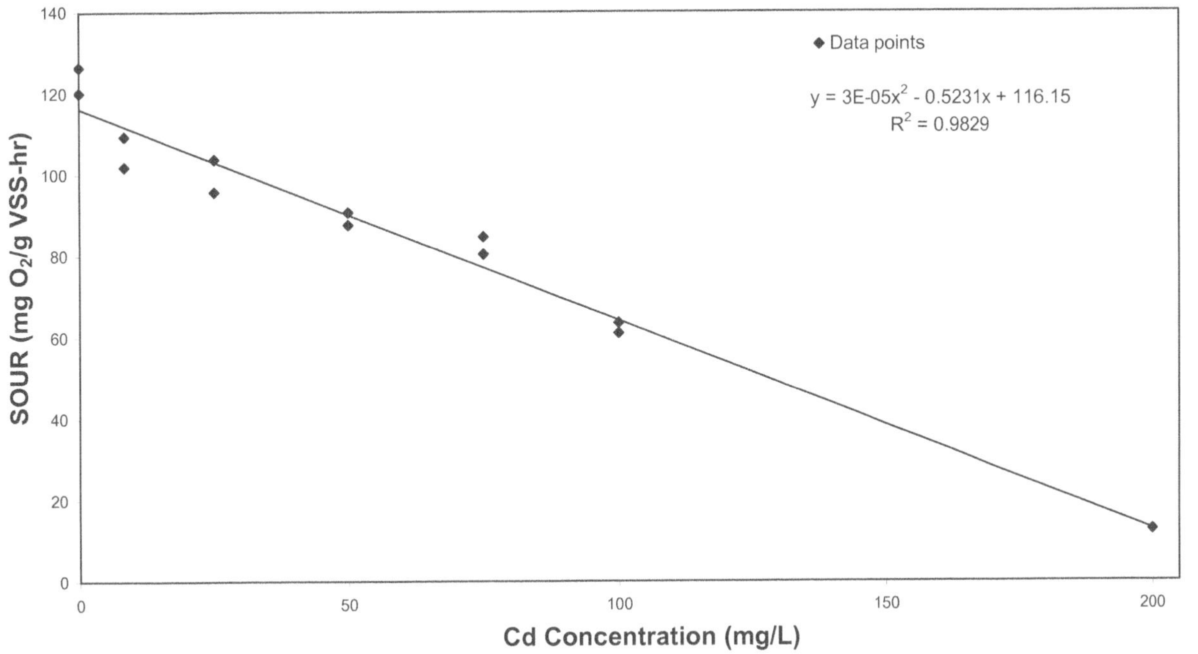

Figure B-1. ICxx Results for the 2-Day SRT Biomass Exposed to Cadmium Shock. The Equation Represents the Best Fit Polynomial to the Data.

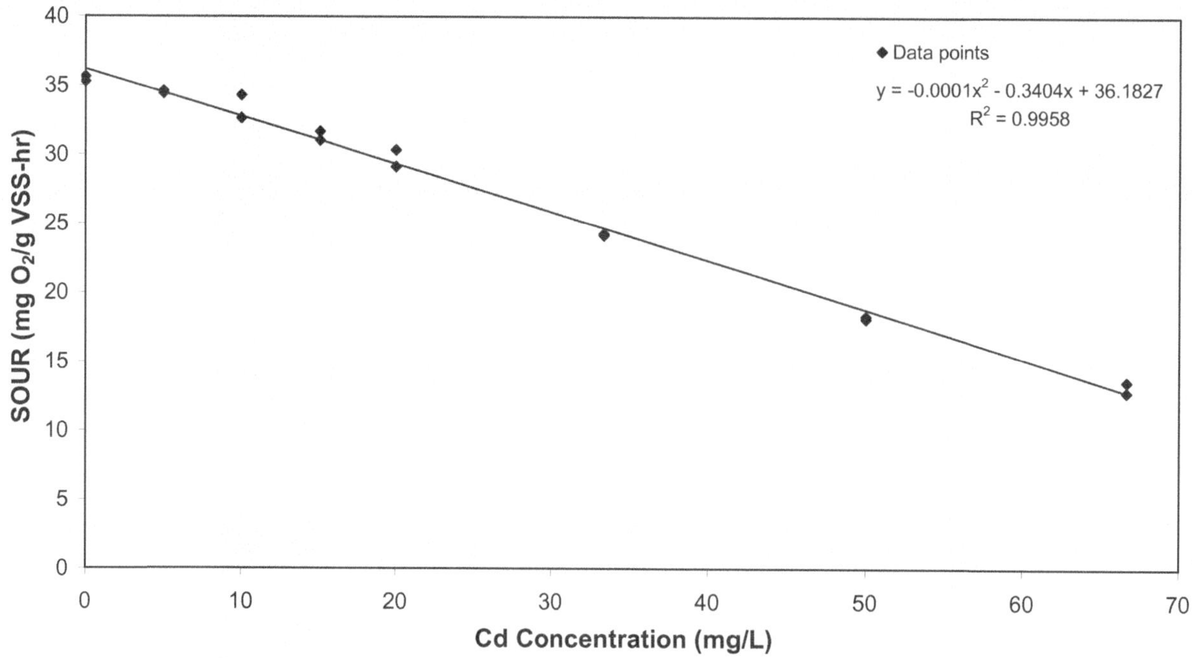

Figure B-2. ICxx Results for the 10-Day SRT Biomass Exposed to Cadmium Shock. The Equation Represents the Best Fit Polynomial to the Data.

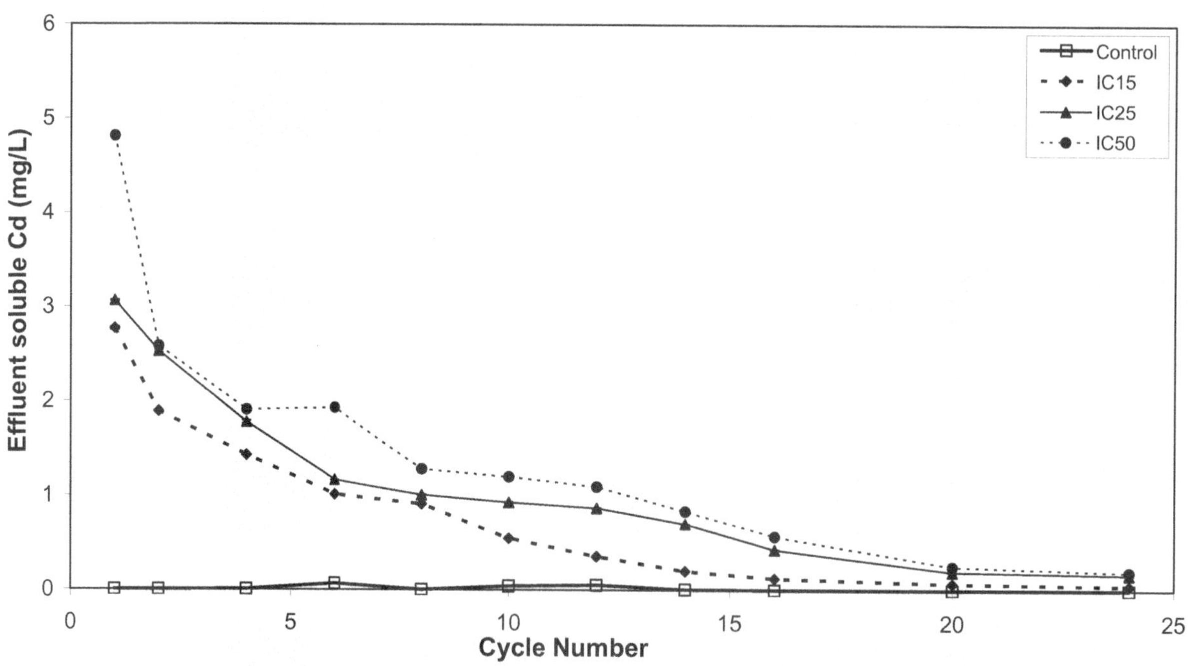

Figure B-3. Effluent Soluble Cadmium for the 2-Day SRT Reactors Exposed to Cadmium Shock.

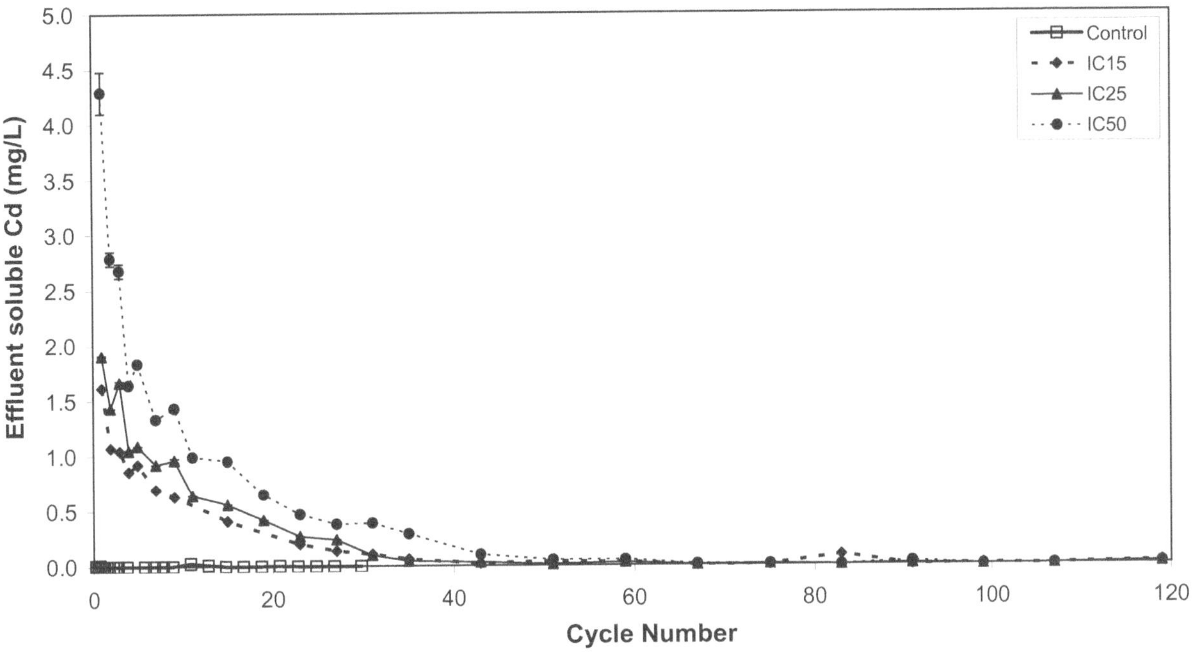

Figure B-4. Effluent Soluble Cadmium for the 10-Day SRT Reactors Exposed to Cadmium Shock.

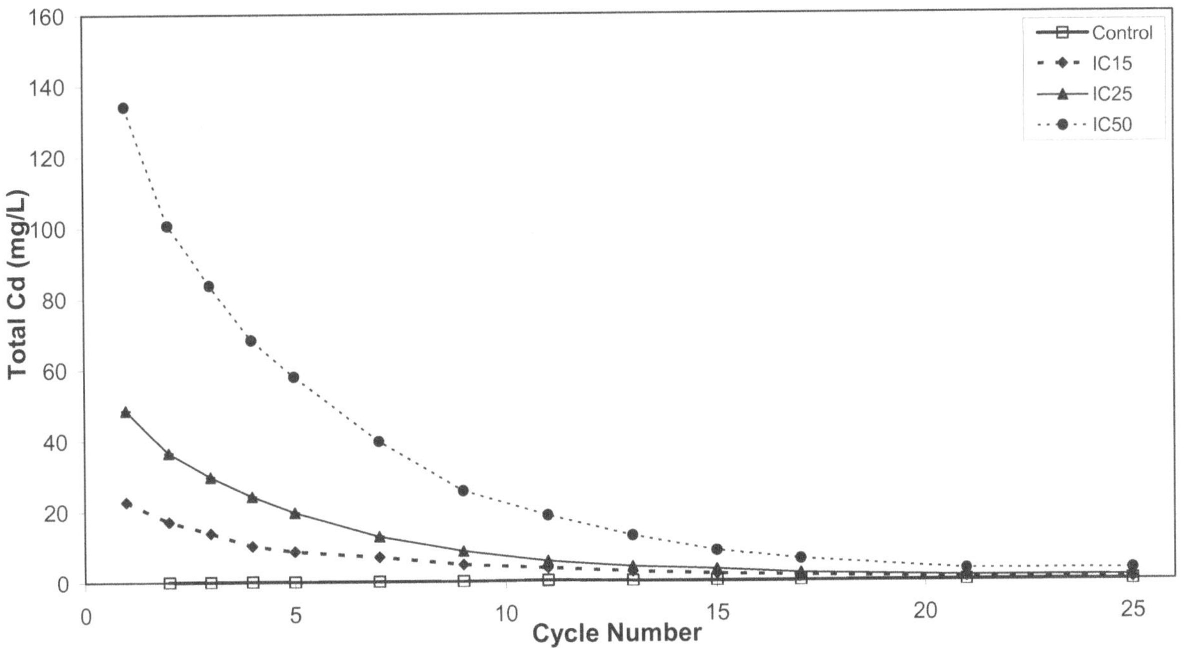

Figure B-5. Total Cadmium in Mixed Liquor for the 2-Day SRT Reactors Exposed to Cadmium Shock; the First Point of Each Line is an Estimate of the Concentration in the Mixed Liquor During Cycle 1 and Assuming no Reactions.

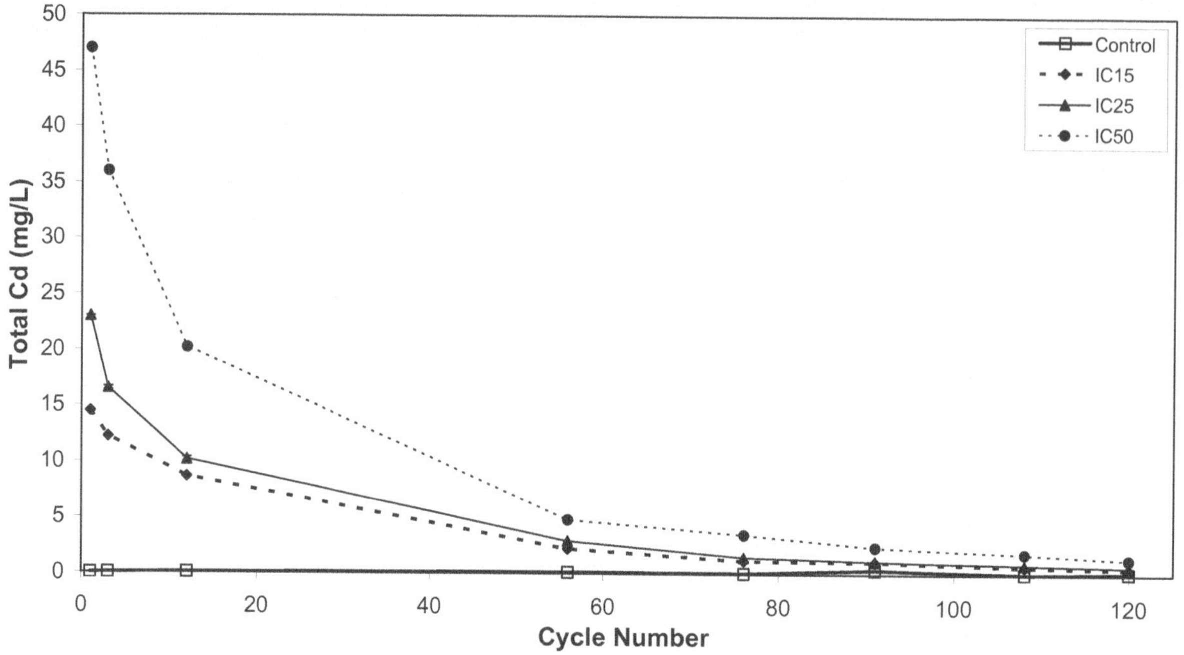

Figure B-6. Total Cadmium in Mixed Liquor for the 10-Day SRT Reactors Exposed to Cadmium Shock.

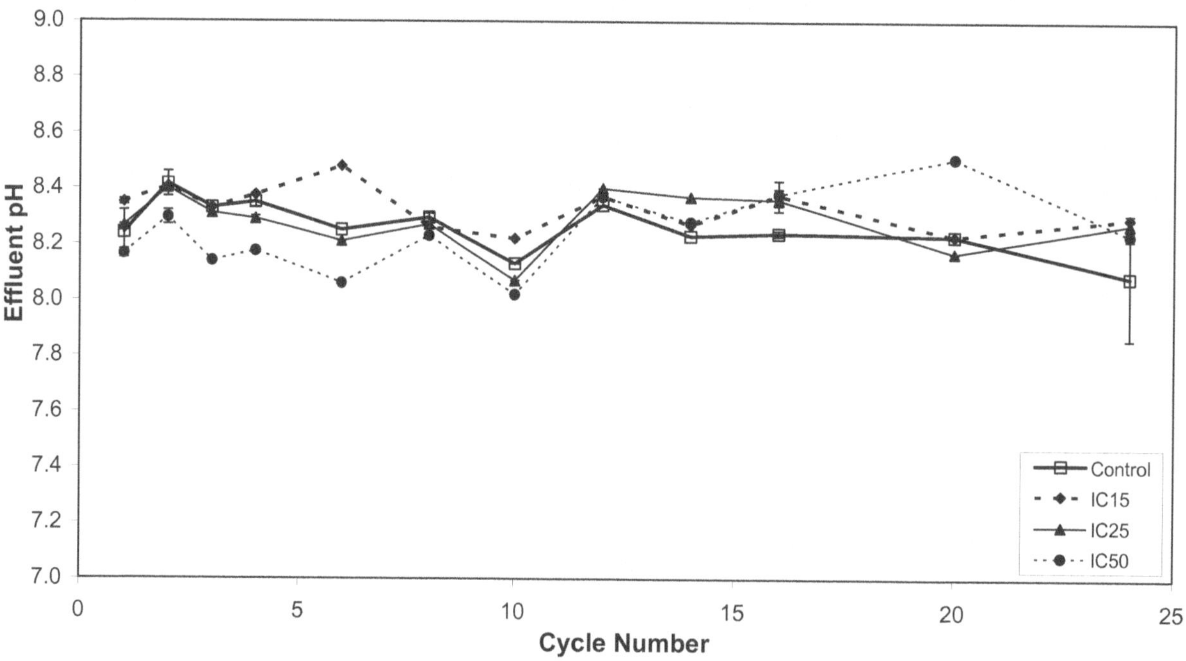

Figure B-7. Effluent pH for the 2-Day SRT Reactors Exposed to Cadmium Shock.

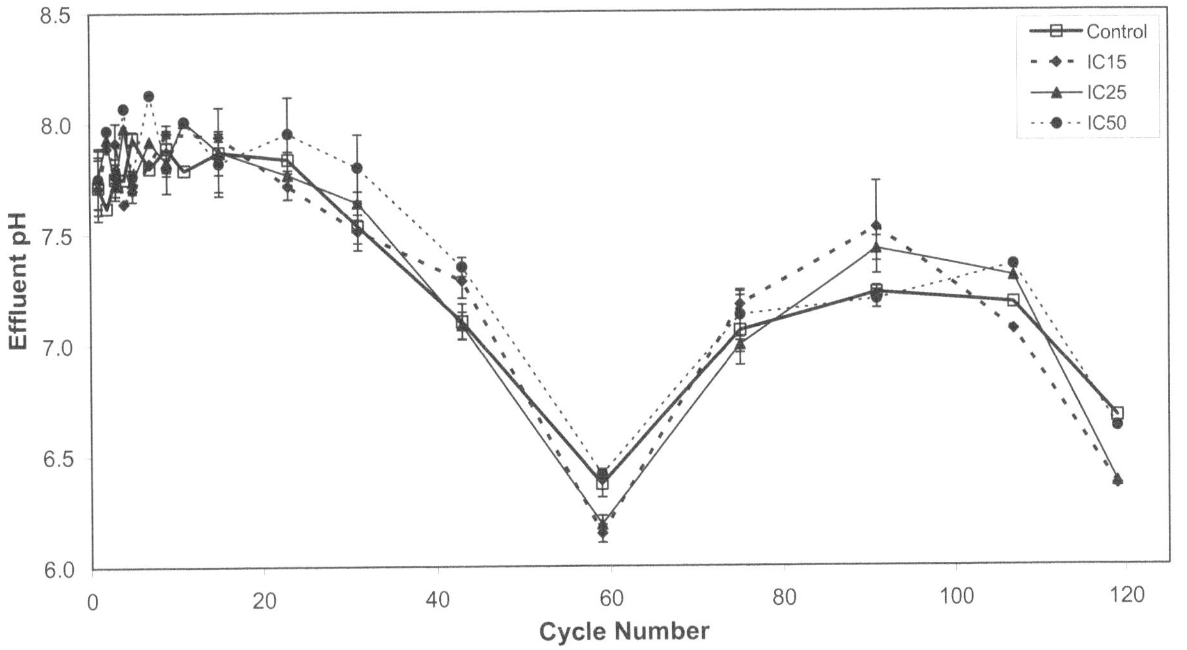

Figure B-8. Effluent pH for the 10-Day SRT Reactors Exposed to Cadmium Shock.

APPENDIX C

ADDITIONAL DATA FOR 1-OCTANOL SHOCK EXPERIMENTS

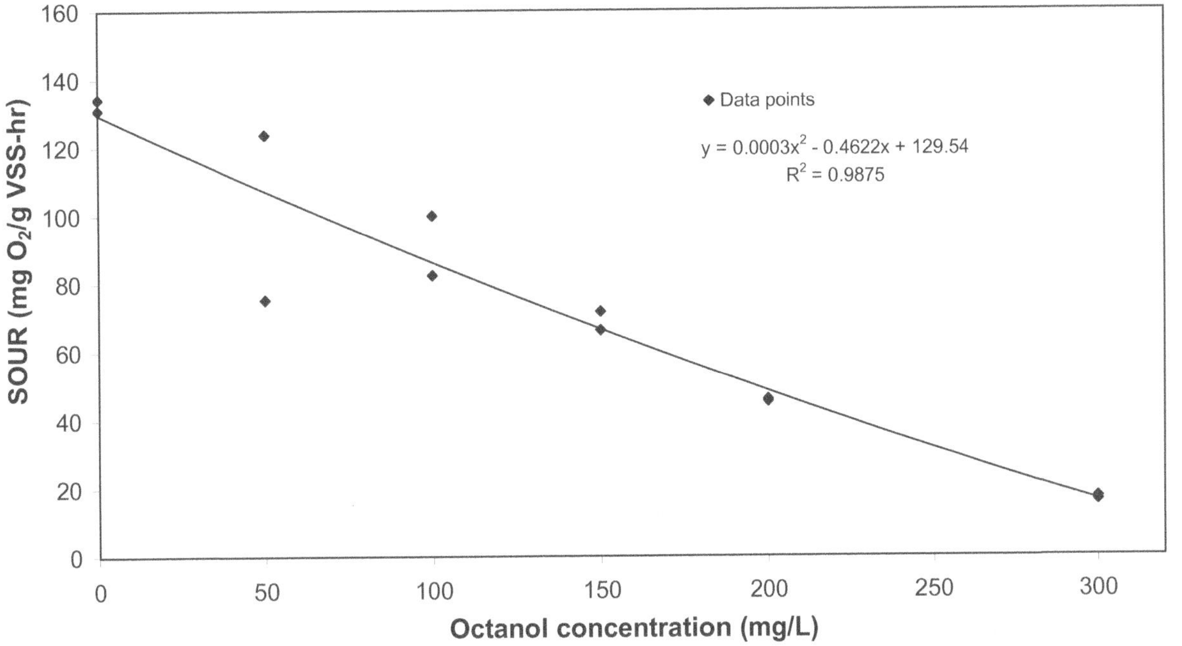

Figure C-1. ICxx Results for the 2-Day SRT Biomass Exposed to Octanol Shock. The Equation Represents the Best Fit Polynomial to the Data.

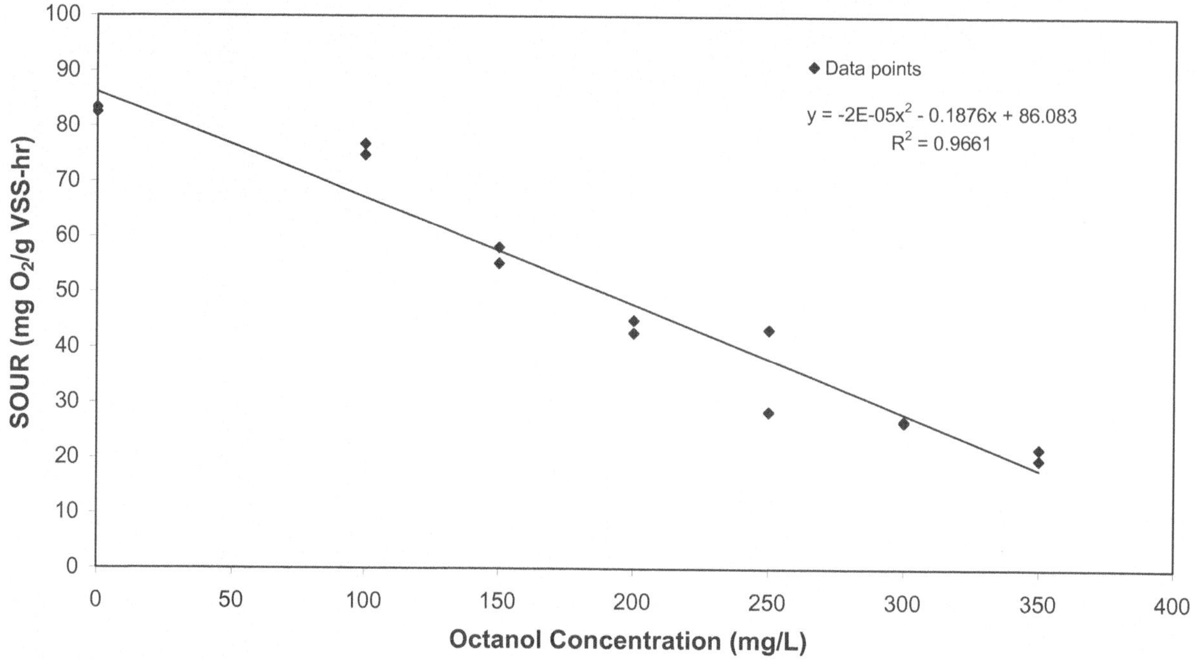

Figure C-2. ICxx Results for the 10-Day SRT Biomass Exposed to Octanol Shock. The Equation Represents the Best Fit Polynomial to the Data.

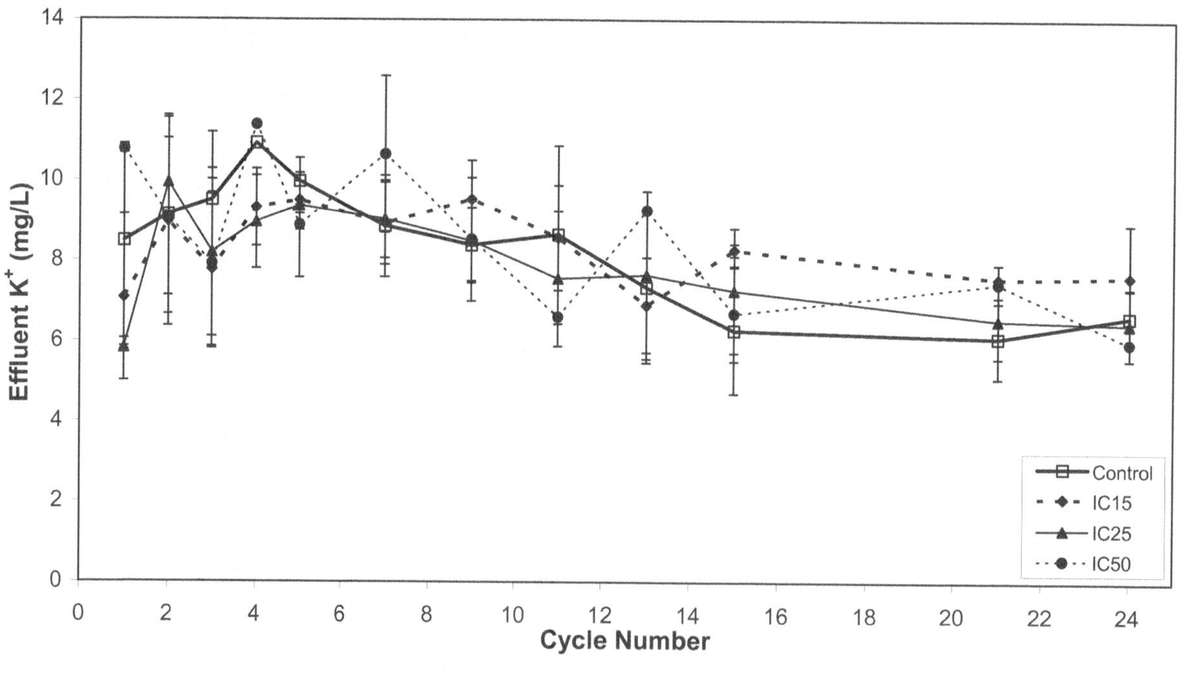

Figure C-3. Effluent Potassium for the 2-Day SRT Reactors Exposed to Octanol Shock.

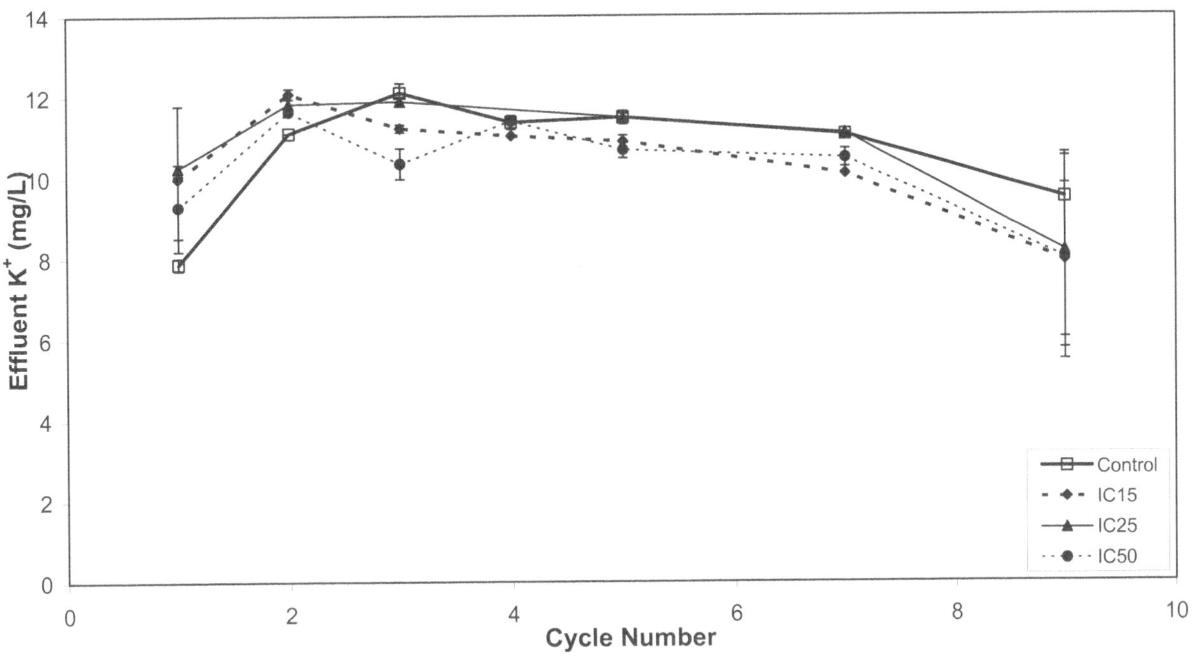

Figure C-4. Effluent Potassium for the 10-Day SRT Reactors Exposed to Octanol Shock. Data Through Cycle 9 Are Shown.

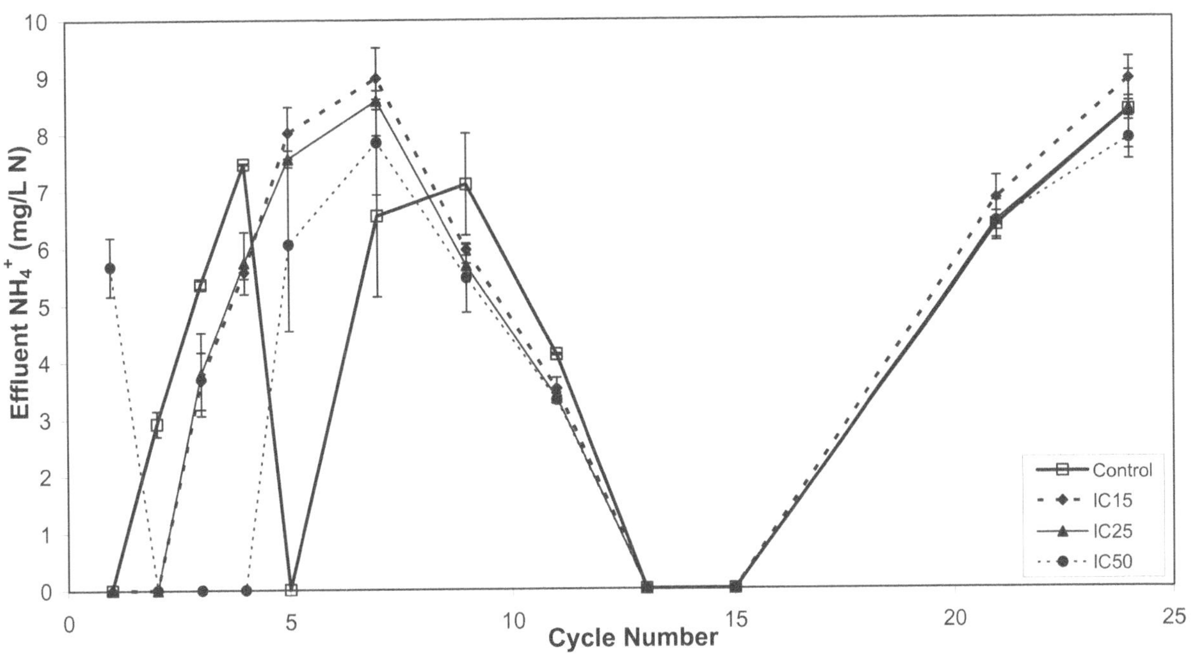

Figure C-5. Effluent Ammonia-Nitrogen for the 2-Day SRT Reactors Exposed to Octanol Shock.

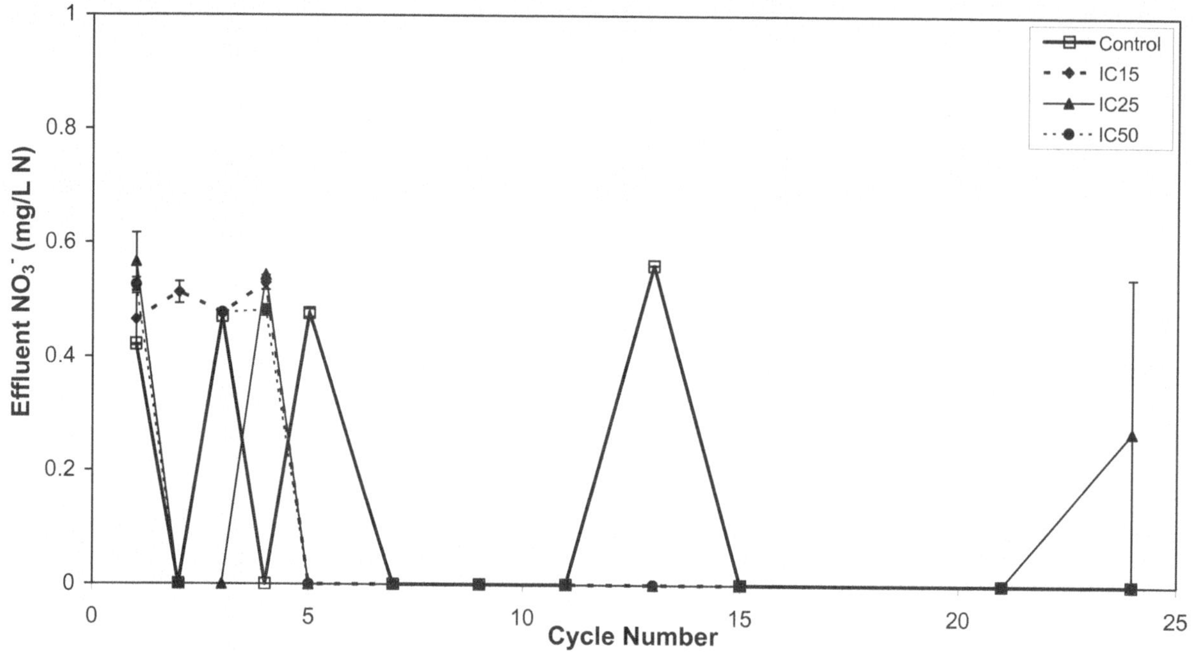

Figure C-6. Effluent Nitrate-Nitrogen for the 2-Day SRT Reactors Exposed to Octanol Shock.

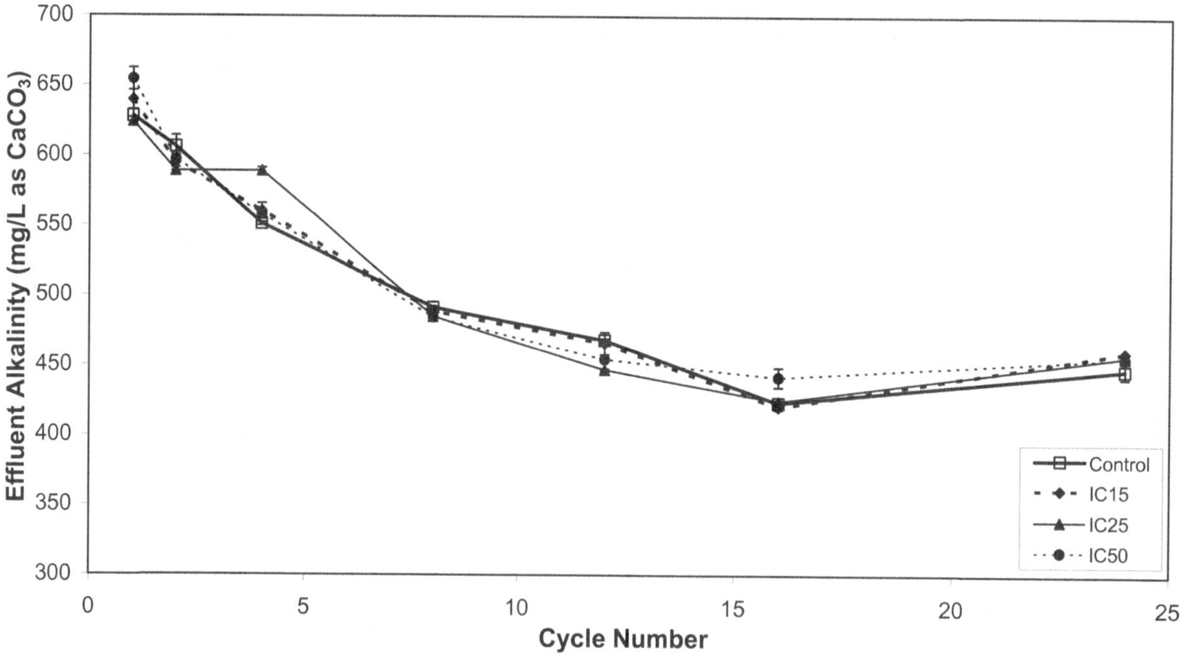

Figure C-7. Effluent Alkalinity for the 2-Day SRT Reactors Exposed to Octanol Shock.

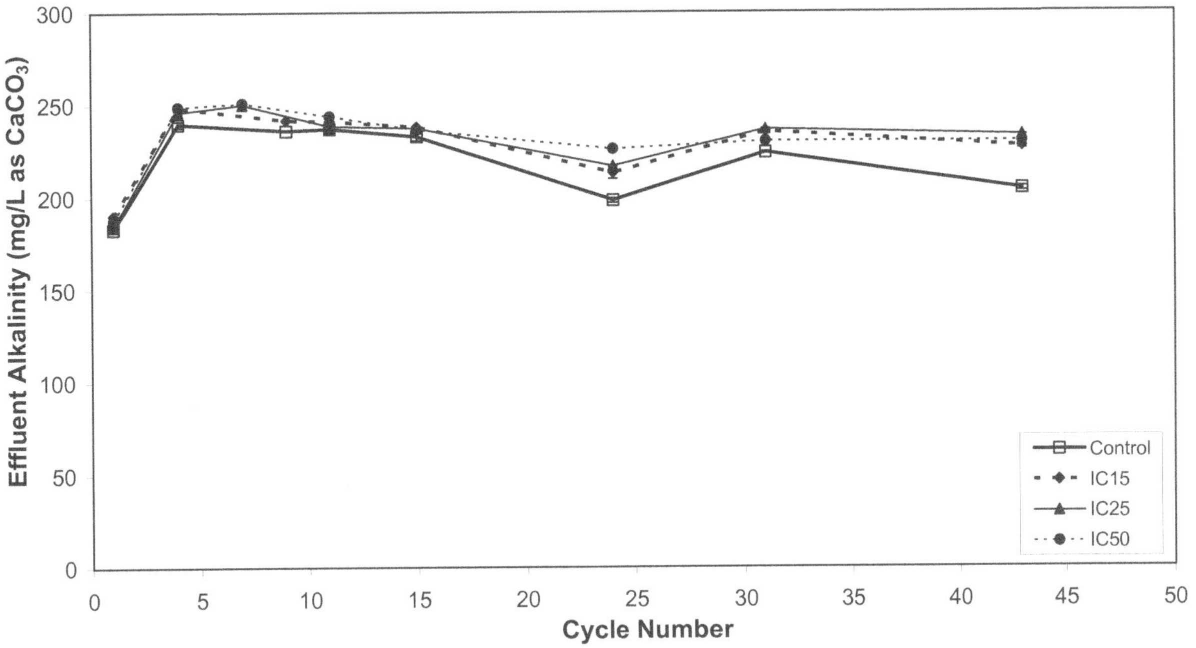

Figure C-8. Effluent Alkalinity for the 10-Day SRT Reactors Exposed to Octanol Shock.

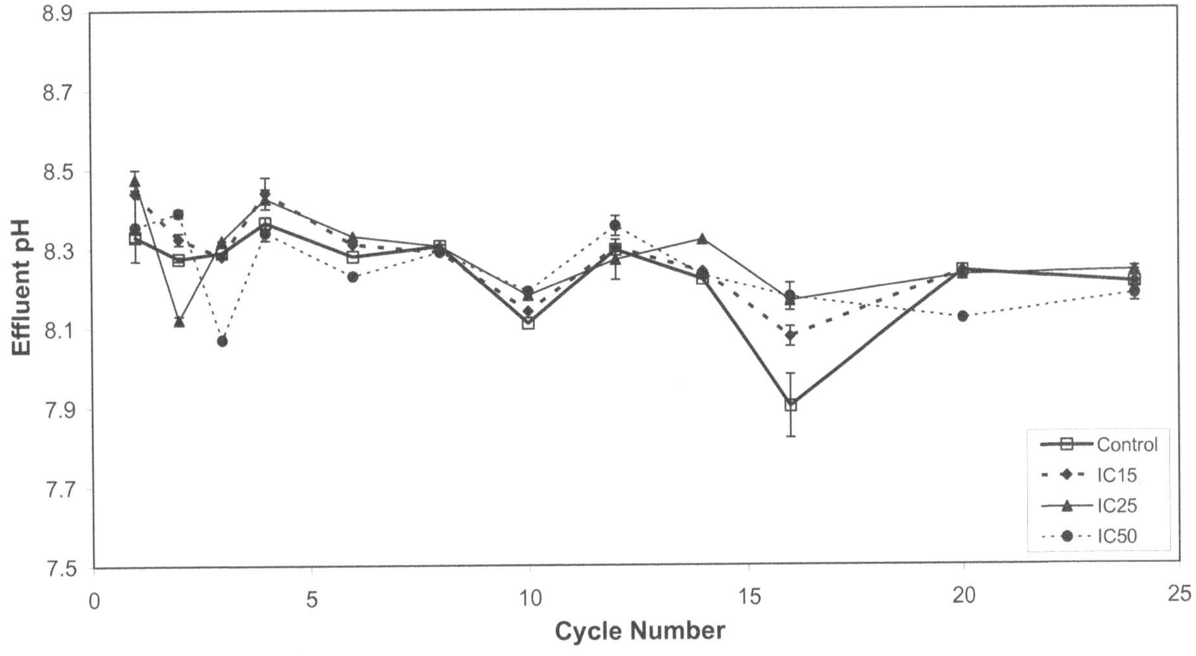

Figure C-9. Effluent pH for the 2-Day SRT Reactors Exposed to Octanol Shock.

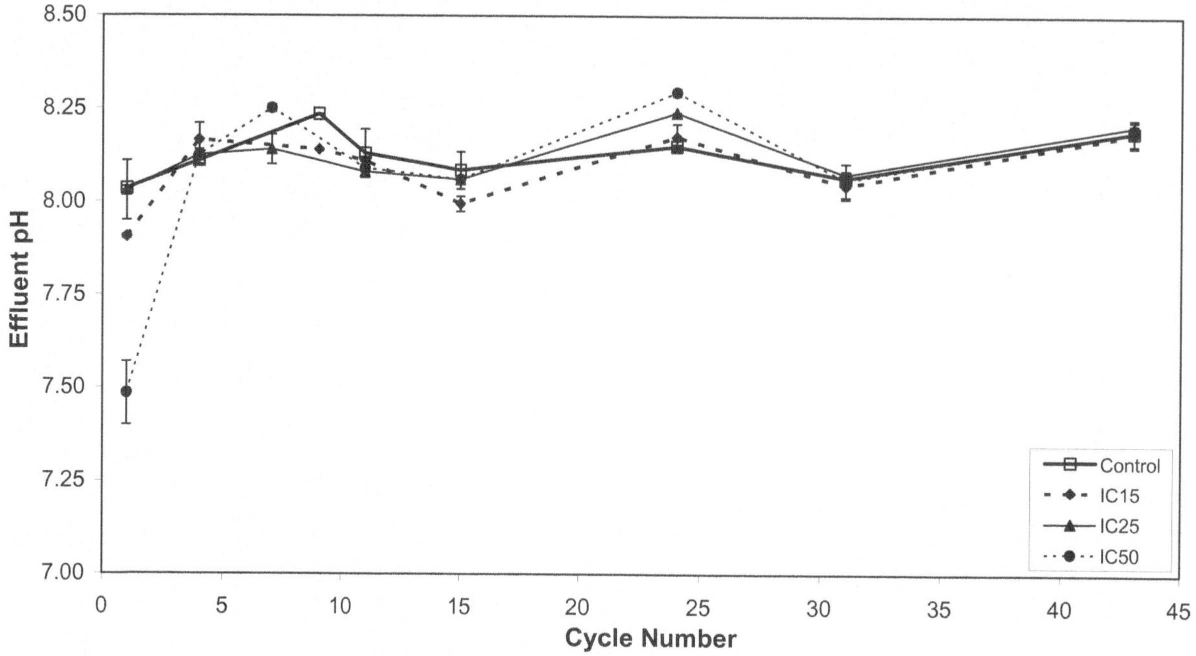

Figure C-10. Effluent pH for the 10-Day SRT Reactors Exposed to Octanol Shock.

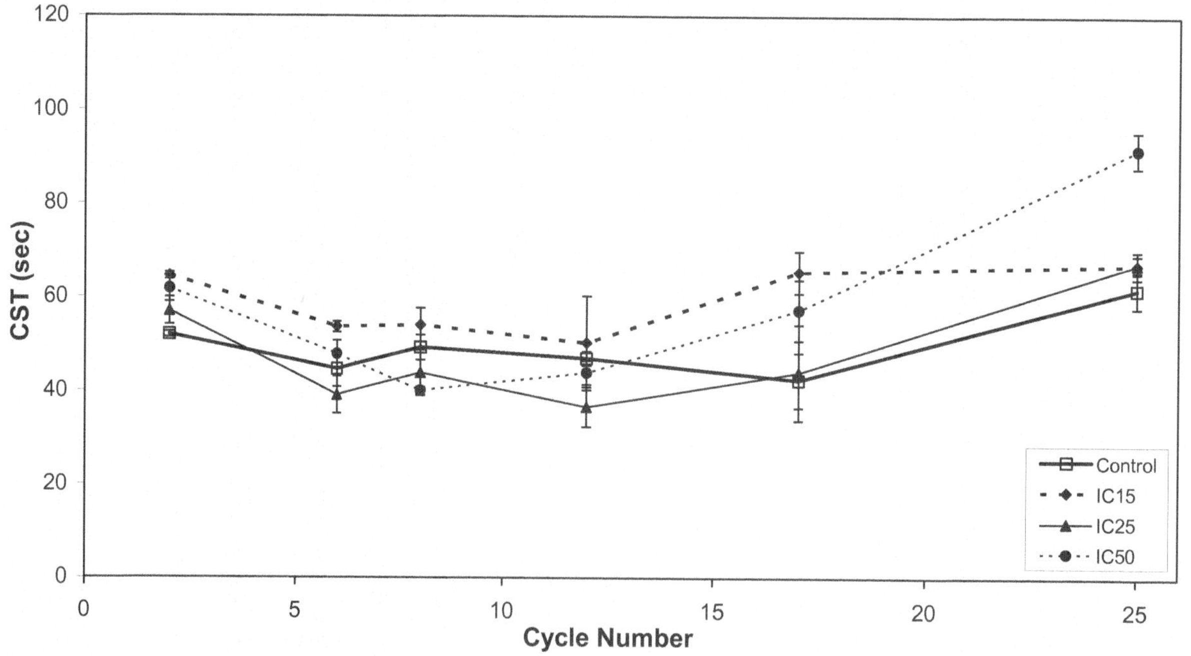

Figure C-11. Mixed Liquor CST for the 2-Day SRT Reactors Exposed to Octanol Shock.

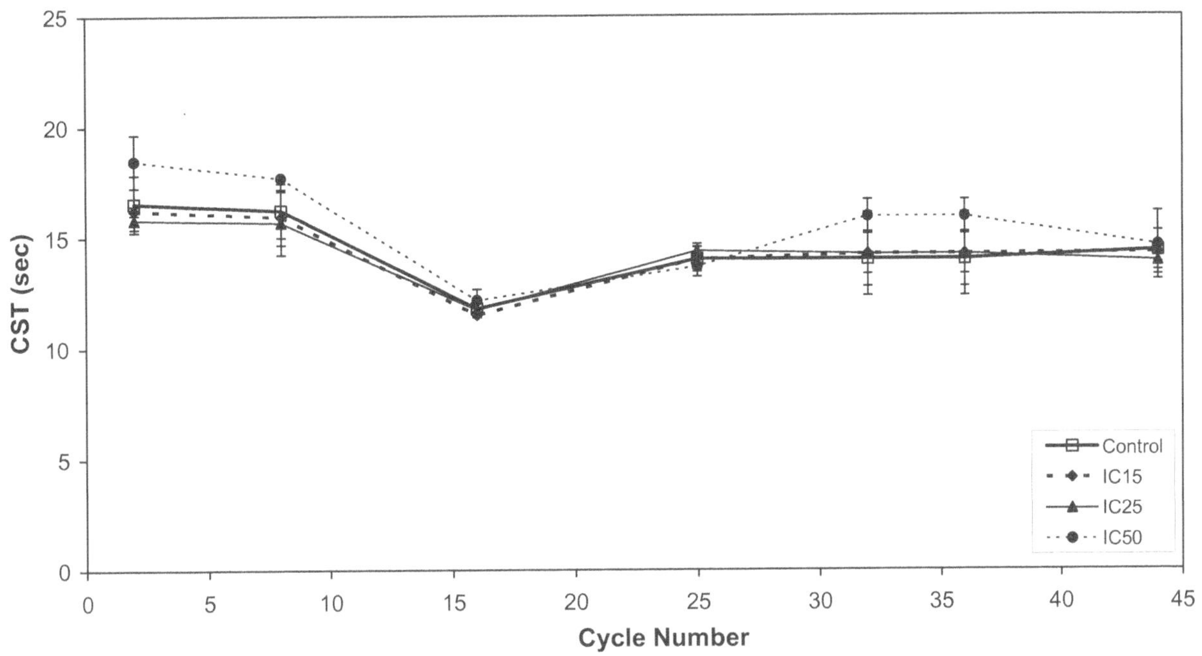

Figure C-12. Mixed Liquor CST for the 10-Day SRT Reactors Exposed to Octanol Shock.

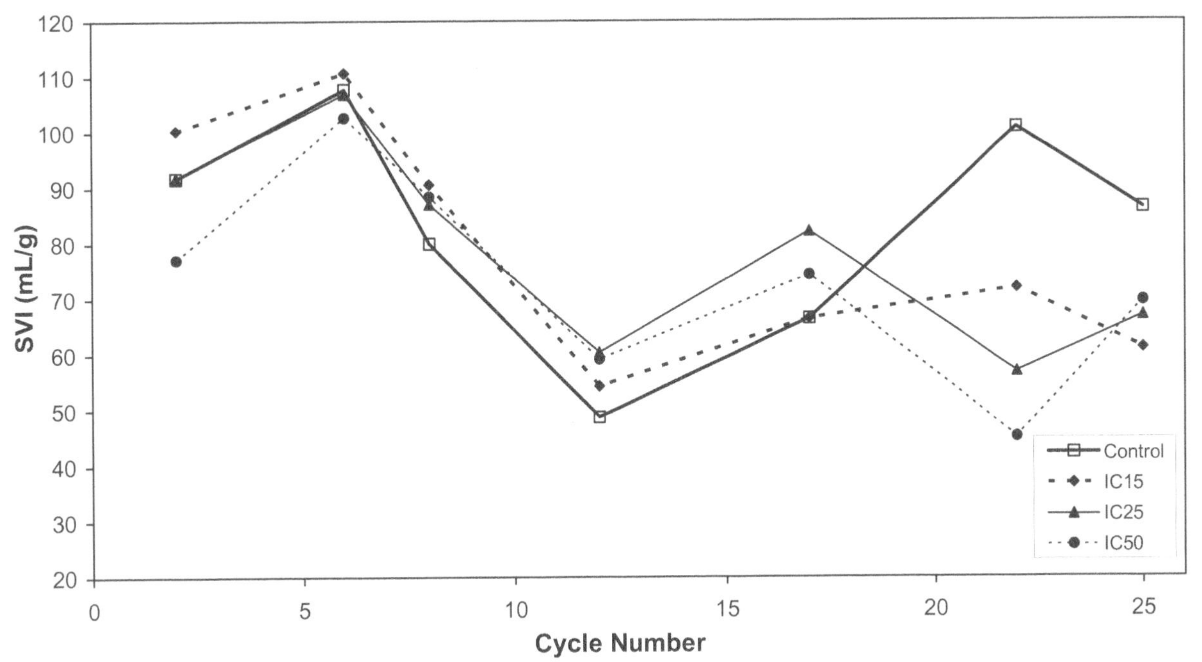

Figure C-13. SVI for the 2-Day SRT Reactors Exposed to Octanol Shock.

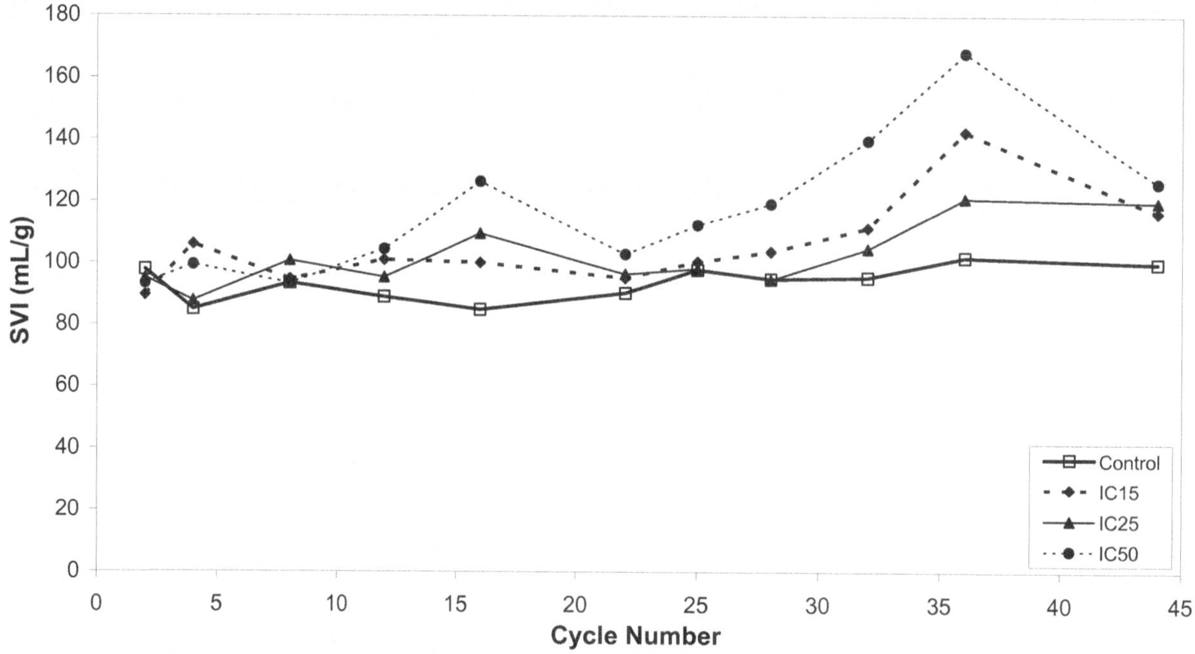

Figure C-14. SVI for the 10-Day SRT Reactors Exposed to Octanol Shock.

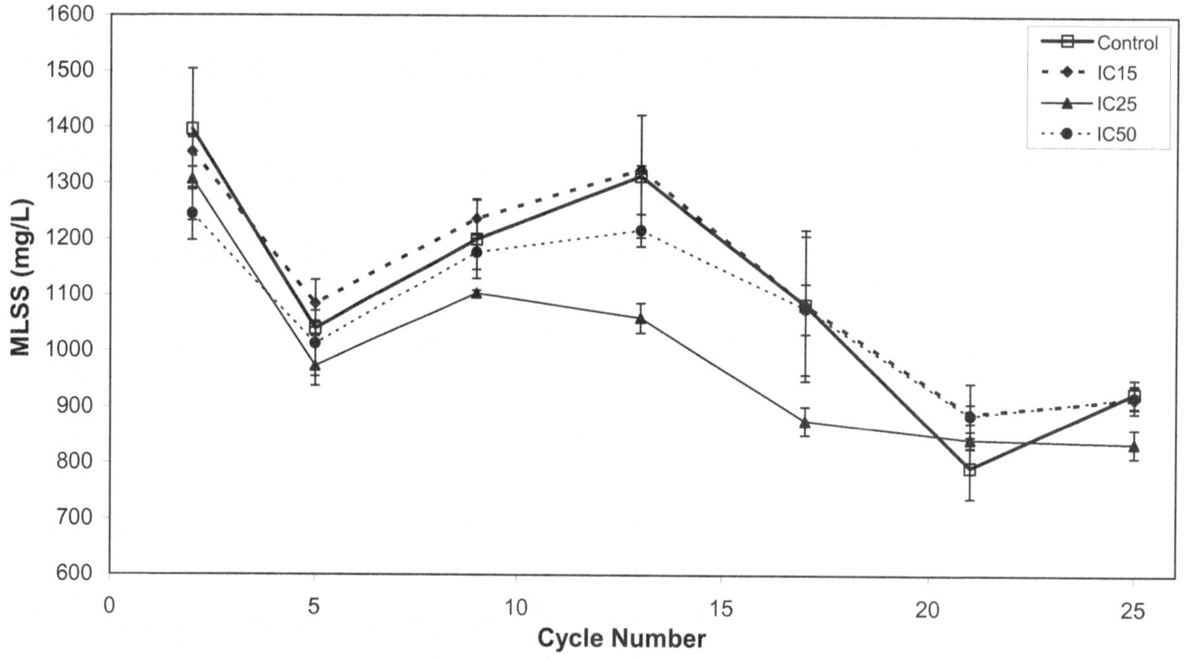

Figure C-15. MLSS for the 2-Day SRT Reactors Exposed to Octanol Shock.

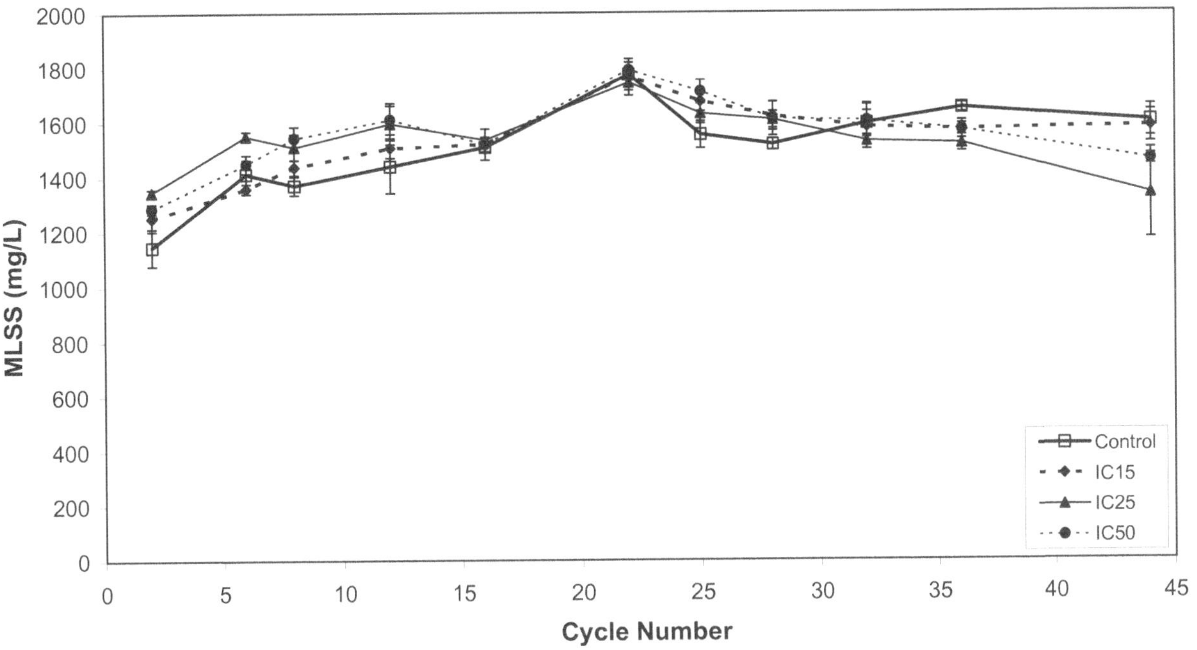

Figure C-16. MLSS for the 10-Day SRT Reactors Exposed to Octanol Shock.

APPENDIX D

ADDITIONAL DATA FOR AMMONIA SHOCK EXPERIMENTS

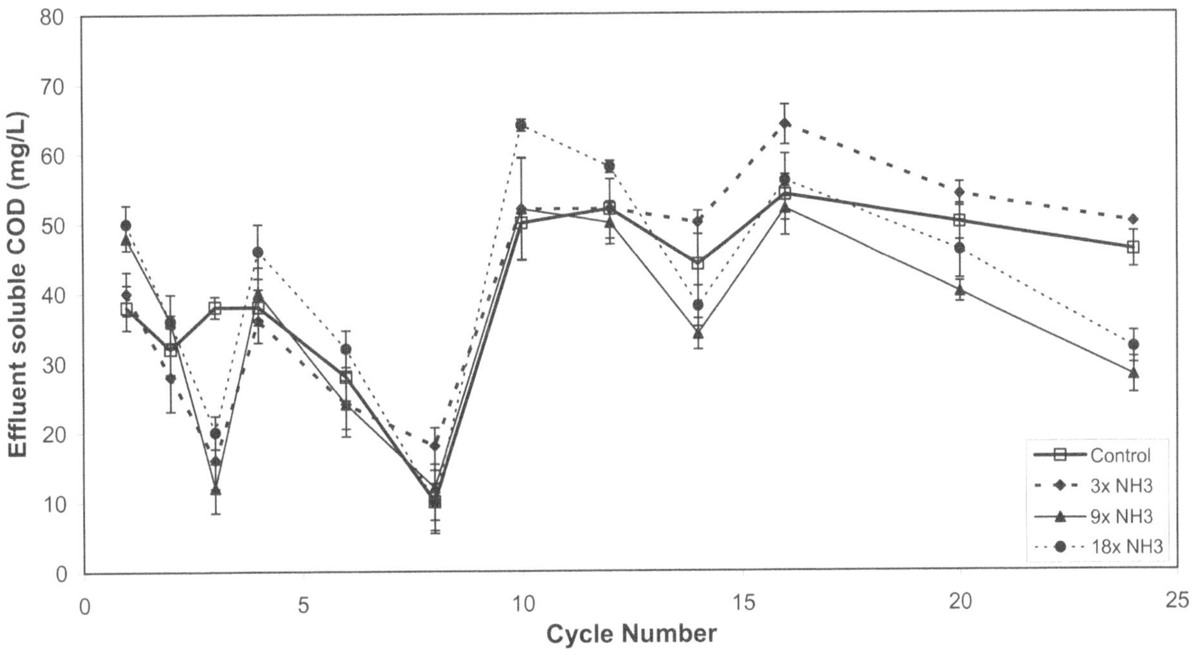

Figure D-1. Effluent Soluble COD for the 2-Day SRT Reactors Exposed to Ammonia Shock.

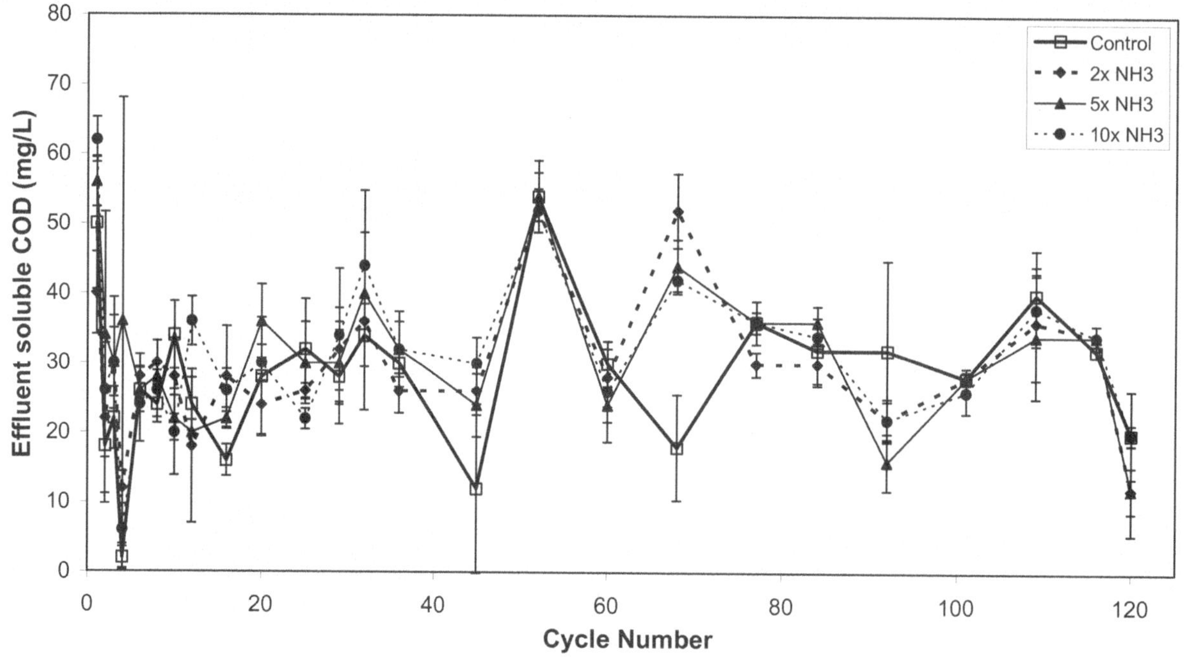

Figure D-2. Effluent Soluble COD for the 10-Day SRT Reactors Exposed to Ammonia Shock.

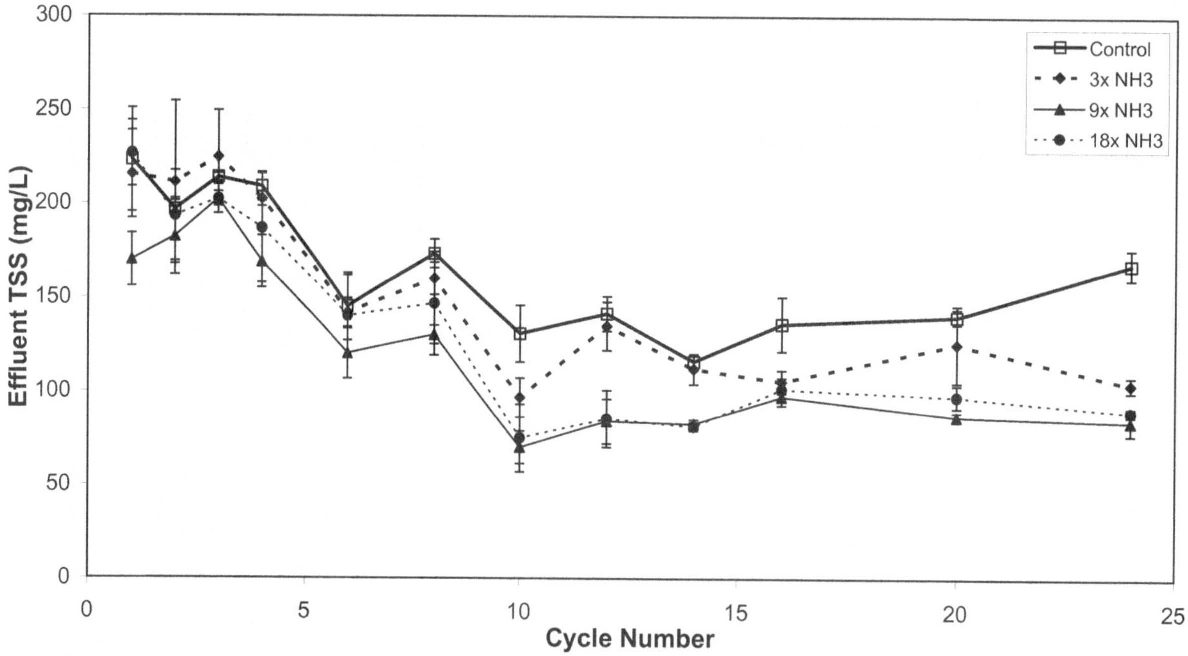

Figure D-3. Effluent TSS for the 2-Day SRT Reactors Exposed to Ammonia Shock.

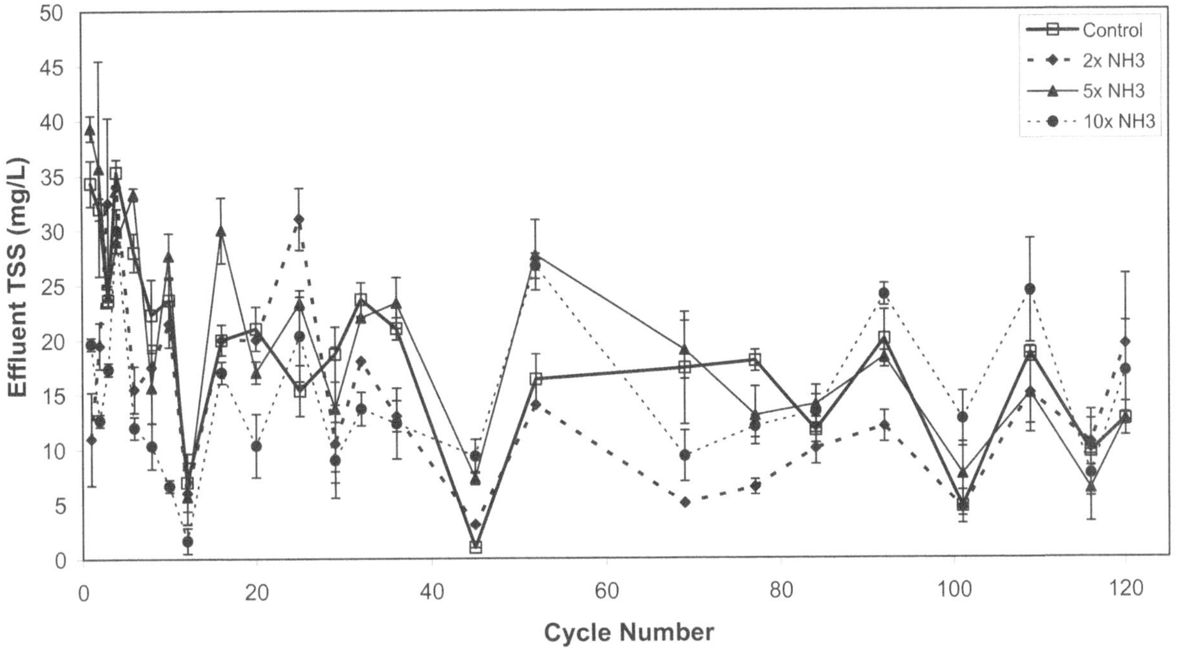

Figure D-4. Effluent TSS for the 10-Day SRT Reactors Exposed to Ammonia Shock.

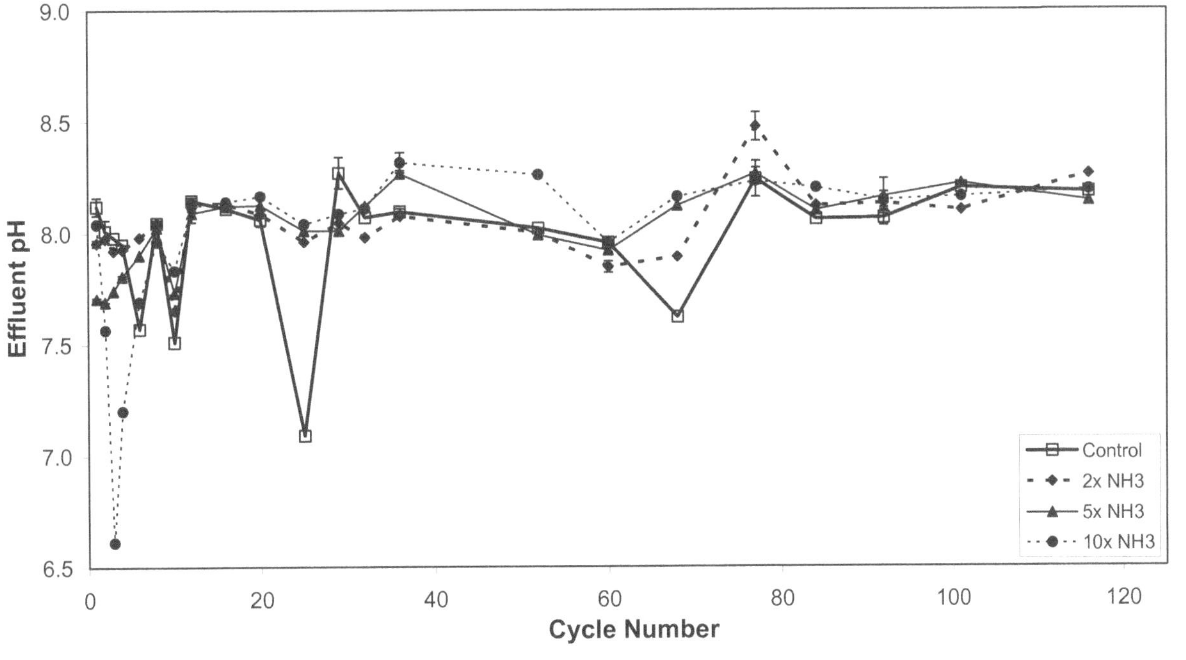

Figure D-5. Effluent pH for the 10-Day SRT Reactors Exposed to Ammonia Shock.

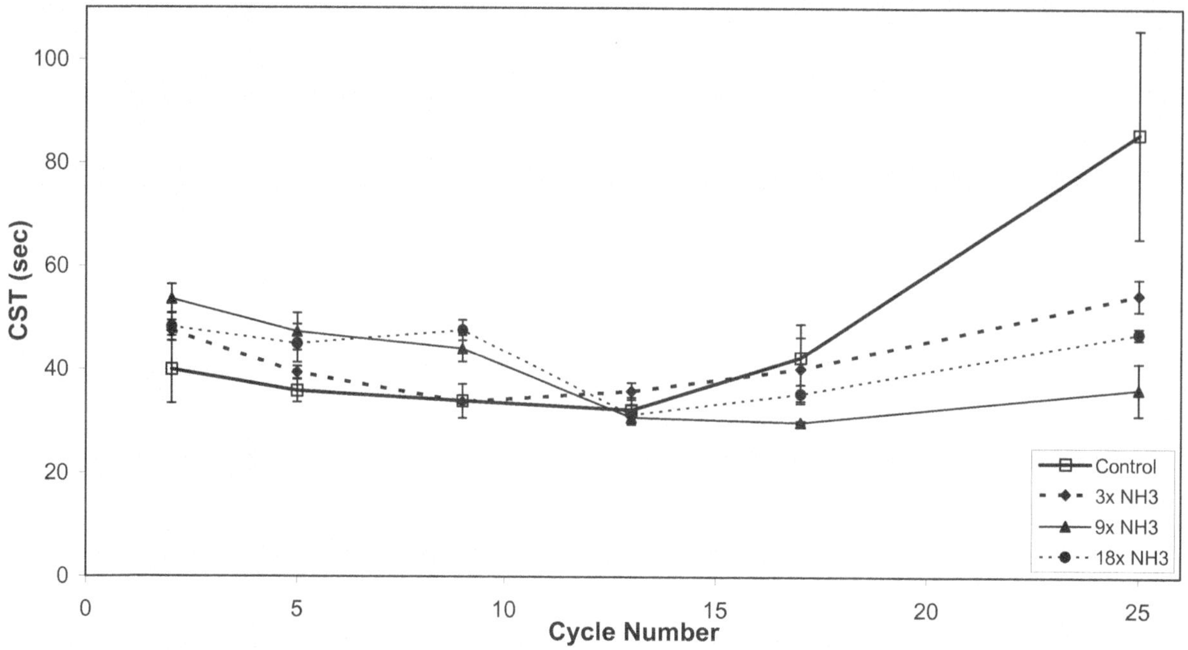

Figure D-6. Mixed Liquor CST for the 2-Day SRT Reactors Exposed to Ammonia Shock.

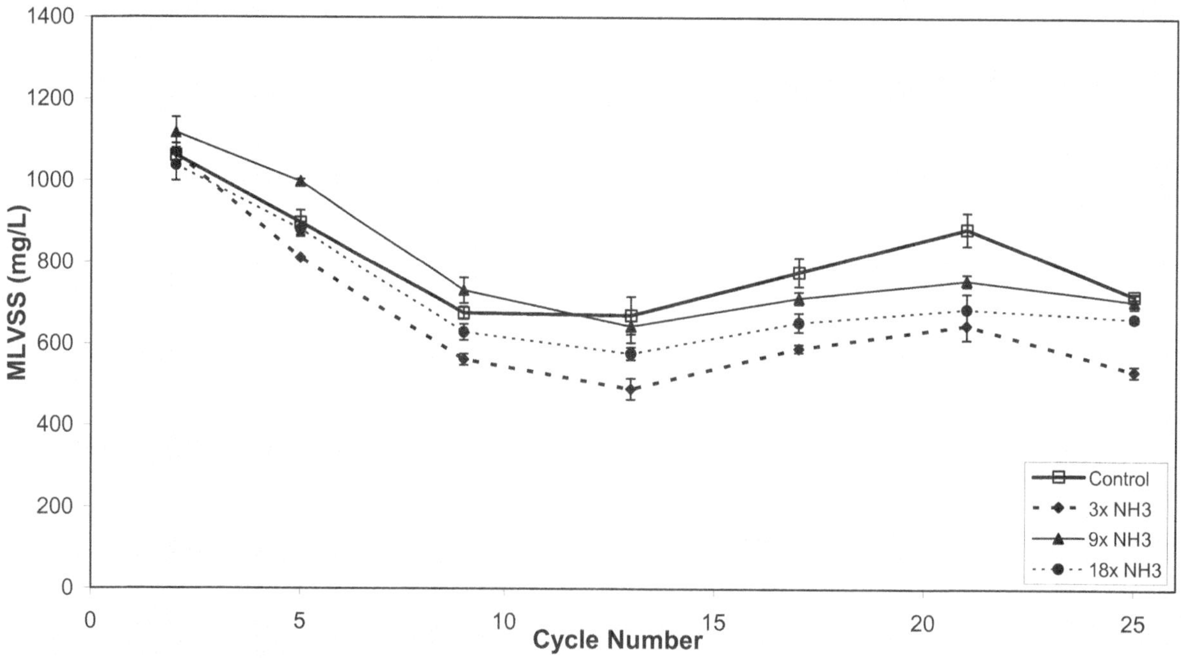

Figure D-7. MLVSS for the 2-Day SRT Reactors Exposed to Ammonia Shock.

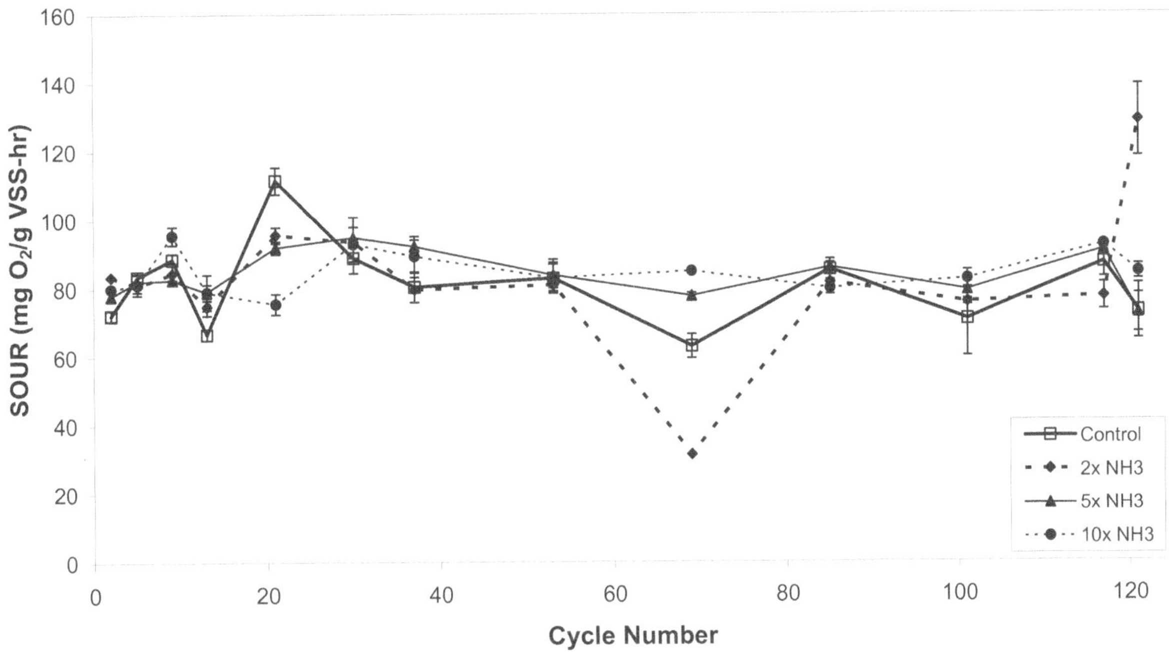

Figure D-8. SOURs for the 10-Day SRT Reactors Exposed to Ammonia Shock.

APPENDIX E

ADDITIONAL DATA FOR ZINC-COMPLEXED CYANIDE SHOCK EXPERIMENTS

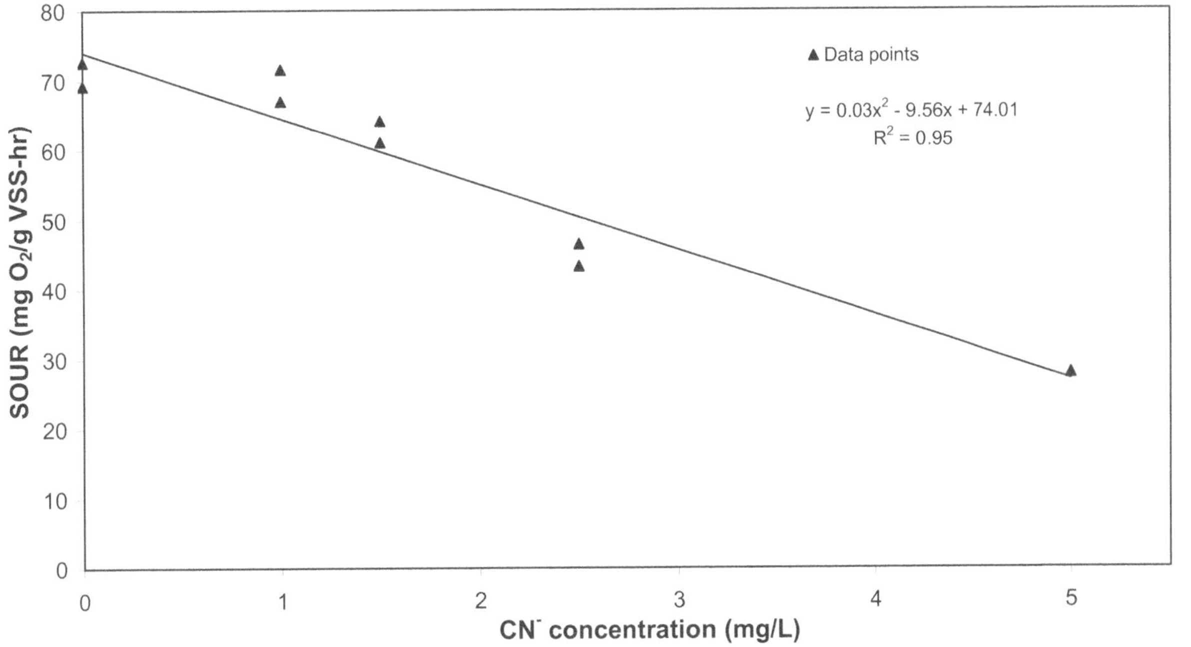

Figure E-1. ICxx Results for the 2-Day SRT Biomass Exposed to Zinc-Complexed Cyanide Shock. The Equation Represents the Best Fit Polynomial to the Data.

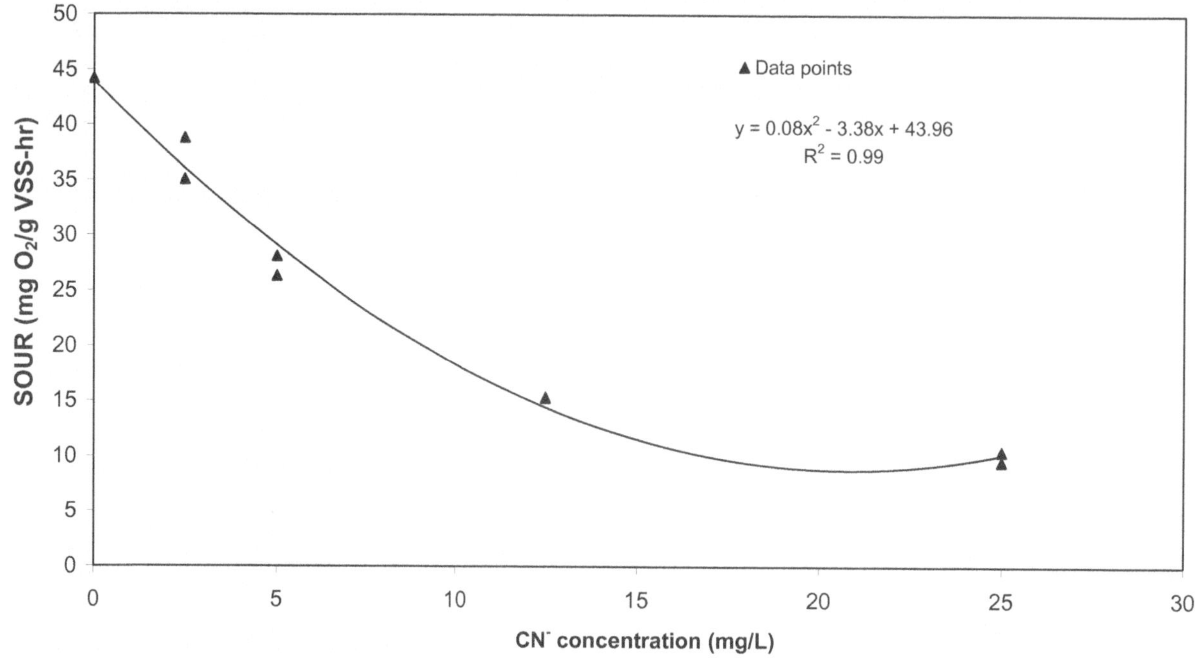

Figure E-2. ICxx Results for the 10-Day SRT Biomass Exposed to Cyanide Shock. The Equation Represents the Best Fit Polynomial to the Data.

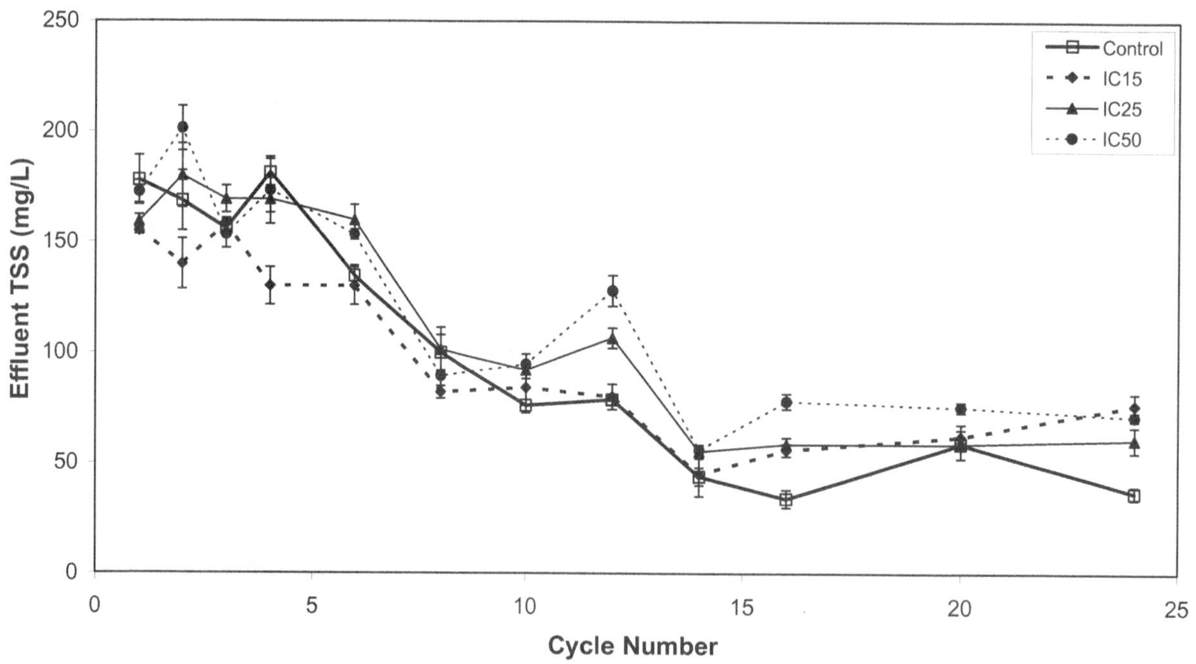

Figure E-3. Effluent TSS for the 2-Day SRT Reactors Exposed to Cyanide Shock.

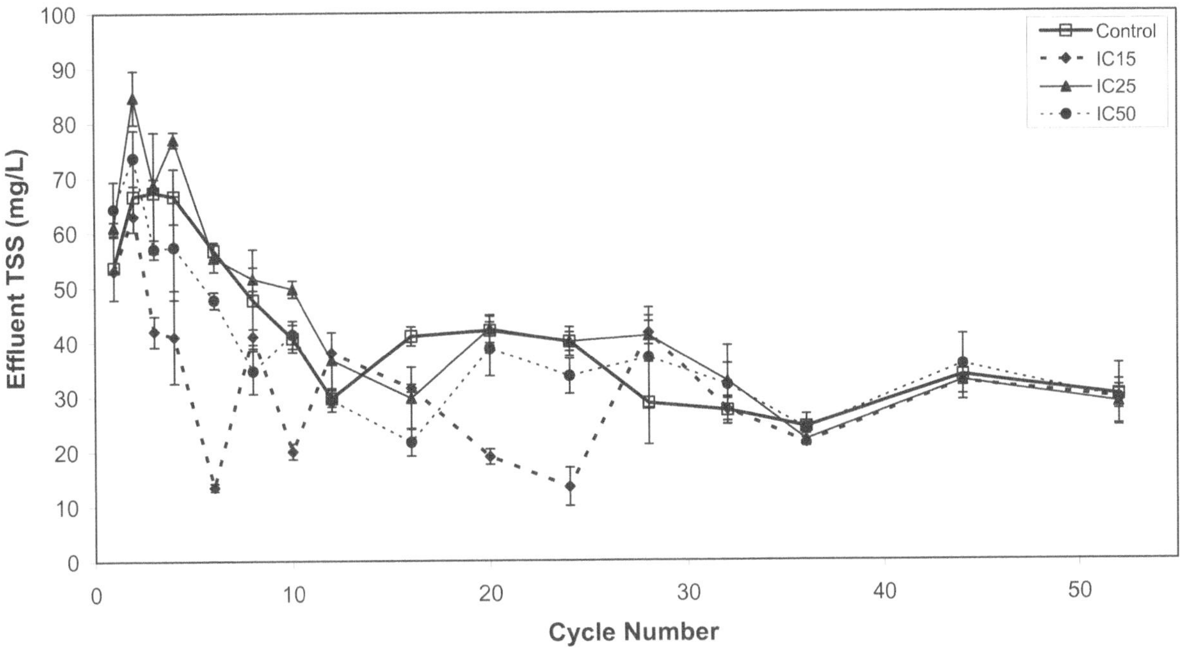

Figure E-4. Effluent TSS for the 10-Day SRT Reactors Exposed to Cyanide Shock.

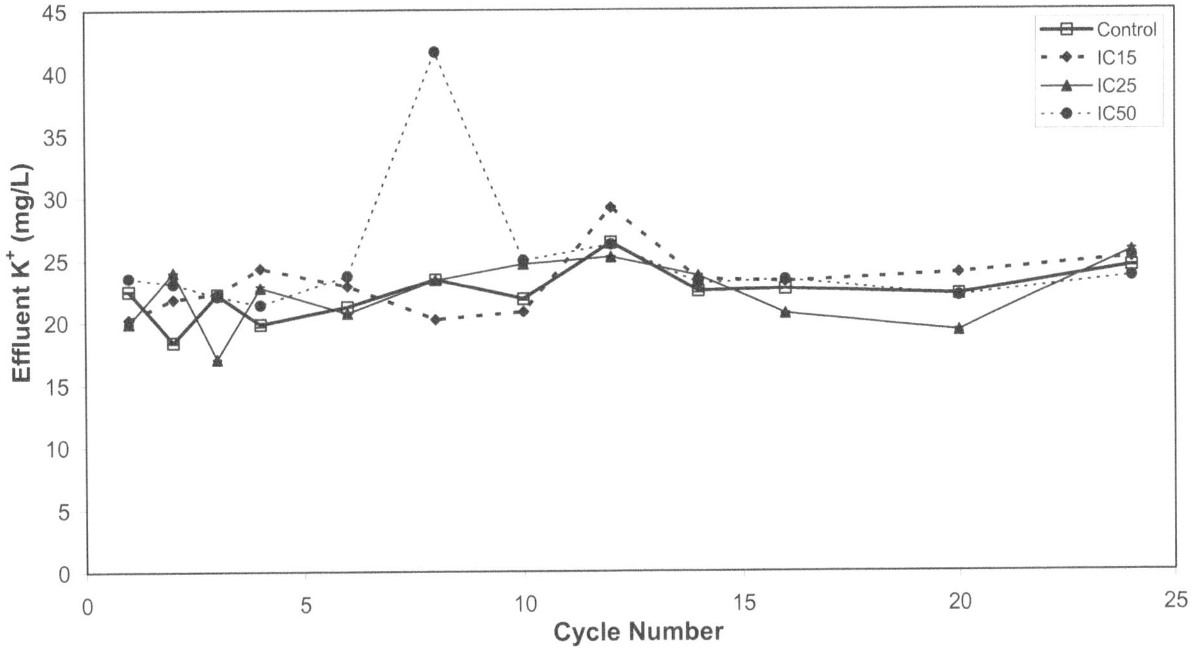

Figure E-5. Effluent Potassium for the 2-Day SRT Reactors Exposed to Cyanide Shock.

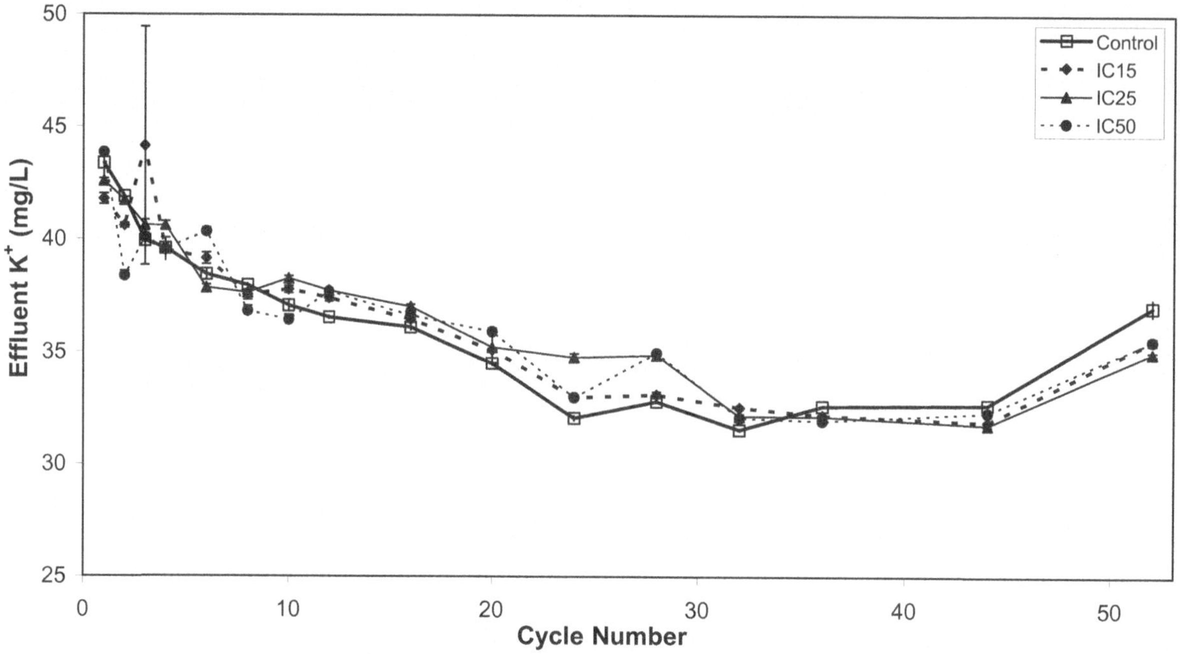

Figure E-6. Effluent Potassium for the 10-Day SRT Reactors Exposed to Cyanide Shock.

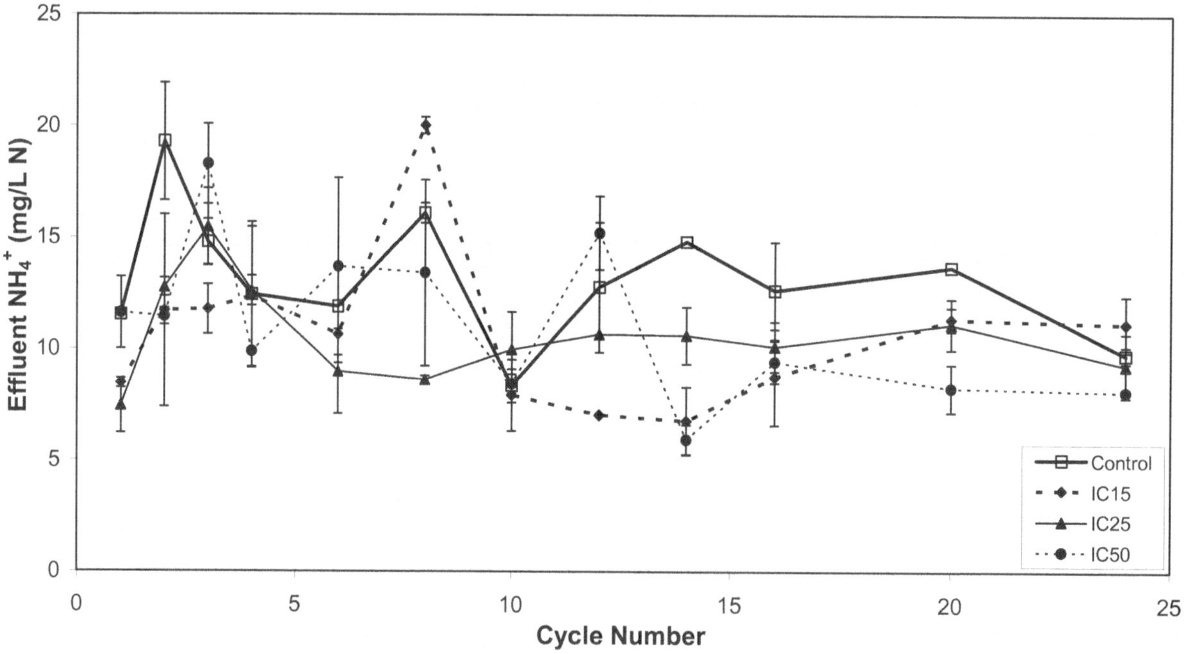

Figure E-7. Effluent Ammonia-Nitrogen for the 2-Day SRT Reactors Exposed to Cyanide Shock.

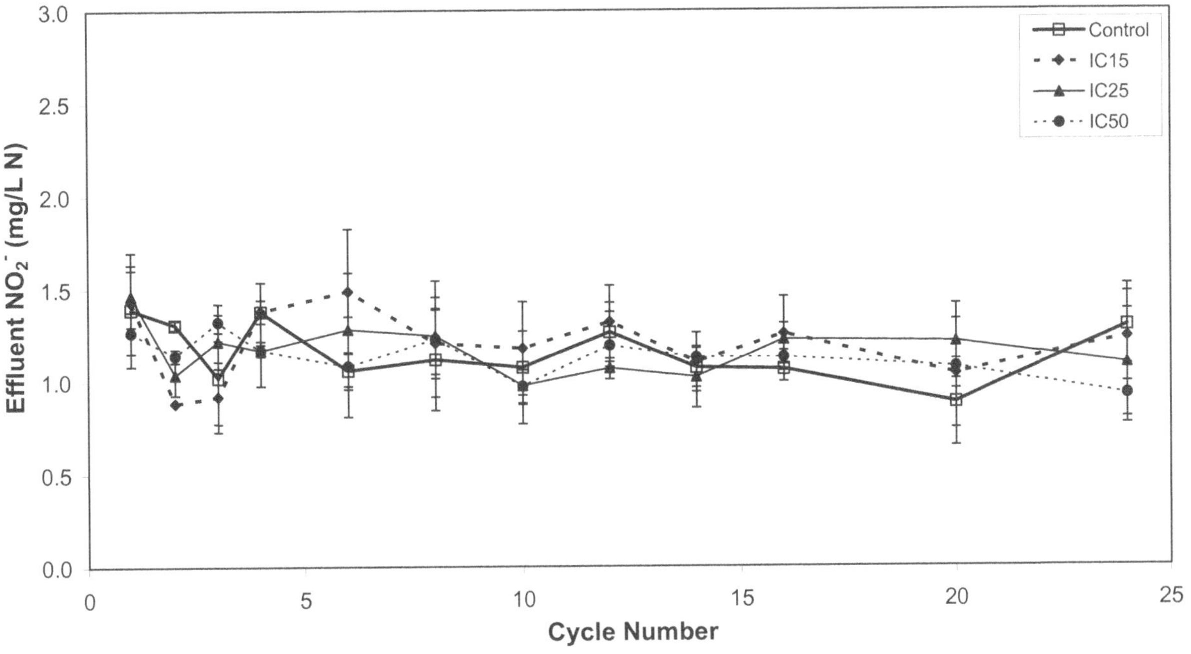

Figure E-8. Effluent Nitrite-Nitrogen for the 2-Day SRT Reactors Exposed to Cyanide Shock.

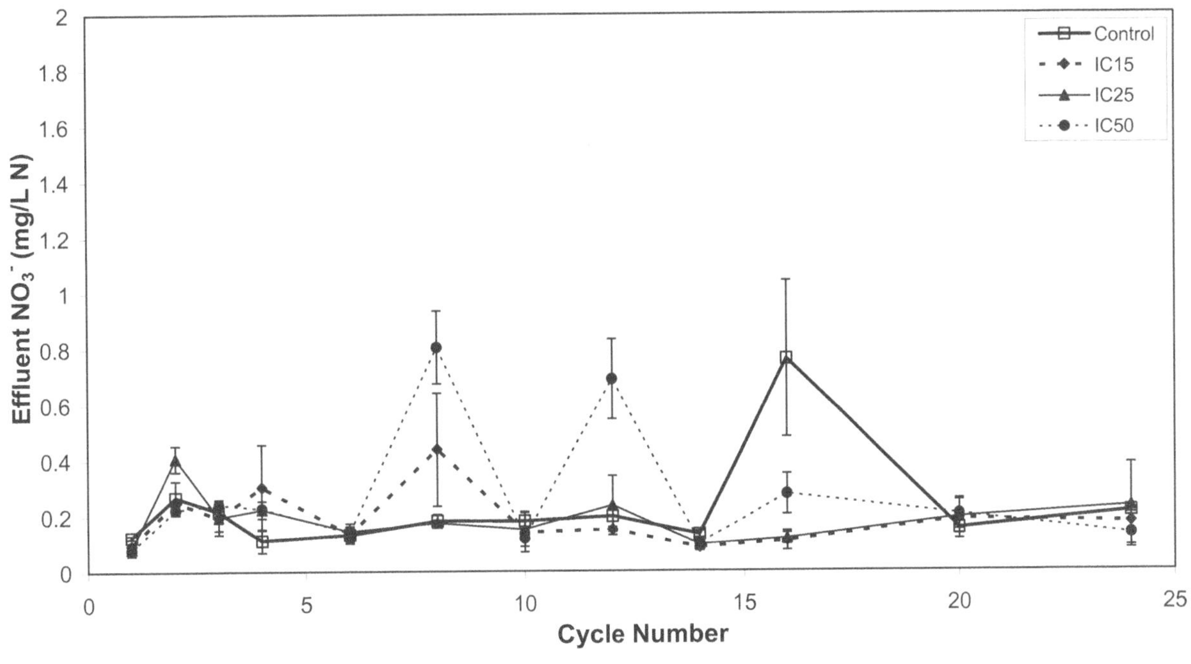

Figure E-9. Effluent Nitrate-Nitrogen for the 2-Day SRT Reactors Exposed to Cyanide Shock.

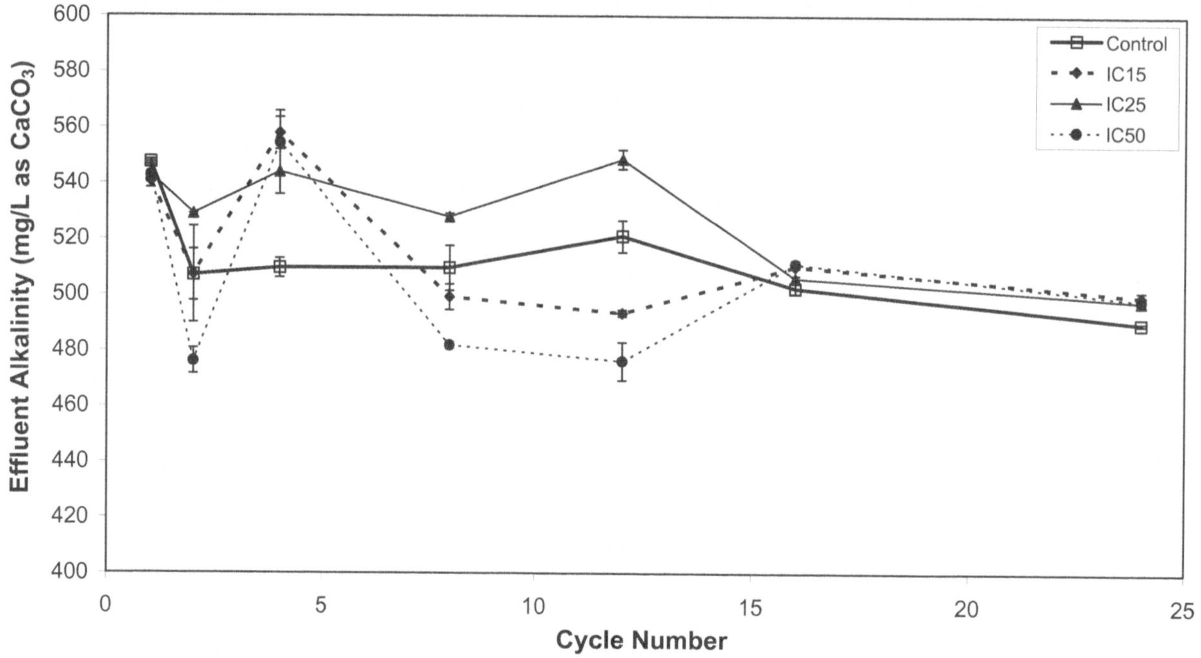

Figure E-10. Effluent Alkalinity for the 2-Day SRT Reactors Exposed to Cyanide Shock.

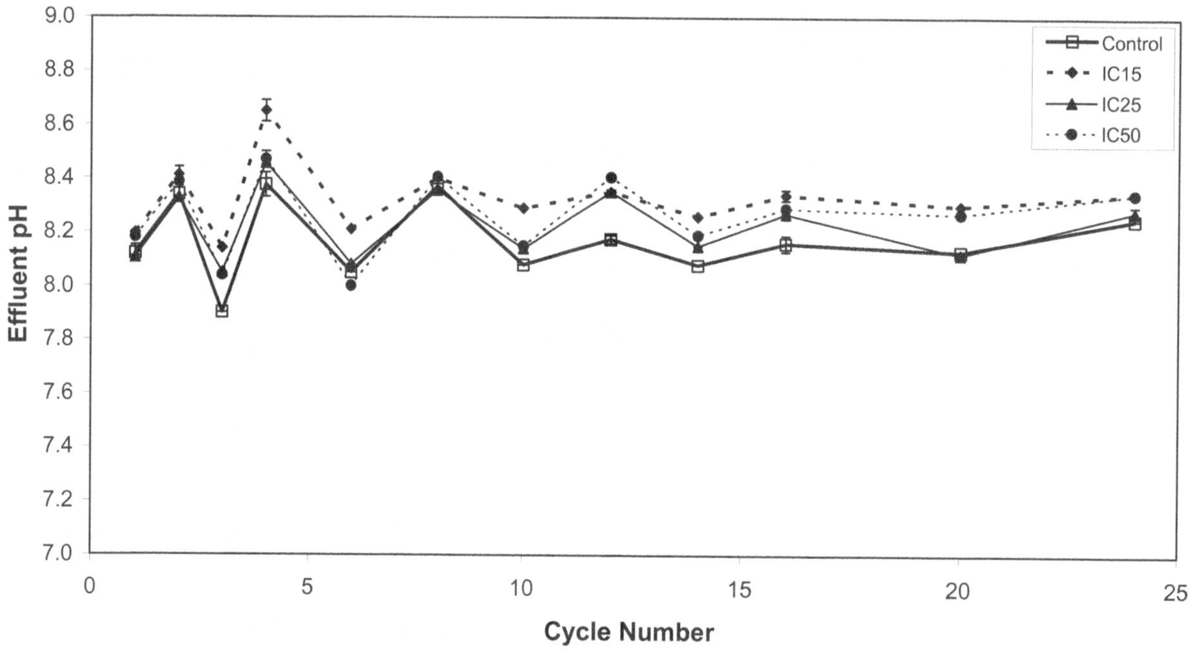

Figure E-11. Effluent pH for the 2-Day SRT Reactors Exposed to Cyanide Shock.

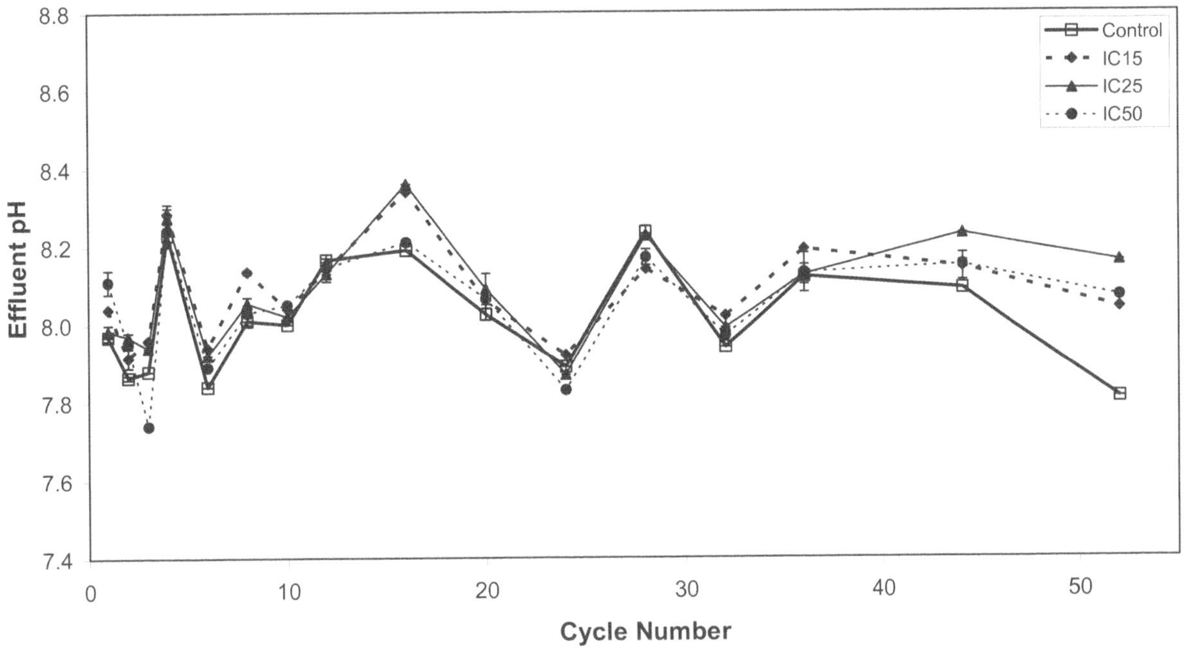

Figure E-12. Effluent pH for the 10-Day SRT Reactors Exposed to Cyanide Shock.

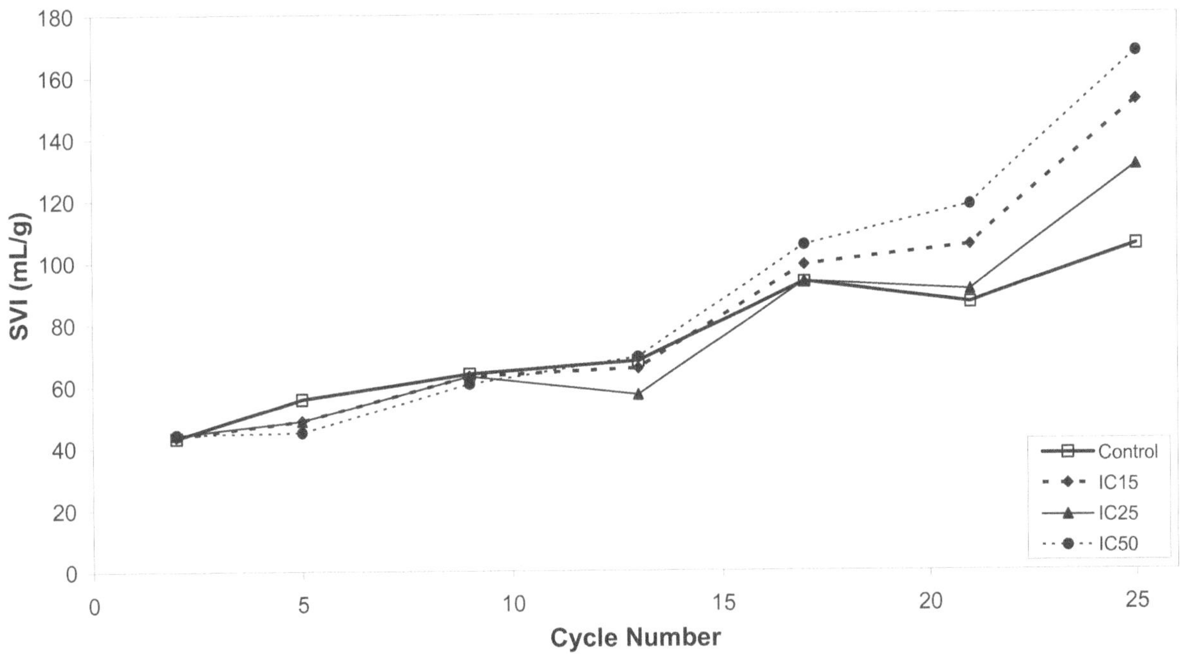

Figure E-13. SVI for the 2-Day SRT Reactors Exposed to Cyanide Shock.

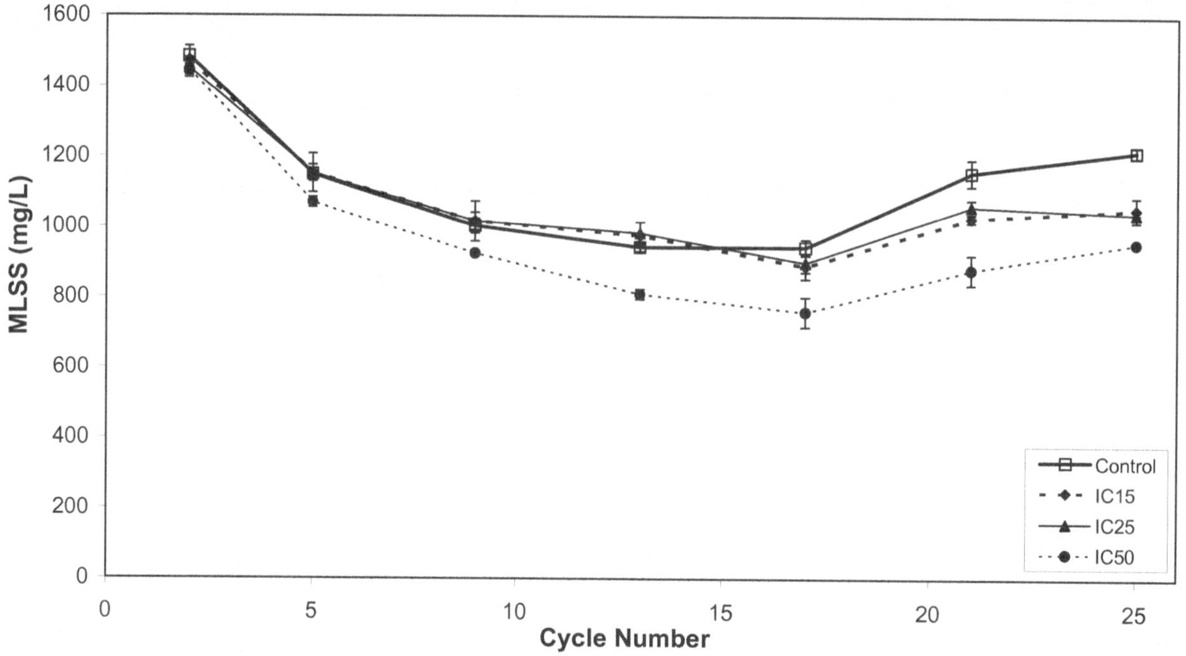

Figure E-14. MLSS for the 2-Day SRT Reactors Exposed to Cyanide Shock.

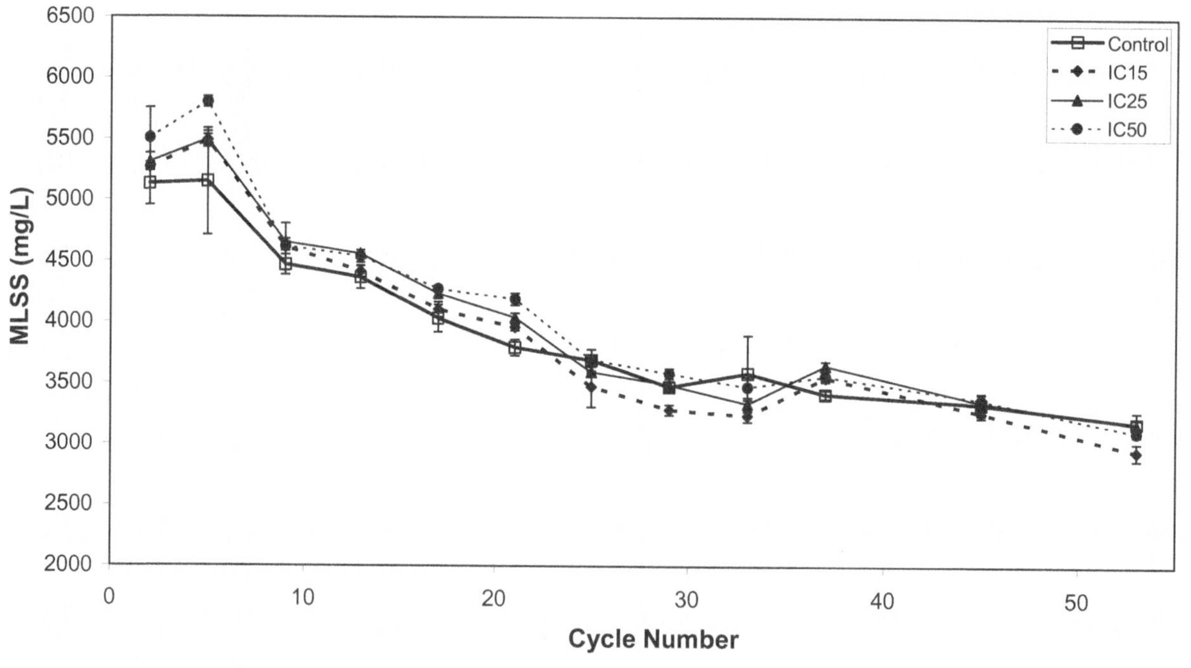

Figure E-15. MLSS for the 10-Day SRT Reactors Exposed to Cyanide Shock.

APPENDIX F

ADDITIONAL DATA FOR 1-CHLORO-2,4-DINITROBENZENE SHOCK EXPERIMENTS

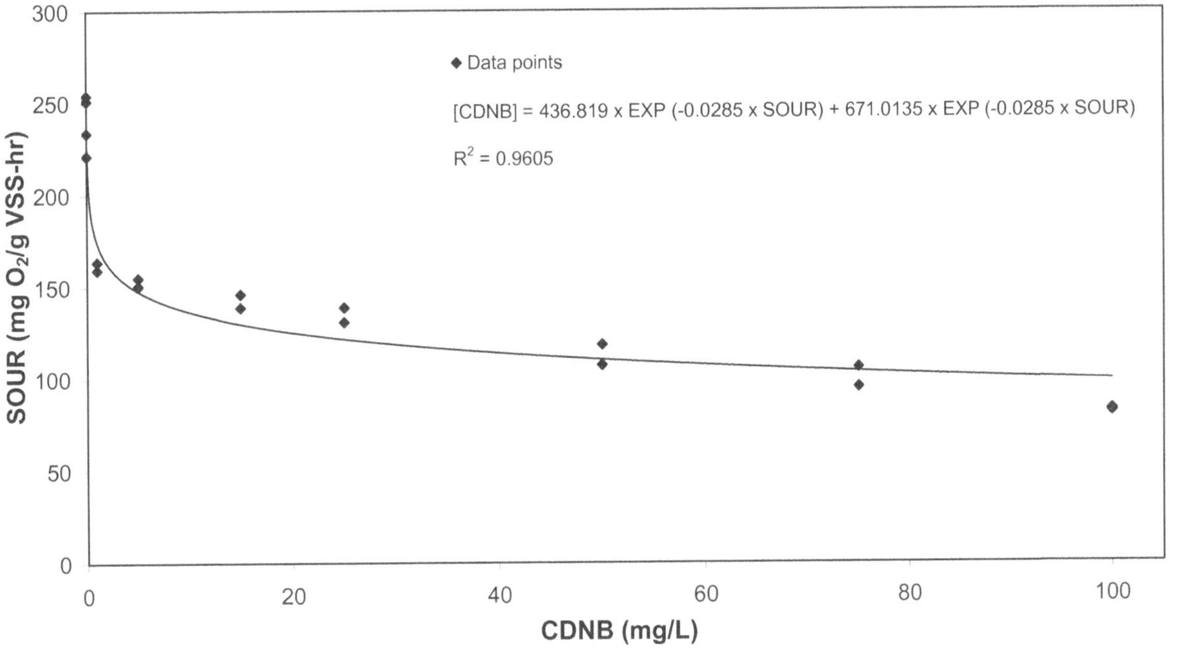

Figure F-1. IC_{XX} Results for the 2-Day SRT Biomass Exposed to CDNB Shock. Equation Represents the Best Fit Logarithmic Curve to the Data.

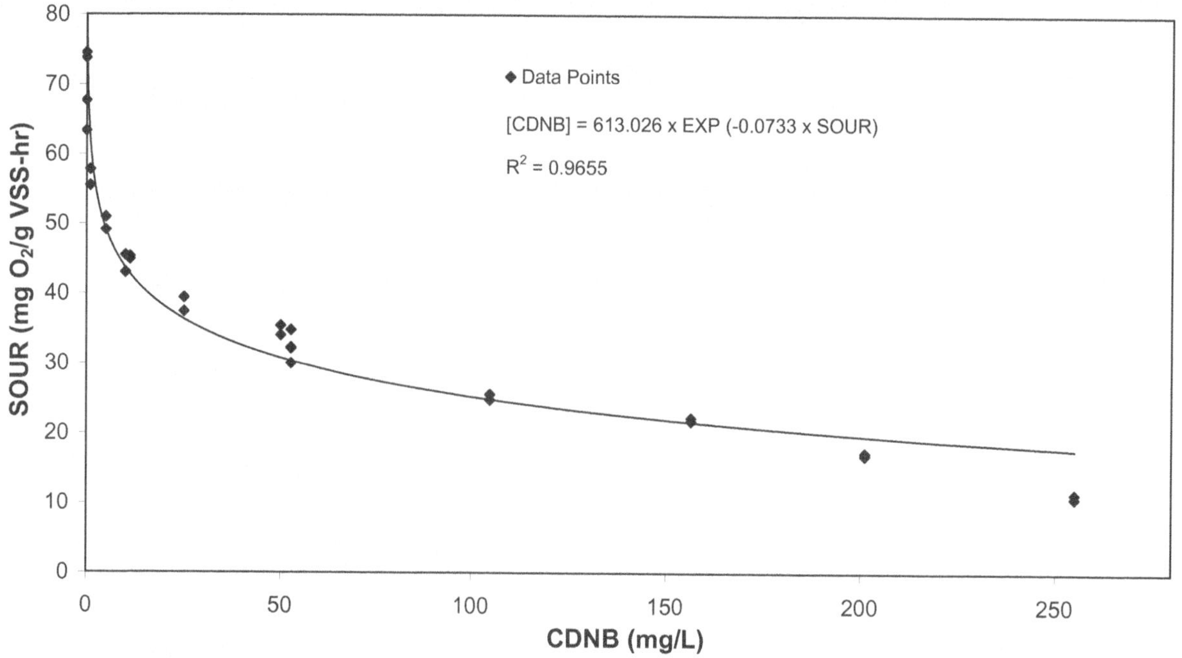

Figure F-2. ICxx Results for the 10-Day SRT Biomass Exposed to CDNB Shock. Equation Represents the Best Fit Logarithmic Curve to the Data.

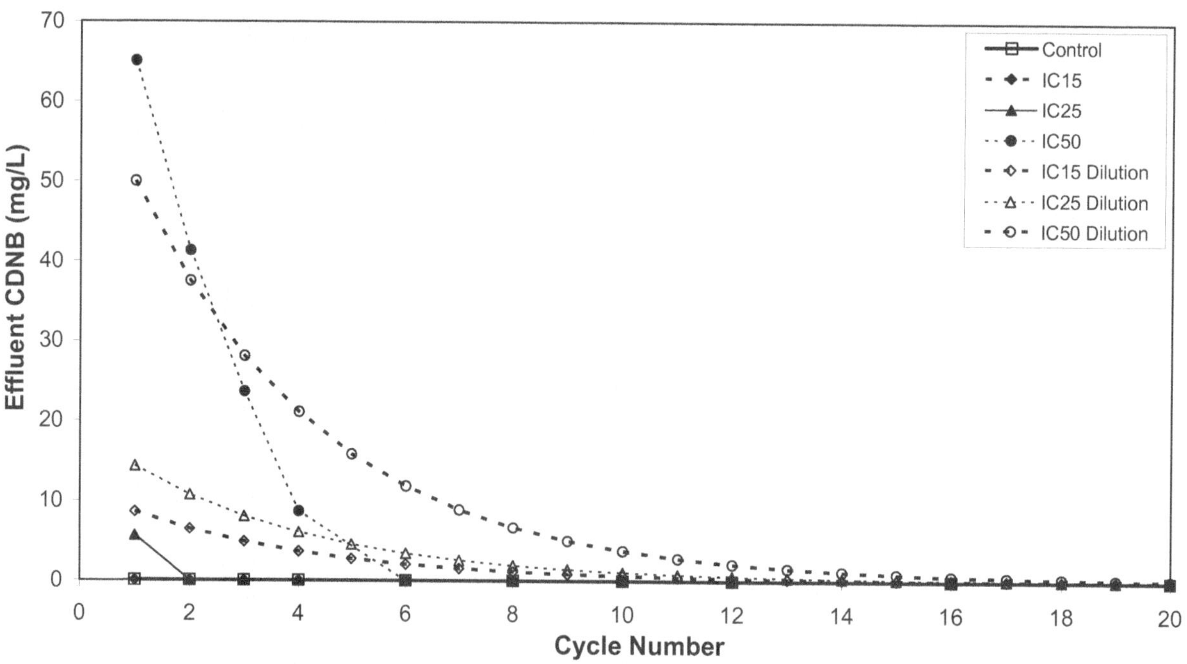

Figure F-3. Effluent Soluble CDNB for the 10-Day SRT Reactors Exposed to CDNB Shock with Theoretical Dilution Models. Data through Cycle 20 Are Shown.

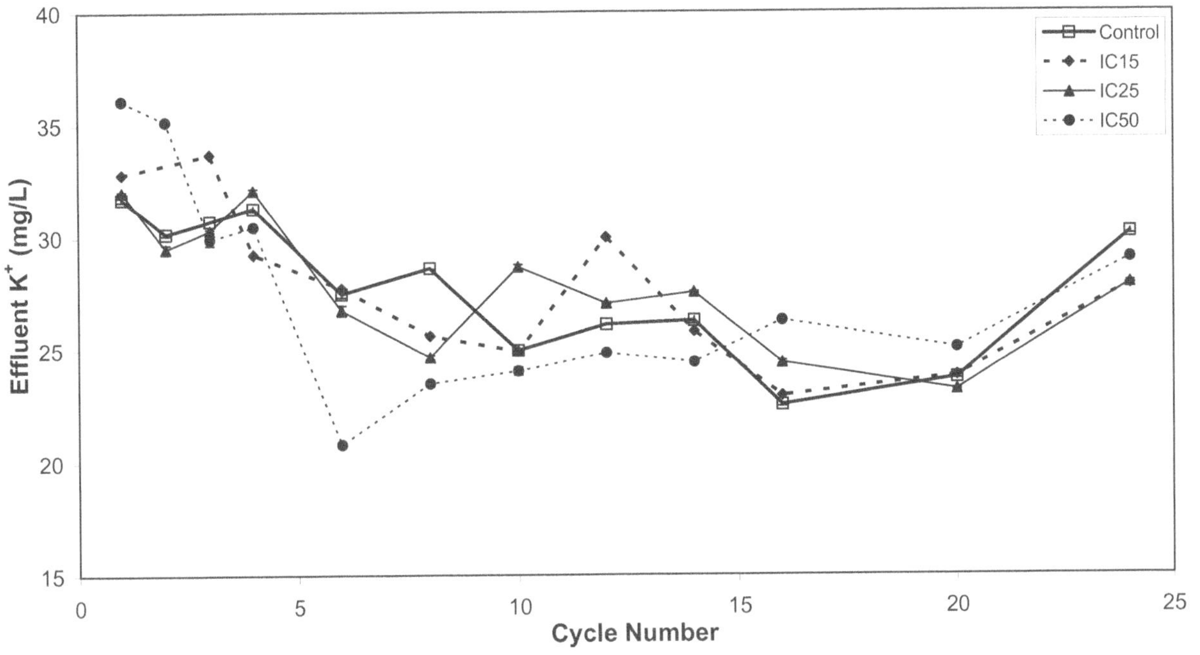

Figure F-4. Effluent K⁺ for the 2-Day SRT Reactors Exposed to CDNB Shock.

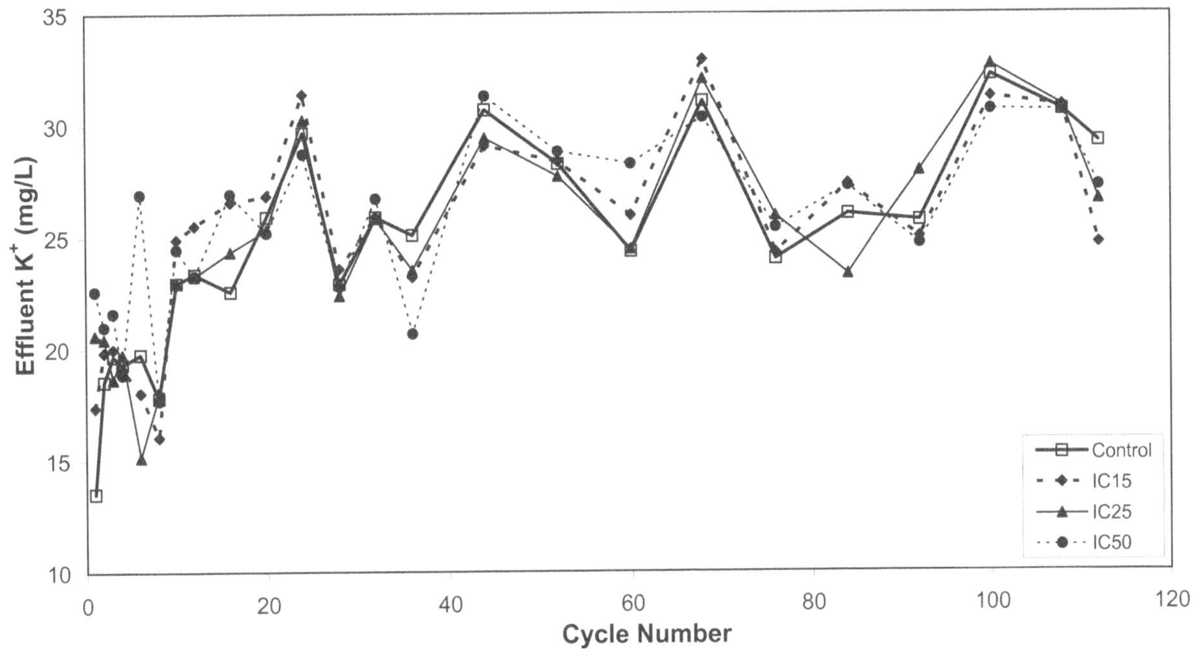

Figure F-5. Effluent K⁺ for the 10-Day SRT Reactors Exposed to CDNB Shock.

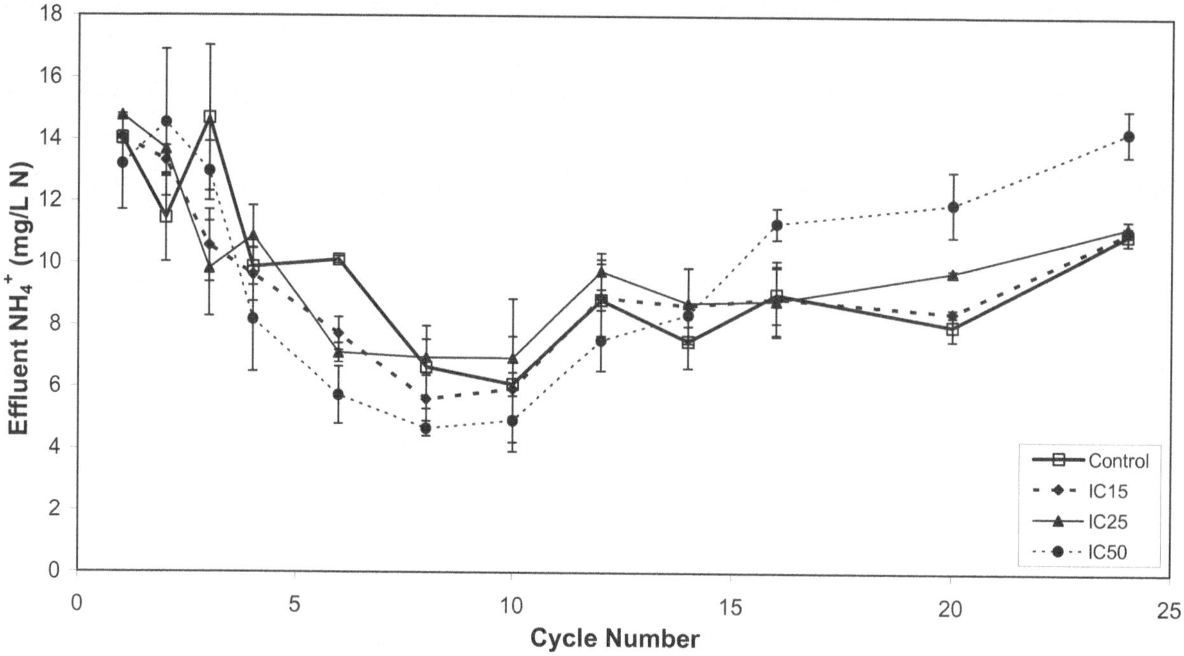

Figure F-6. Effluent Ammonia-Nitrogen for the 2-Day SRT Reactors Exposed to CDNB Shock.

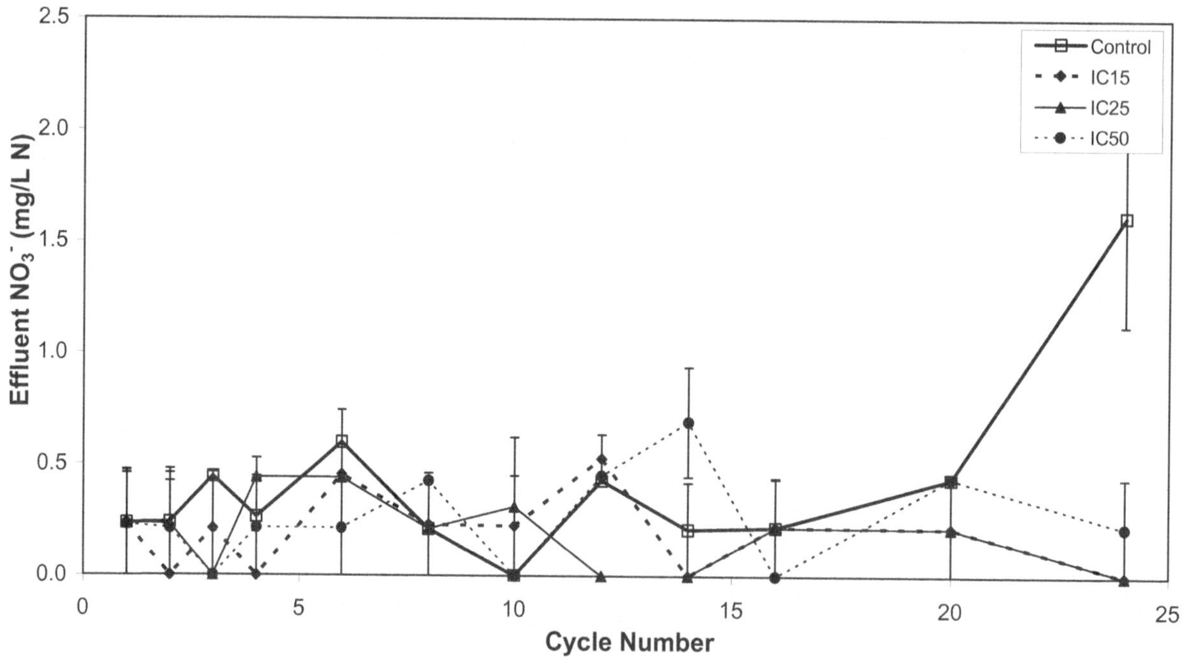

Figure F-7. Effluent Nitrate-Nitrogen for the 2-Day SRT Reactors Exposed to CDNB Shock.

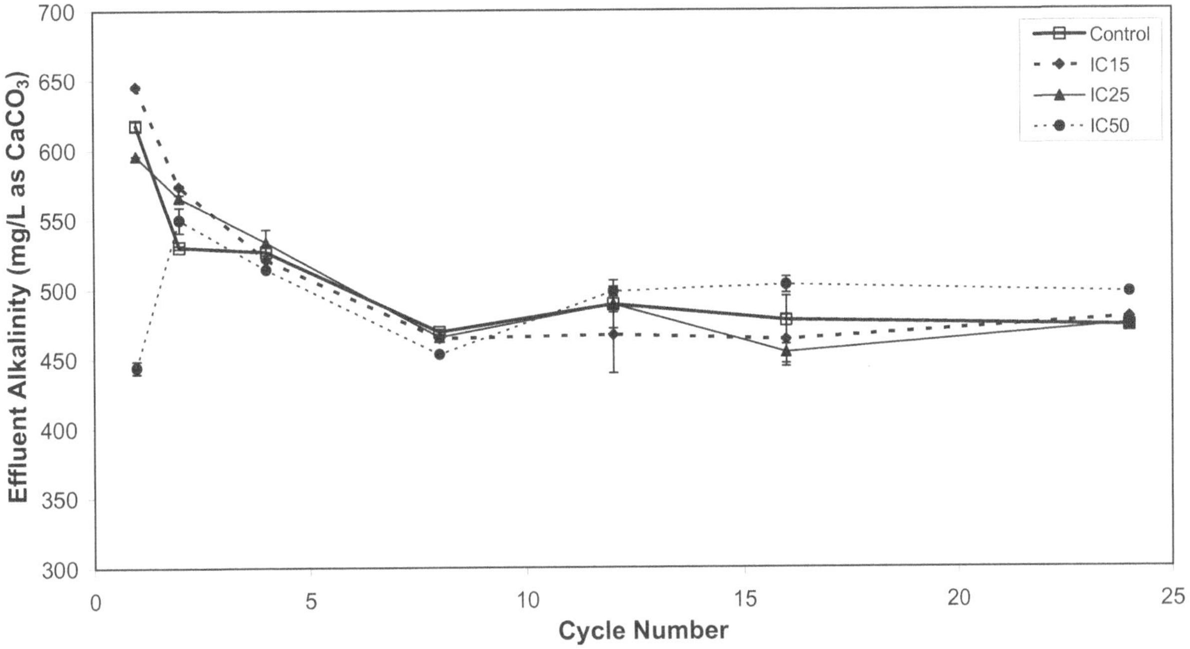

Figure F-8. Effluent Alkalinity for the 2-Day SRT Reactors Exposed to CDNB Shock.

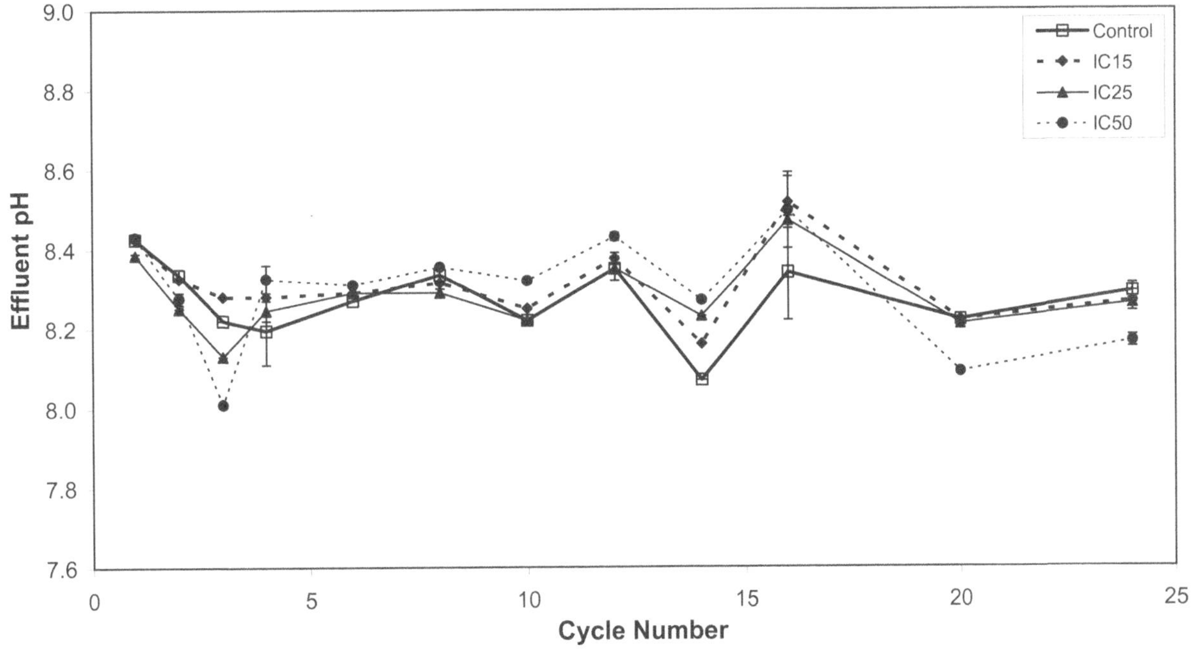

Figure F-9. Effluent pH for the 2-Day SRT Reactors Exposed to CDNB Shock.

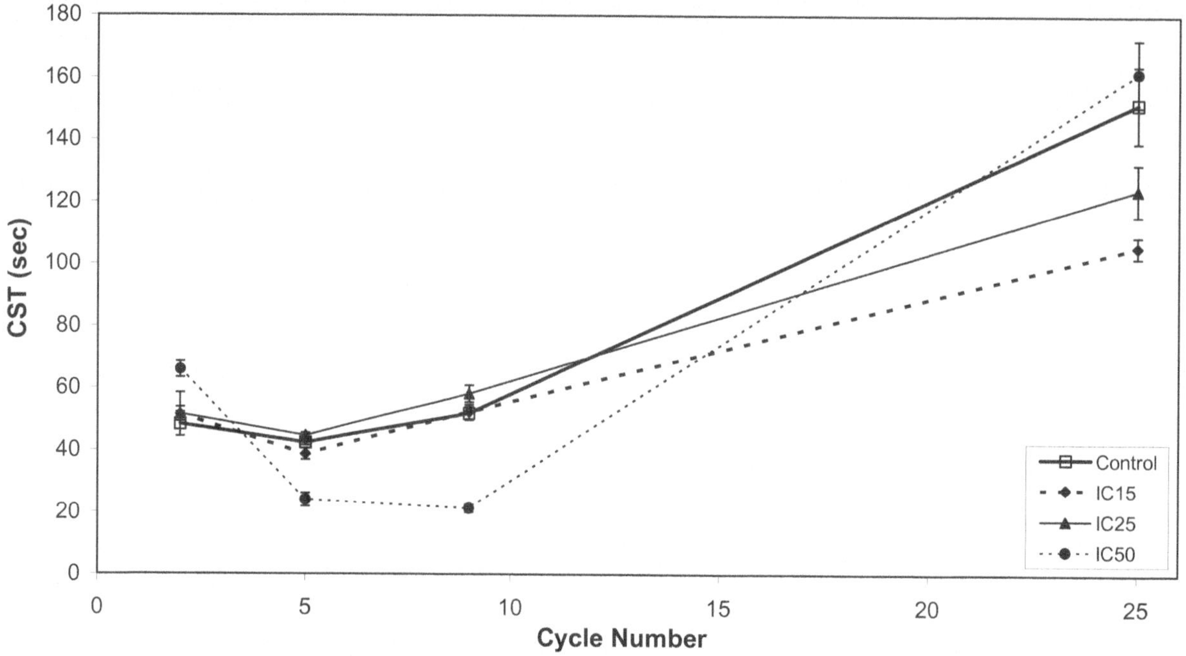

Figure F-10. Mixed Liquor CST for the 2-Day SRT Reactors Exposed to CDNB Shock.

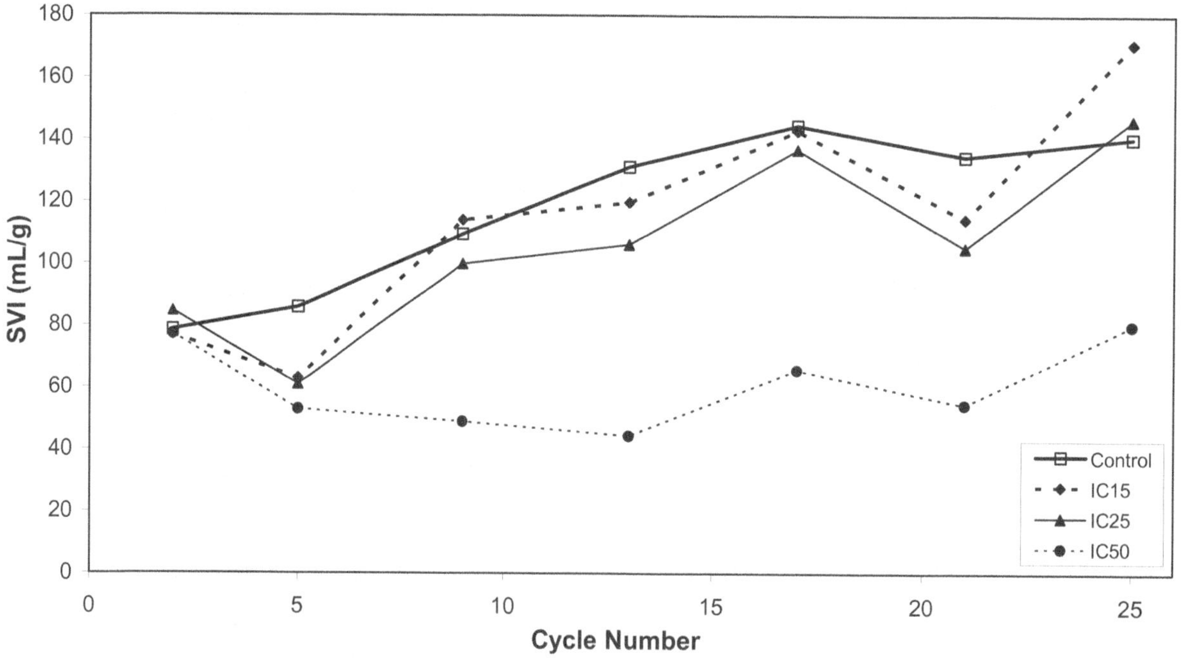

Figure F-11. SVI for the 2-Day SRT Reactors Exposed to CDNB Shock.

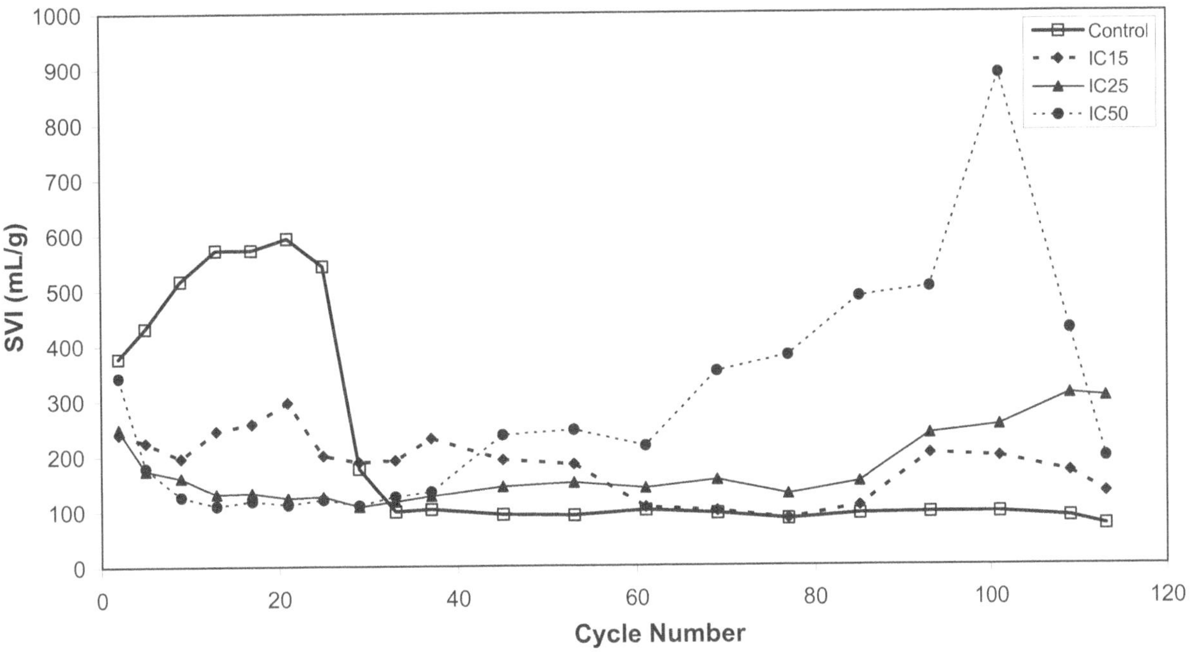

Figure F-12. SVI for the 10-Day SRT Reactors Exposed to CDNB Shock.

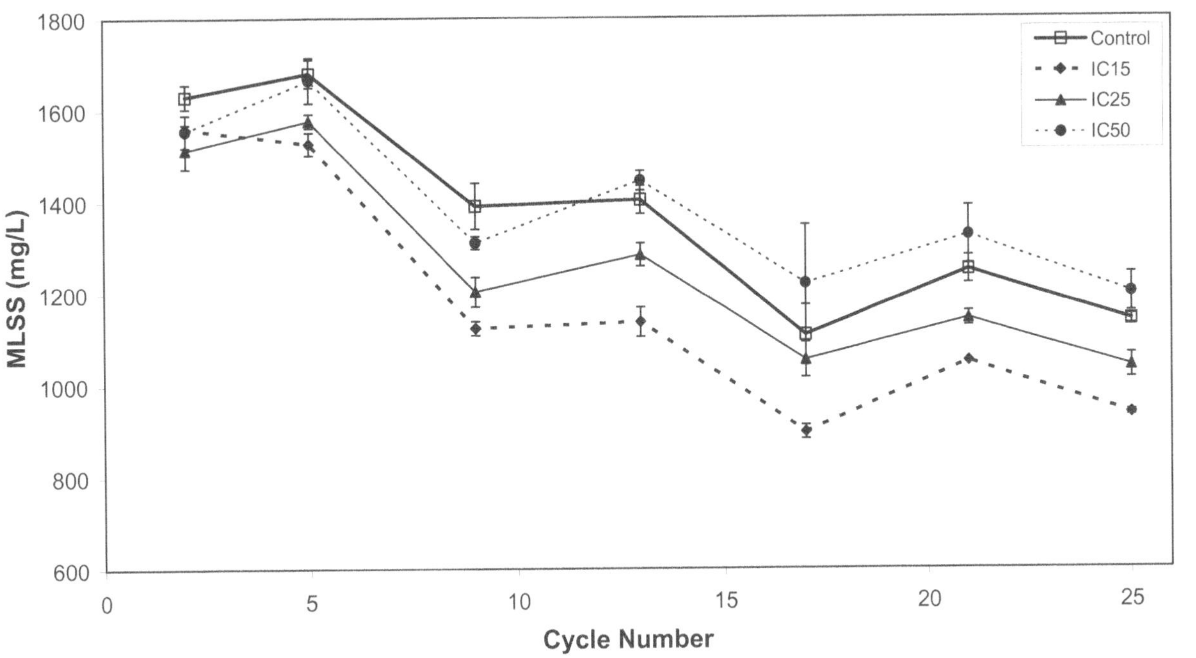

Figure F-13. MLSS for the 2-Day SRT Reactors Exposed to CDNB Shock.

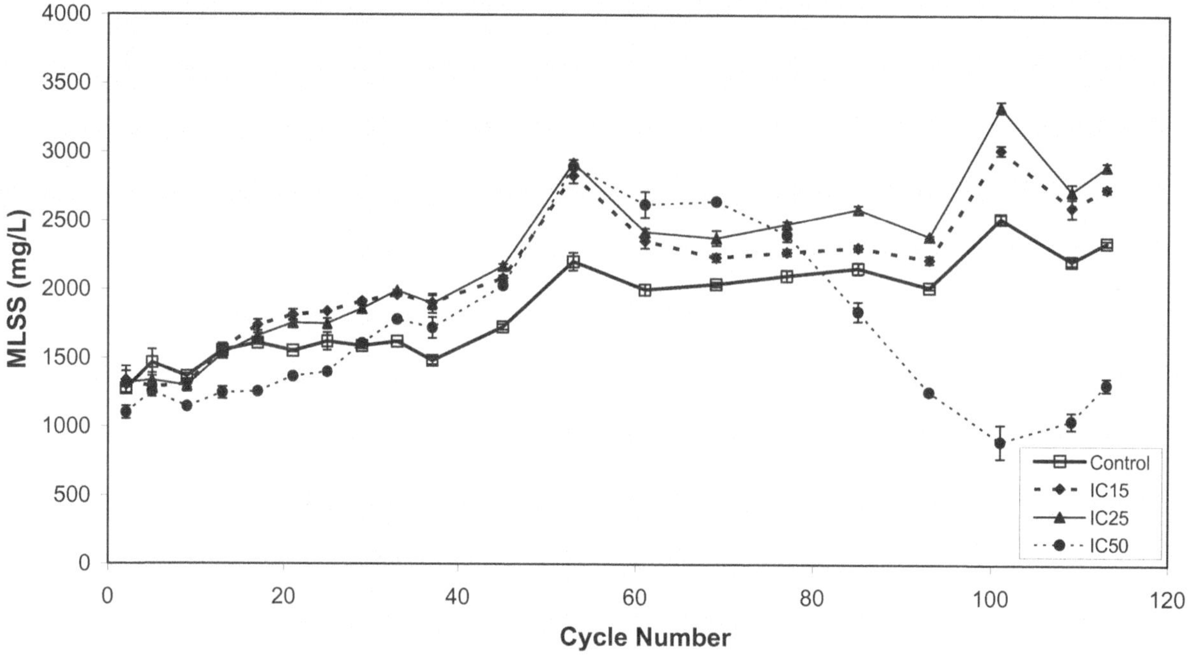

Figure F-14. MLSS for the 10-Day SRT Reactors Exposed to CDNB Shock.

APPENDIX G

ADDITIONAL DATA FOR 2,4-DINITROPHENOL SHOCK EXPERIMENTS

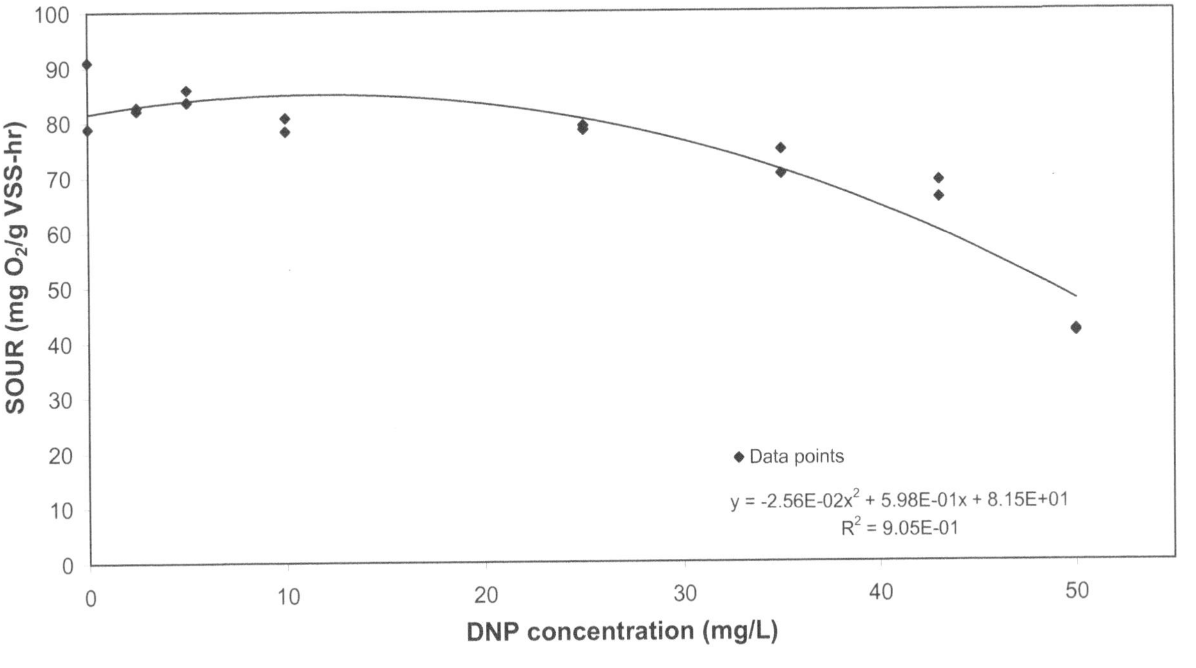

Figure G-1. IC_{xx} Results for the 2-Day SRT Biomass Exposed to DNP Shock. SOUR vs. DNP Concentration.

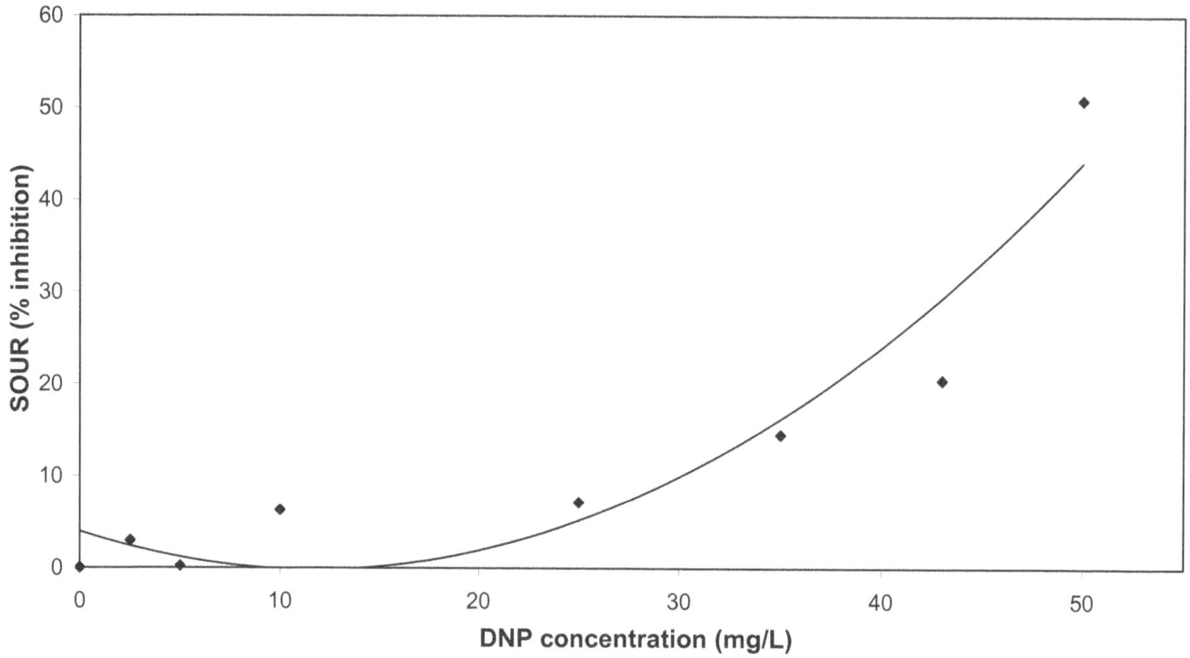

Figure G-2. IC_{xx} Results for the 2-Day SRT Biomass Exposed to DNP Shock. Percent Inhibition vs. DNP Concentration.

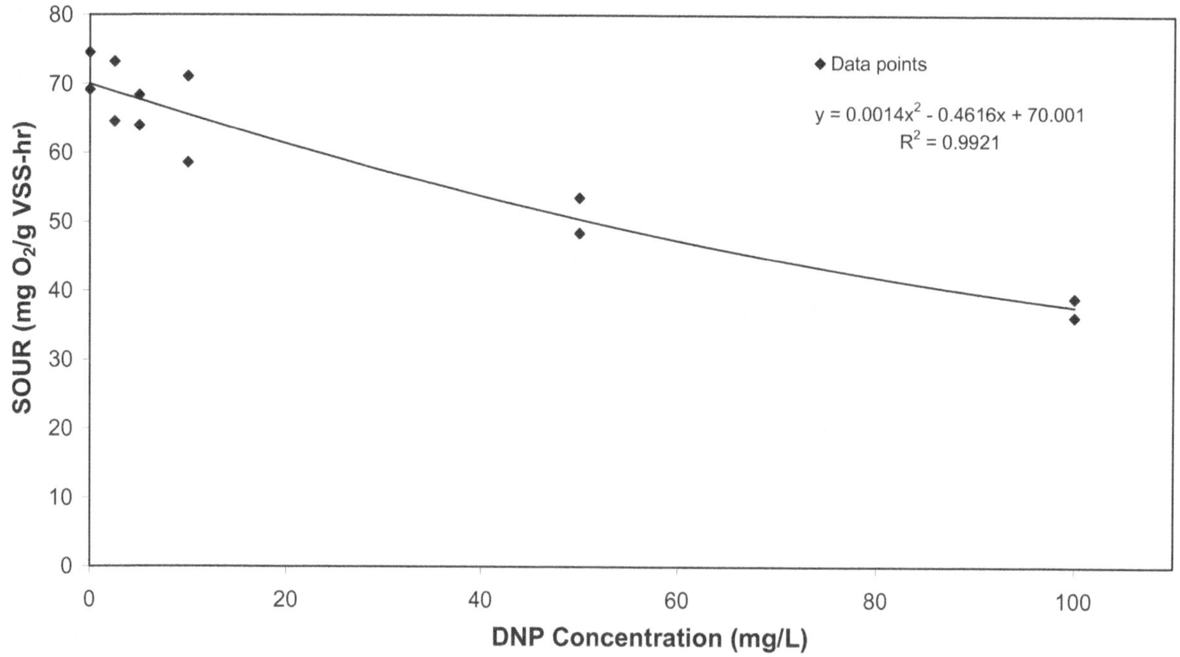

Figure G-3. IC_{xx} Results for the 10-Day SRT Biomass Exposed to DNP Shock. SOUR vs. DNP Concentration.

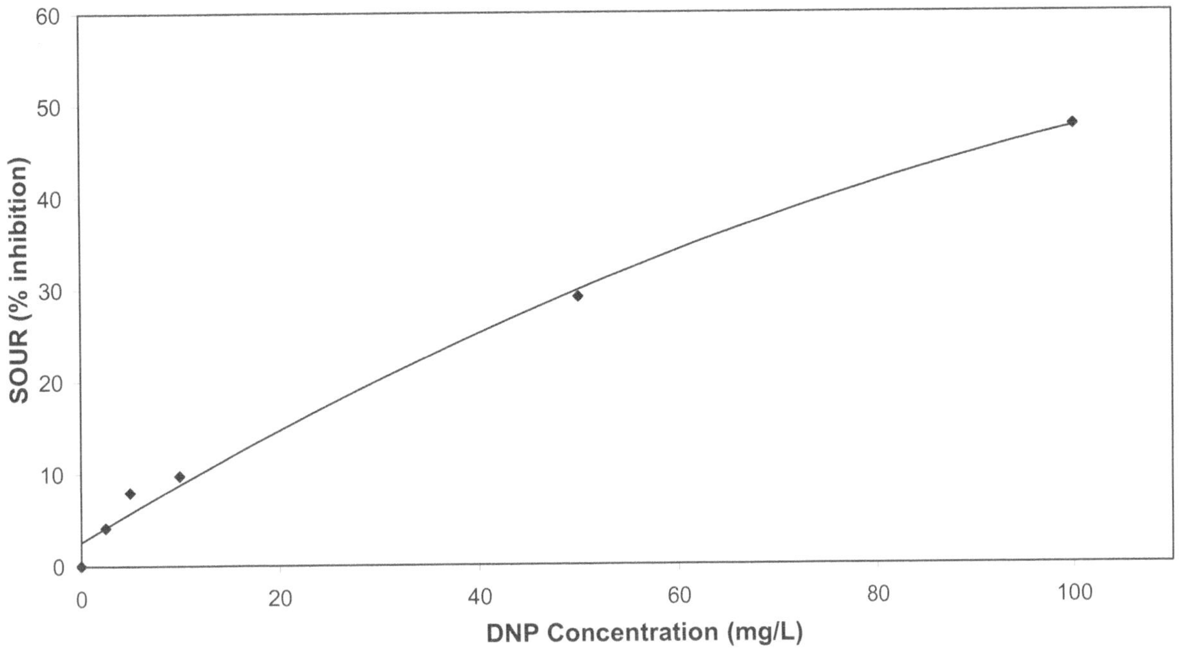

Figure G-4. IC_{xx} Results for the 10-Day SRT Biomass Exposed to DNP Shock. Percent Inhibition vs. DNP Concentration.

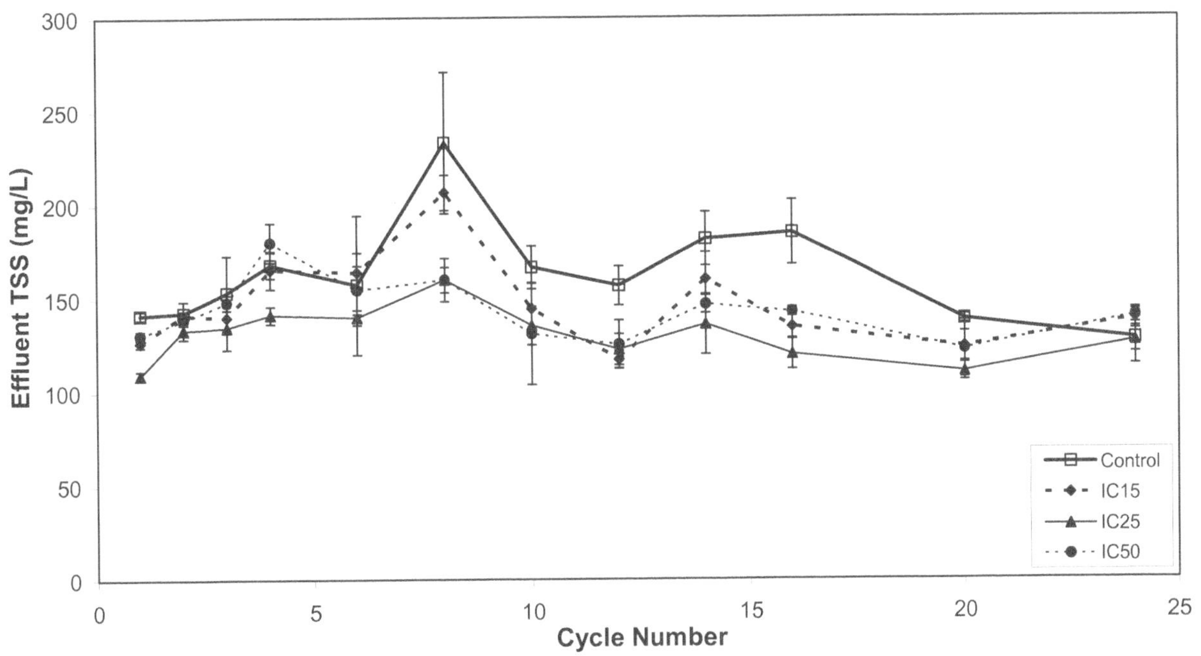

Figure G-5. Effluent TSS for the 2-Day SRT Reactors Exposed to DNP Shock.

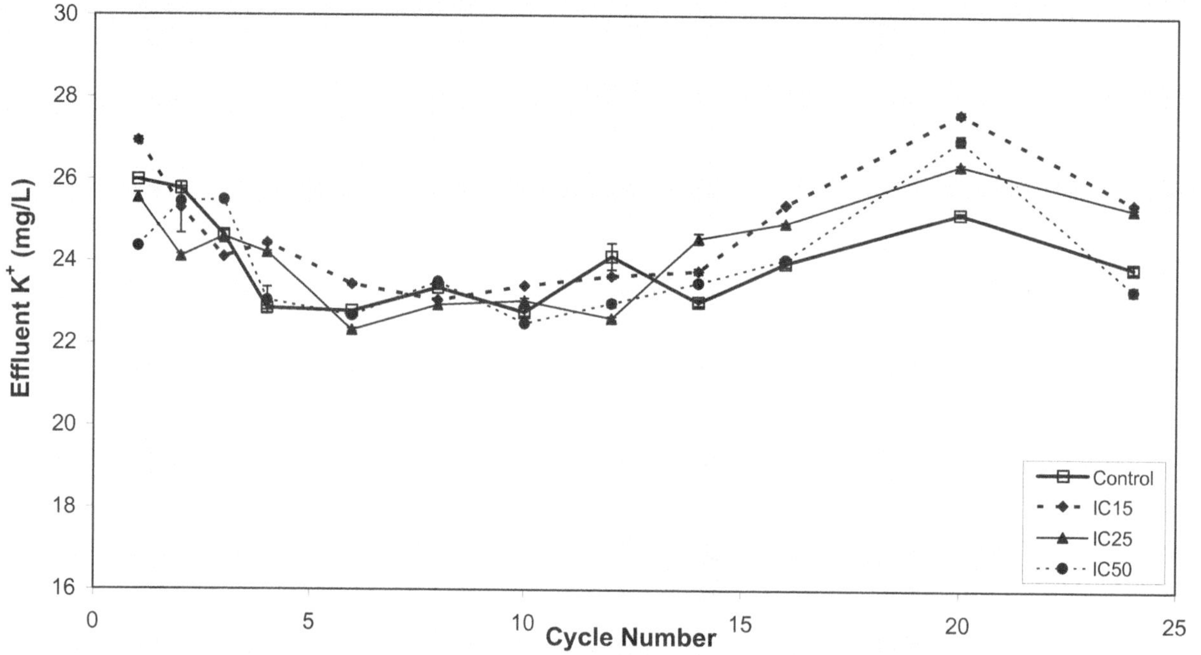

Figure G-6. Effluent Potassium for the 2-Day SRT Reactors Exposed to DNP Shock.

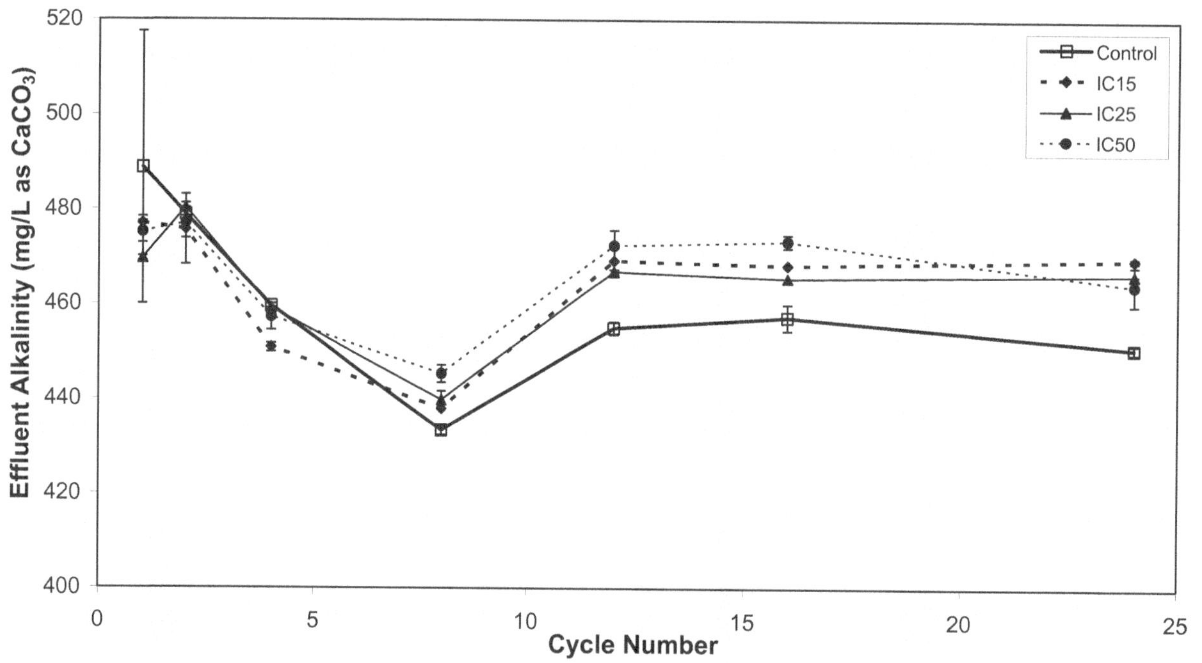

Figure G-7. Effluent Alkalinity for the 2-Day SRT Reactors Exposed to DNP Shock.

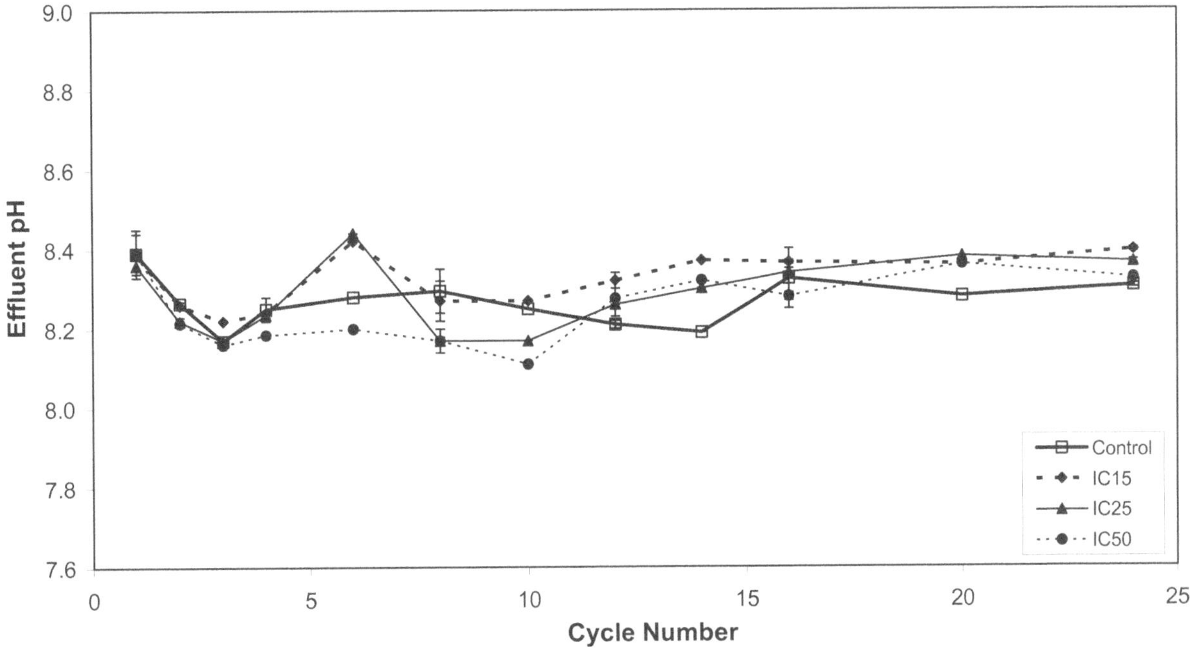

Figure G-8. Effluent pH for the 2-Day SRT Reactors Exposed to DNP Shock.

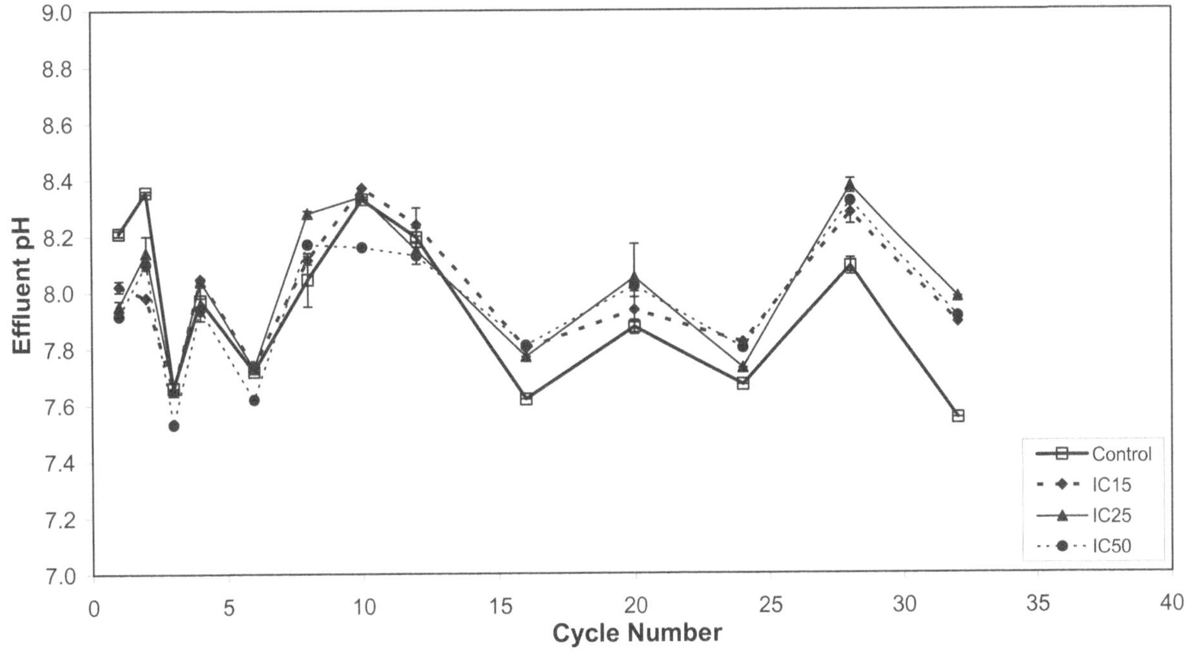

Figure G-9. Effluent pH for the 10-Day SRT Reactors Exposed to DNP Shock.

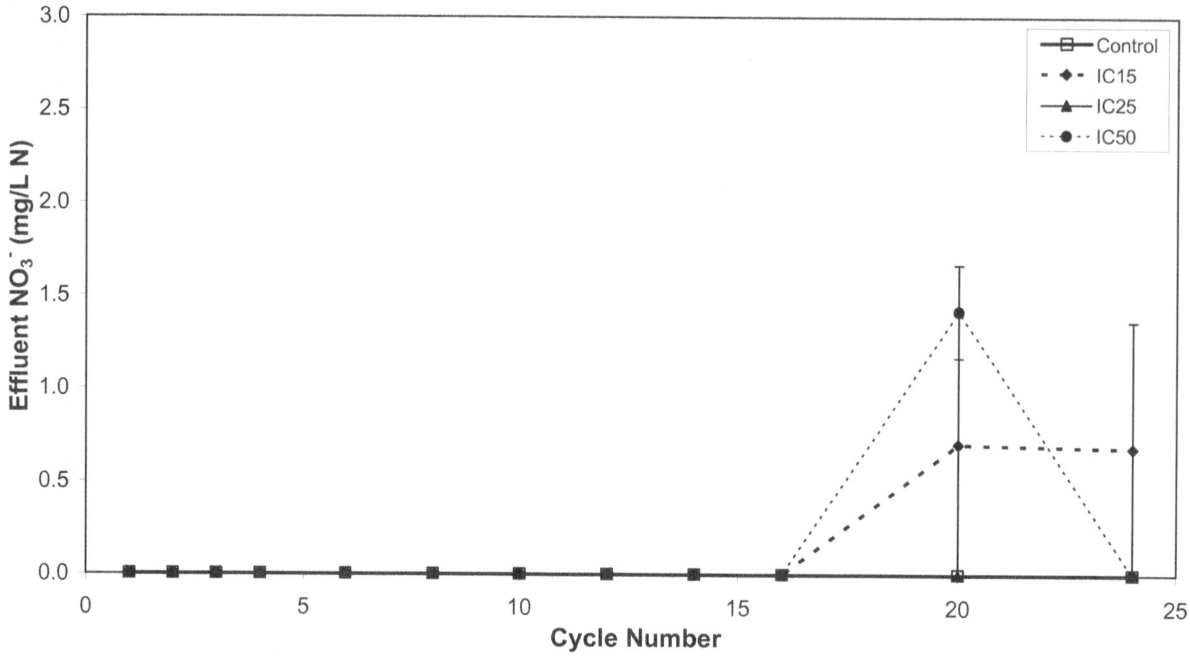

Figure G-10. Effluent Nitrate-Nitrogen for the 2-Day SRT Reactors Exposed to DNP Shock.

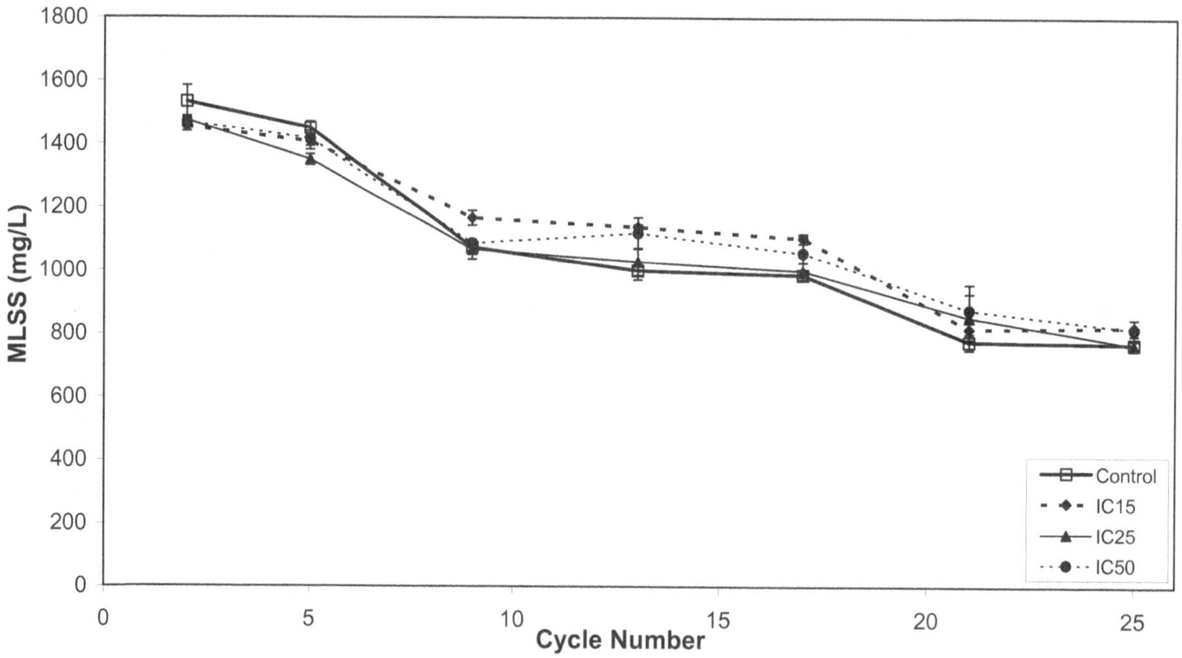

Figure G-11. MLSS for the 2-Day SRT Reactors Exposed to DNP Shock.

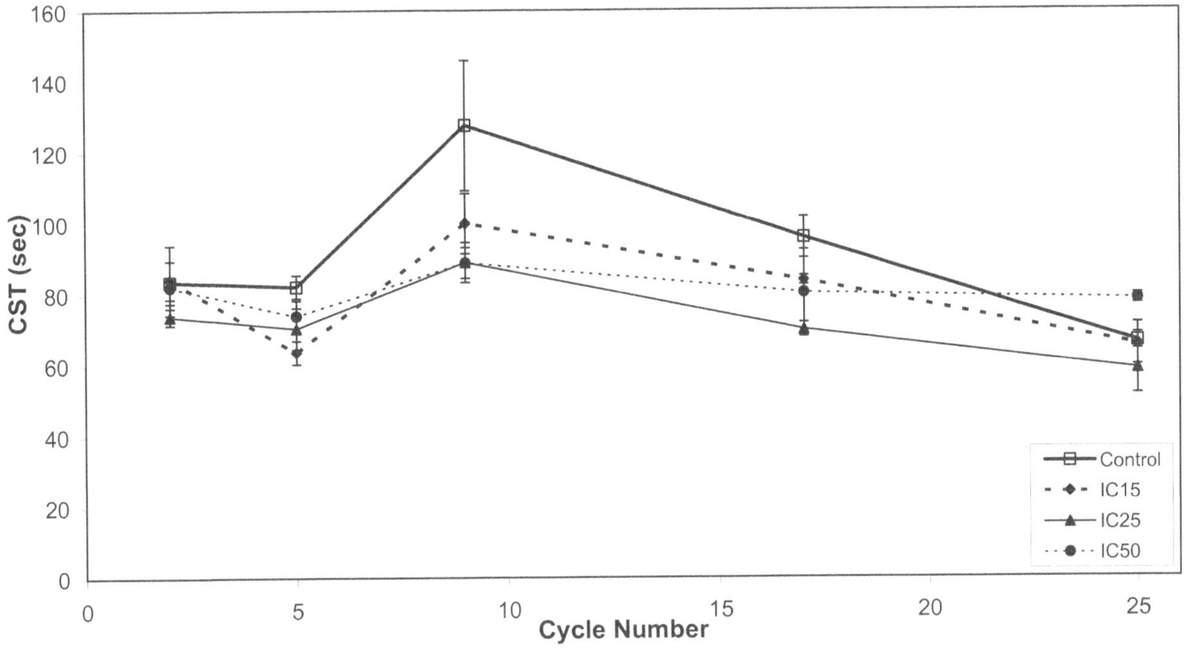

Figure G-12. Mixed Liquor CST for the 2-Day SRT Reactors Exposed to DNP Shock.

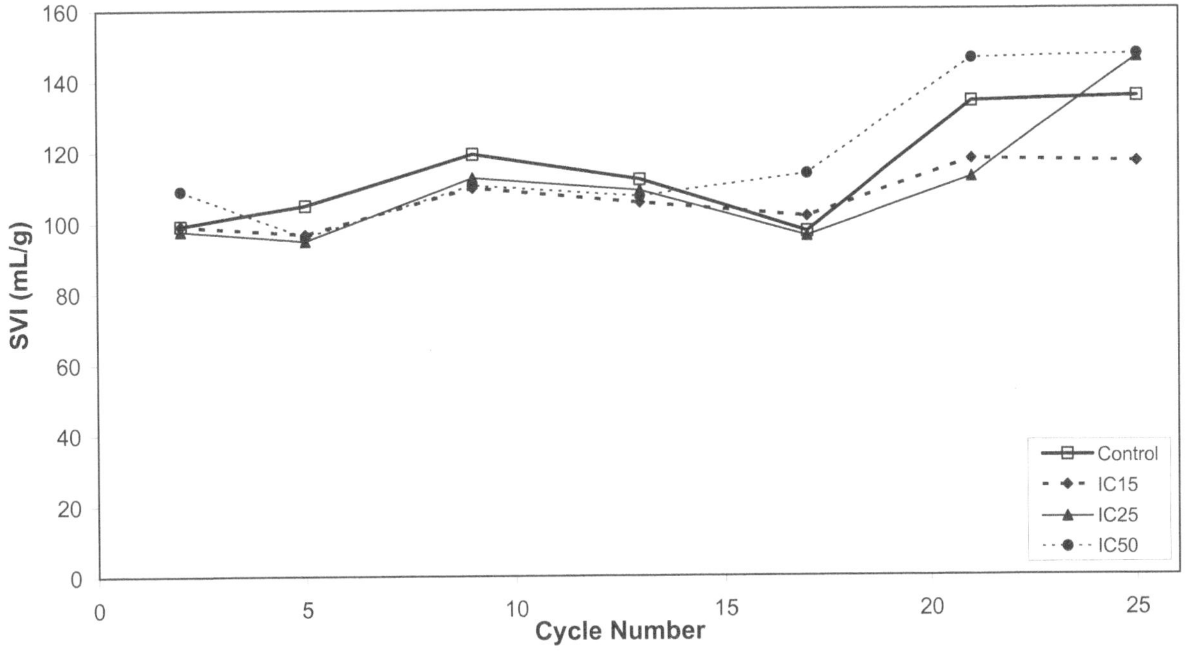

Figure G-13. SVI for the 2-Day SRT Reactors Exposed to DNP Shock.

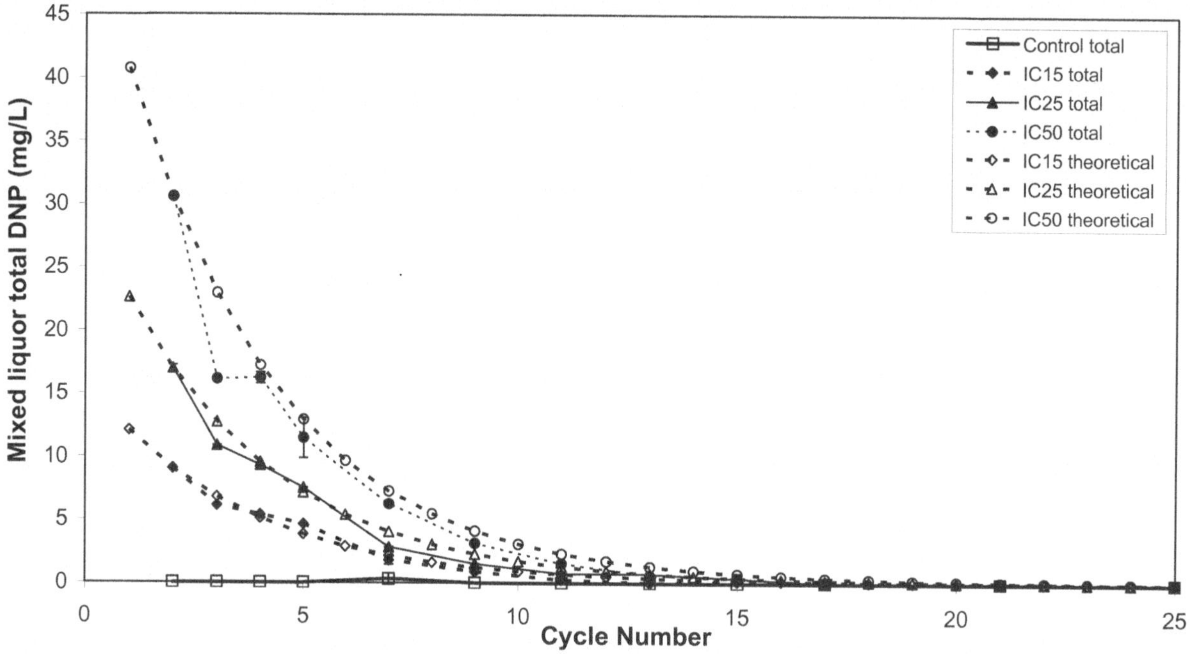

Figure G-14. Mixed Liquor Total DNP for the 2-Day SRT Reactors Exposed to DNP Shock. The Dashed Lines Represent Theoretical Dilution Curves.

APPENDIX H

ADDITIONAL DATA FOR pH SHOCK EXPERIMENTS

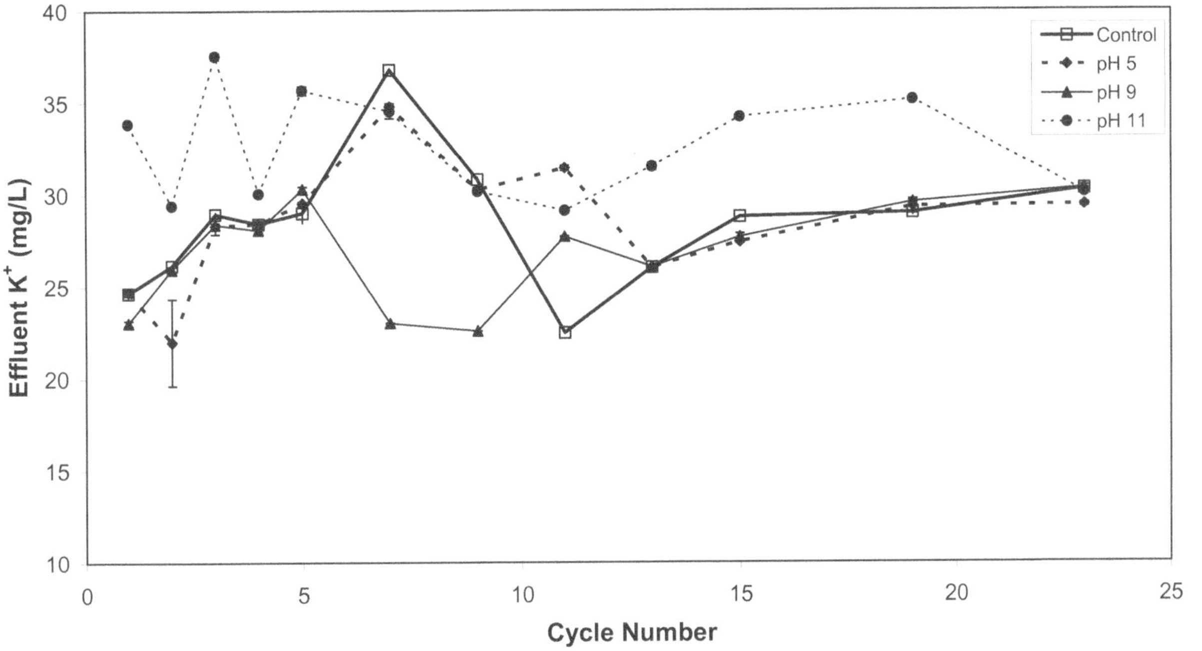

Figure H-1. Effluent Potassium for the 2-Day SRT Reactors Exposed to Various pH Values.

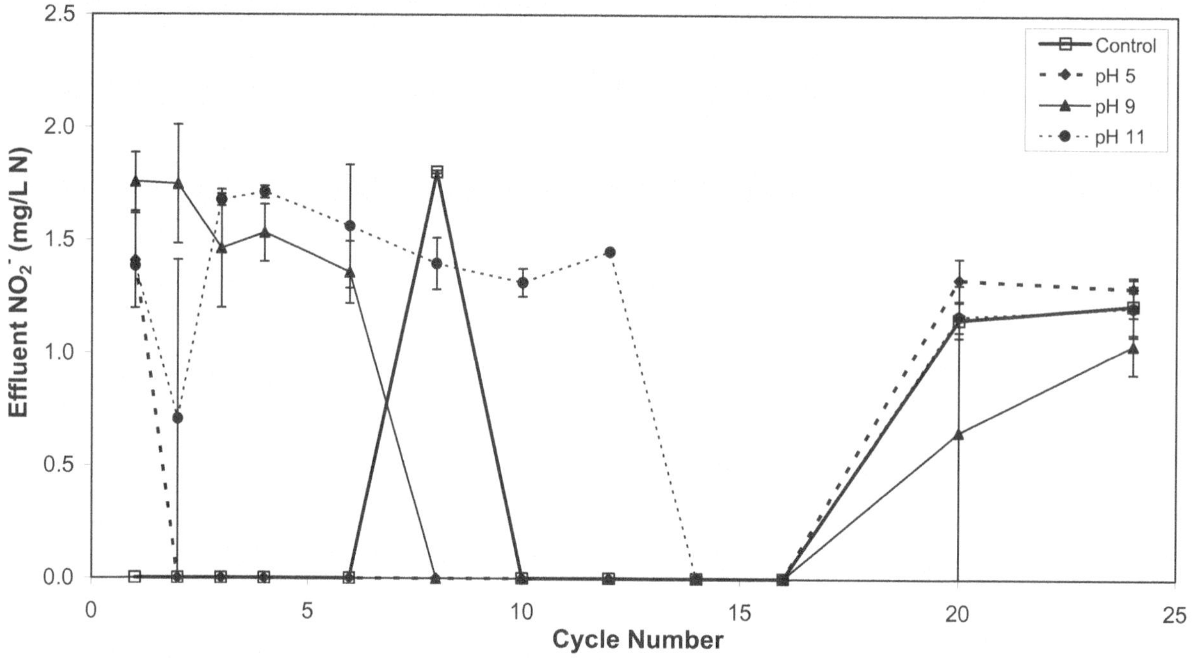

Figure H-2. Effluent Nitrite-Nitrogen for the 2-Day SRT Reactors Exposed to Various pH Values.

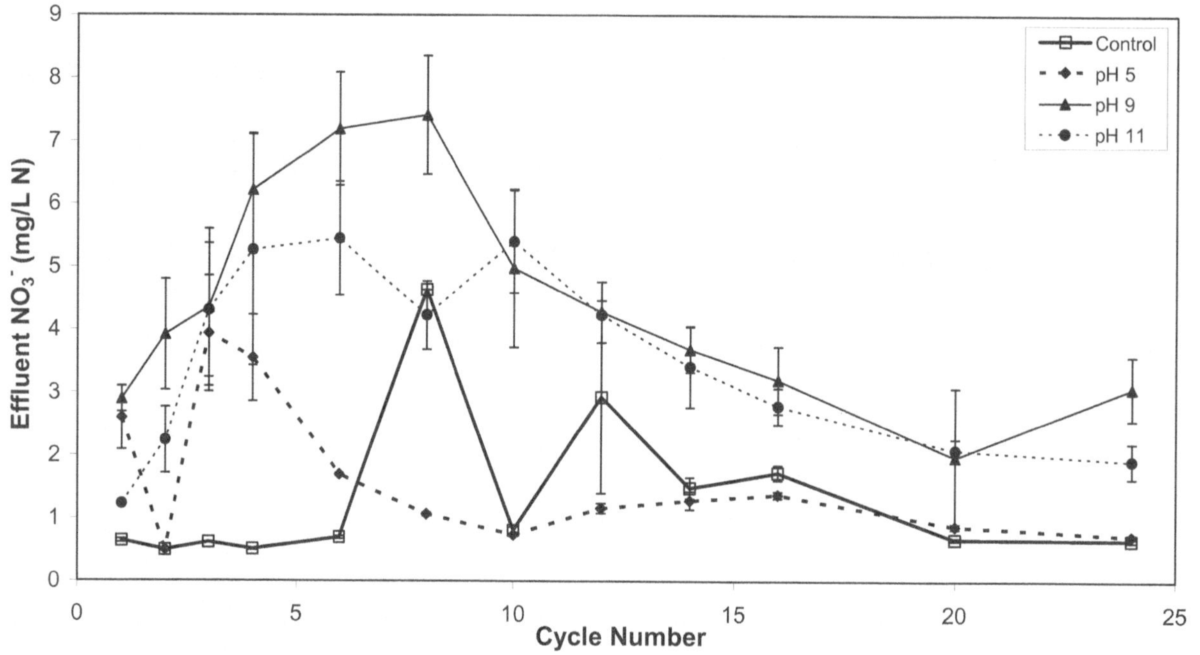

Figure H-3. Effluent Nitrate-Nitrogen for the 2-Day SRT Reactors Exposed to Various pH Values.

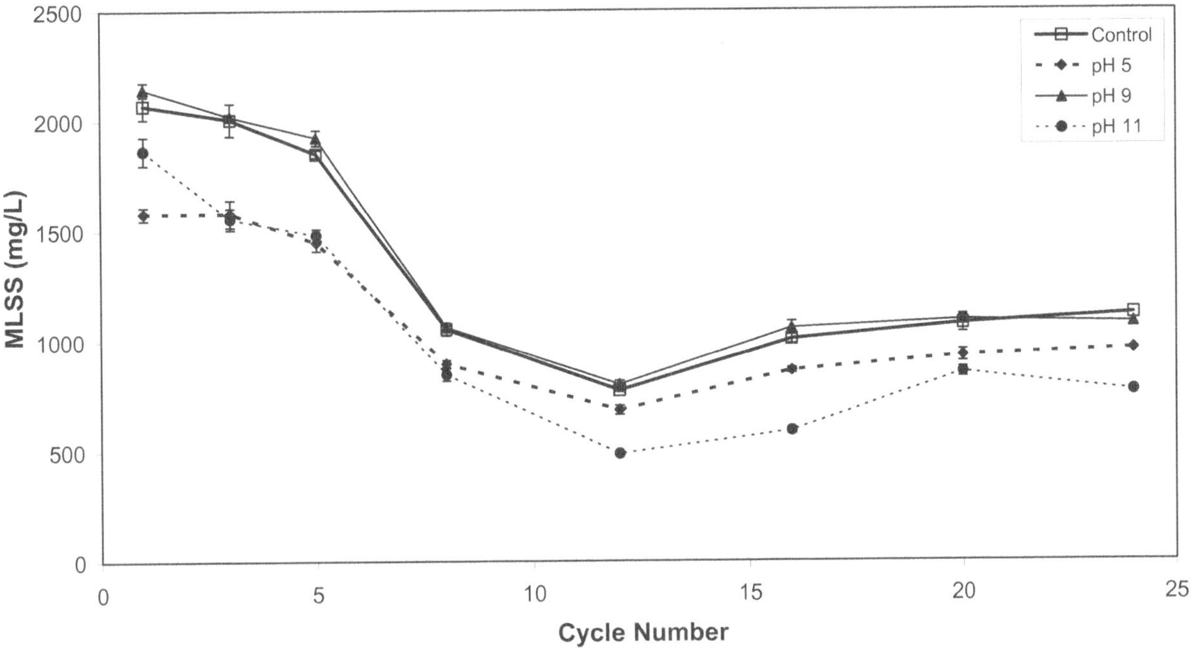

Figure H-4. MLSS for the 2-Day SRT Reactors Exposed to Various pH Values.

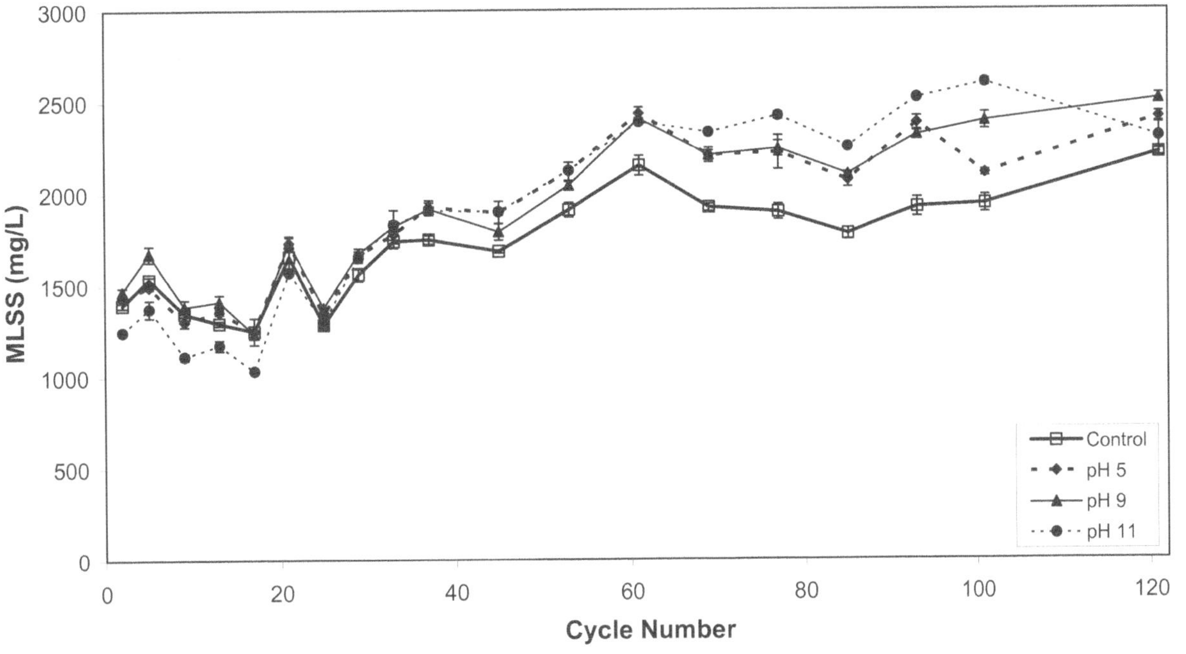

Figure H-5. MLSS for the 10-Day SRT Reactors Exposed to Various pH Values.

REFERENCES

Agency for Toxic Substances and Disease Registry (1995) *Toxicological Profile for Dinitrophenols*, U.S. Department of Health and Human Services, Public Health Service, Agency for Toxic Substances and Disease Registry.

Aleem, M.I. (1966) Generation of reducing power in chemosynthesis. II. Energy-linked reduction of pyridine nucleotides in the chemoautotroph *Nitrosomonas europaea*. *Biochim. Biophys. Acta,* **113**(2), 216-224.

Anthonisen, A.C., Loehr, R.C., Prakasam, T.S., and Srinath, E.G. (1976) Inhibition of nitrification by ammonia and nitrous acid. *J. Water Pollut. Con. Fed.*, **48**(5), 835-852.

APHA (1998) *Standard Methods for the Examination of Water and Wastewater*. 20th Edition. American Public Health Association, Washington, D.C.

Apontoweil, P. and Berends, W. (1975a) Glutathione biosynthesis in *Escherichia coli* K-12: properties of the enzymes and regulation. *Biochim. Biophys. Acta,* **399**(1), 1-9.

Apontoweil, P. and Berends, W. (1975b) Isolation and initial characterization of glutathione-deficient mutants of *Escherichia coli* K-12. *Biochim. Biophys. Acta,* **399**(1), 10-22.

Arden, S.R., Sinor, J.D., Potthoff, W.K., and Aizenman, E. (1998) Subunit-specific interactions of cyanide with the N-methyl-D-aspartate receptor. *J. Biol. Chem.,* **273**(34), 21505-21511.

Bagby, M.M. and Sherrard, J.H. (1981) Combined effects of cadmium and nickel on activated sludge process. *J. Water Pollut. Con. Fed.*, **53**(11), 1609-1619.

Bakker, E.P. and Randall, L.L. (1984) The requirement for energy during export of beta-lactamase in *Escherichia coli* is fulfilled by the total proton motive force. *EMBO J.,* **3**(4), 895-900.

Baldwin, D.D. and Campbell, C.E. (2001) Short-term effects of low pH on the microfauna of an activated sludge wastewater treatment system. *Water Qual. Res. J. Can.*, **36**(3), 519-535.

Barnes, E.M., Jr. and Jayakumar, A. (1993) NH_4^+ transport systems in *Escherichia coli.* In E.P. Bakker (ed.), *Alkali Cation Transport Systems in Prokaryotes*, vol 1. CRC Press, Inc., Boca Raton, FL, p. 397-409.

Bearden, A., Sinks, G., Vaes, W., Ramous, E., Herens, J., and Schults, T. (1999) Bioavailability, biodegradation, and acclimation of *Tetrahymena pyriformis* to 1-octanol. *Ecotox. Environ. Safety,* **44**, 86-91.

Blaha, J. (1976) Mathematical analysis of the chemical system "cyanide-heavy metals" in water-determination of components and toxicity of the system-I: The theoretical solution. *Water Res.*, **10,** 815-819.

Blum, D. and Speece, R. (1991) A database of chemical toxicity to environmental bacteria and its use in interspecies comparisons and correlations. *J Water Pollut. Con. Fed.,* **63**, 198-207.

Bochner, B.R., Lee, P.C., Wilson, S.W., Cutler, C.W., and Ames, B.N. (1984) AppppA and related adenylylated nucleotides are synthesized as a consequence of oxidation stress. *Cell,* **37**, 225-232.

Booth, I.R. (1999) The regulation of intracellular pH in bacteria. Wiley, Chichester (*Novartis Foundation Symposium 221).* London.

Booth, I.R. (1985) Regulation of cytoplasmic pH in bacteria. *Microbiol. Rev.,* **49**(2), 359-378.

Bott, C.B. (2001) Elucidating the role of toxin-induced microbial stress response in wastewater treatment process upset. Ph.D. Dissertation. Virginia Polytechnic Institute and State University, Blacksburg, VA.

Bott, C.B. and Love, N.G. (2001) The immunochemical detection of stress proteins in activated sludge exposed to toxic chemicals. *Water Res.,* **35**(1), 91-100.

Bott, C.B. and Love, N.G. (2002) Investigating a mechanistic cause for activated sludge deflocculation in response to shock loads of toxic electrophilic chemicals, *Water Environ. Res.,* **74**(3), 306-315.

Bott, C.B. and Love, N.G. (2004) Implicating the glutathione-gated potassium efflux system as a cause of electrophile-induced activated sludge deflocculation. *Appl. Environ. Microbiol.*, **70**(9), 5569-5578.

Bradbury, S. and Lipnick, R. (1990) Introduction: structural properties for determining mechanisms of toxic action. *Environ. Health Perspect.,* **87**, 181-182.

Brummett, T.B. and Ordal, G.W. (1977) Inhibition of amino acid transport in *Bacillus subtilis* by uncouplers of oxidative phosphorylation. *Arch. Biochem. Biophys.,* **178**(2), 368-72.

Budavari, S. (1989) *The Merck Index – Encyclopedia of chemicals, drugs, and biologicals.* Merck and Co., Inc. Rahway, NJ.

Campos, J.L., Garrido-Fernandez, J.M., Mendez, R., and Lema, J.M. (1999) Nitrification at high ammonia loading rates in an activated sludge unit. *Bioresource Technol.,* **68**, 141-148.

Chain, P., Lamerdin, J., Larimer, F., Regala, W., Lao, V., Land, M., Hauser, L., Hooper, A., Klotz, M., Norton, J., Sayavedra-Soto, L., Arciero, D., Hommes, N., Whittaker, M., and Arp, D. (2003). Complete genome sequence of the ammonia-oxidizing bacterium and obligate chemolithoautotroph *Nitrosomonas europaea*. *J Bacteriol.*, **185**, 2759-2773.

Chappell, M.A. and Evangelou, V.P. (2000) Influence of added K^+ on ammonium selectivity/mobility by soils with vermiculitic behavior. *Soil Sci.*, **165**(11), 858-868.

Choe, J.W., Kim, J.H., and Chung, T.H. (1999) Effect of competing cations on ammonium ion exchange: characteristics of zeolite in a bi-solute solution. *J. Environ. Sci. Health – Part A,* **34**(7), 1553-1567.

Chong, N.M., Pai, S.L., and Chen, C.H. (1997) Bioaugmentation of an activated sludge receiving pH shock loadings. *Bioresource. Technol.,* **59**, 235-240.

Clayton, G. and Clayton, F. (1982) *Patty's industrial hygiene and toxicology: Volume 2A, 2B, 2C: Toxicology. 3rd ed.* John Wiley Sons, New York, NY.

Collins, Y.E. and Stotzky, G. (1989) Factors affecting the toxicity of heavy metals to microbes. In *Metal Ions and Bacteria*, eds T.J. Beveridge and R.J. Doyle, John Wiley & Sons, Inc., New York, NY, pp. 31-90.

Cooper, G. (2000) *The Cell - A Molecular Approach.* ASM Press, Washington, D.C.

CRC Handbook of Chemistry and Physics, 3rd Electronic Edition (07/02) http://www.hbcpnetbase.com/

Daigger, G.T. and Sadick, T.E. (1998) Evaluation of methods to detect and control nitrification inhibition with specific application to incinerator flue-gas scrubber water. *Water Environ. Res.,* **70**(7), 1248-1257.

Decker, S.J. and Lang, D.R. (1977) Mutants of *Bacillus megaterium* resistant to uncouplers of oxidative phosphorylation. *J Biol Chem,* **252**(17), 5936-5938.

Dilworth, M.J. and Glenn, A.R. (1999) Problems of adverse pH and bacterial strategies to combat it. Wiley, Chichester (*Novartis Foundation Symposium 221).* London.

Eckenfelder, W.W. Jr. (2000) *Industrial Wastewater Pollution Control.* McGraw Hill, Inc., Boston, MA.

Enequist, H.G., Hirst, T.R., Harayama, S., Hardy, S.J., and Randall, L.L. (1981) Energy is required for maturation of exported proteins in *Escherichia coli. Eur. J. Biochem.,* **116**, 227-233.

Environmental Defense (2004) Chemical profile for 1-chloro-2,4-dinitrobenzene (CAS Number: 97-00-07), http://www.scorecard.org/chemical-profiles/summary.tcl?edf_substance_id=97-00-7

Fallon, R.D., Cooper, D.A., and Speece, R. (1991) Anaerobic biodegradation of cyanide under methanogenic conditions, *Appl. Environ. Microbiol.*, **57**(6), 1656-1662.

Ferguson, G.P., Nikolaev, Y., McLaggan, D., Maclean, M., and Booth, I.R. (1997) Survival during exposure to the electrophilic reagent N-ethylmaleimide in *Escherichia coli*: role of KefB and KefC potassium channels. *J. Bacteriol.,* **179**(4), 1007-1012.

Ferguson, G.P. and Booth, I.R. (1998) Importance of glutathione for growth and survival of *Escherichia coli* cells: detoxification of methylglyoxal and maintenance of intracellular K^+, *J. Bacteriol.*, **180**, 4314-4318.

Ferguson, G.P., Creighton, R.I., Nikolaev, Y., and Booth, I.R. (1998) Importance of RpoS and Dps in survival of exposure of both exponential- and stationary-phase *Escherichia coli* cells to the electrophile N-ethylmaleimide. *J. Bacteriol.*, **180**, 1030-1036.

Ferguson, G.P., McLaggan, D., and Booth, I.R. (1995) Potassium channel activation by glutathione-S-conjugates in *Escherichia coli* – Protection against methylglyoxal is mediated by cytoplasmic acidification, *Mol. Microbiol.*, **17**, 1025-1033.

Gage, D.J. and Neidhardt, F.C. (1993) Adaptation of *Escherichia coli* to the uncoupler of oxidative phosphorylation 2,4-dinitrophenol. *J. Bacteriol.,* **175**(21), 7105-7108.

Gaudy, A. Jr., Gaudy, E., Feng, Y., and Brueggemann, G. (1982) Treatment of cyanide waste by the extended aeration process. *J. Water Pollut. Con. Fed.,* **54**(2), 153-164.

Geller, B.L. (1991) Energy requirements for protein translocation across the *Escherichia coli* inner membrane. *Mol. Microbiol.,* **5**(9), 2093-2098.

Gilmore, K.R., Husovitz, K.J., Holst, T., and Love, N.G. (1999) Influence of organic and ammonia loading on nitrifier activity and nitrification performance for a two-stage biological aerated filter system. *Water Sci. Technol.,* **39**(7), 227-234.

Grady, C.P.L. Jr., Daigger, G.T., and Lim, H.C. 1999. *Biological Wastewater Treatment*, 2nd Edition. Marcel Dekker, Inc. New York, NY.

Gijzen, H., Bernal, E., and Ferrer, H. (2000) Cyanide toxicity and cyanide degradation in anaerobic wastewater treatment. *Water Res.,* **34**(9), 2447-2454.

Grunditz, C., Dalhamma, G. (2001) Development of nitrification inhibition assays using pure cultures of *Nitrosomonas* and *Nitrobacter*. *Water Res.*, **35**, 433-440.

Gupta, R.L., Saini, B.H., and Juneja, T.R. (1997) Nitroreductase independent mutagenicity of 1-halogenated-2,4- dinitrobenzenes. *Mutat. Res.,* **381**(1), 41-47.

He, N.G., Awasthi, S., Singhal, S.S., Trent, M.B., and Boor, P.J. (1998) The role of glutathione S-transferases as a defense against reactive electrophiles in the blood vessel wall. *Toxicol. Appl. Pharmacol.,* **152**(1), 83-89.

Heipieper, H., Diefenbach, R., and Keweloh, H. (1992) Conversionof *cis* unsaturated fatty acids to *trans* , a possible mechanism for the protection of phenol-degrading *Pseudomonas putida* P8 from substrate toxicity. *Appl. Environ. Microbiol.,* **58,** 1847-1852.

Henriques, I.D.S, Holbrook, R.D., Kelly, R.T., II, and Love, N.G. (In press) The impact of floc size on respiration inhibition by soluble toxicants – a comparative investigation. *Water Res.*

Higgins, M.J. and Novak, J.T. (1997a) The effect of cations on the settling and dewatering of activated sludges: laboratory results. *Water Environ. Res.*, **69**(2), 215-224.

Higgins, M.J. and Novak, J.T. (1997b) Dewatering and settling of activated sludges: The case for using cation analysis. *Water Environ. Res.*, **69**(2), 225-232.

Hooper, A.B. and Terry, K.R. (1973) Specific inhibitors of ammonia oxidation in *Nitrosomonas*. *J Bacteriol,* **115**(2), 480-5.

Hu, Z.Q., Chandran, K., Grasso, D., and Smets, B.F. (2002) Effect of nickel and cadmium speciation on nitrification inhibition. *Environ. Sci. Technol.*, **36**(14), 3074-3078.

Hughes, M.N. and Poole, R.K. (1989) *Metals and Microorganisms*. Chapman and Hall, New York, NY.

Hung, C.H. and Pavlostathis, S.G. (1997) Aerobic biodegradation of thiocyanate. *Water Res.*, **31**(11): 2761-2770.

Iizumi, T., Mizumoto, M., and Nakamura, K. (1998) A bioluminescence assay using *Nitrosomonas europaea* for rapid and sensitive detection of nitrification inhibitors. *Appl Environ Microbiol,* **64**(10), 3656-3662.

Ingram, L. (1977) Changes in lipid composition of *Escherichia coli* resulting from growth with organic solvents and with food additives. *Appl. Environ. Microbiol.,* **35**(5), 1233-1236.

Jenkins, D., Richard, M., and Daigger, G. (1993) *Manual on the Causes and Control of Activated Sludge Bulking and Foaming*. Lewis Publishers, Inc., Chelsea, MI.

Kholdebarin, B., Mozafar, A., Frossard, E. (1998) Differential inhibition of nitrification by three metabolic inhibitors. *J. Plant Nutr.*, **21**(5): 929-936.

Kleiner, D. (1985) Bacterial ammonium transport. *FEMS Microbiol. Rev.*, **32**, 87-100.

Knowles, C. (1988) Cyanide utilization and degradation by microorganisms. *Ciba Foundation Symposium,* **40**(3-15).

Lee, M.W. and Park, J.M. (1998) Biological nitrogen removal from coke plant wastewater with external carbon addition. *Water Environ Res.*, **70**(5): 1090-1095.

Lee, P.C., Bochner, B.R., and Ames, B.N. (1983) App A, Heat-shock stress, and cell oxidation. *P. Natl. Sci. Acad USA,* **80**, 7496-7500.

Lewandowski, Z. (1984) Biological denifrication in the presence of cyanide. *Water Res.,* **18**(3), 289-297.

Lewis, K., Naroditskaya, V., Ferrante, A., and Fokina, I. (1994) Bacterial-resistance to uncouplers. *J. Bioenerg. Biomembr.,* **26**(6), 639-646.

Li, X.Z. and Zhao, Q.L. (1999) Inhibition of microbial activity of activated sludge by ammonia in leachate. *Environ. Int.,* **25**(8), 961-968.

Liu, Y. (2000) Effect of chemical uncoupler on the observed growth yield in batch culture of activated sludge. *Water Res.,* **34**(7), 2025-2030.

Liu, C.H. and Lo, K.V. (2001) Ammonia removal from compost leachate using zeolite III. Regeneration of zeolite columns. *J. Environ. Sci. Health – Part A*, **36**(10), 1825-1843.

Lordi, D., Lue-Hing, C., Whitebloom, S., Kelada, N., and Dennison, S. (1980) Cyanide problems in municipal wastewater treatment plants. *J. Water Pollut. Con. Fed.,* **52**(3), 597-609.

Love, N.G. and Bott, C.B. (2000) *A Review and Needs Survey of Upset Early Warning Devices*. Project 99-WWF-2. Water Environment Research Foundation, Alexandria, VA.

Ludzack, F.J. and Schaffer, R.B. (1962) Activated sludge treatment of cyanide, cyanate and thiocyanate. *J. Water Pollut. Con. Fed.,* **34**, 320-341.

Luthy, R.G. (1981) Treatment of coal coking and coal-gasification wastewaters. *J. Water Pollut. Con. Fed.*, **53**(3), 325-339.

Madigan, M.T., Martinko, J.M., and Parker, J. (2003) *Brock: Biology of Microorganisms*. Prentice Hall, Upper Saddle River, NJ.

Madoni, P., Davoli, D., and Guglielmi, L. (1999) Response of SOUR and AUR to heavy metal contamination in activated sludge. *Water Res.,* **33**, 2459-2464.

Mayhew, M. and Stephenson, T. (1997) Low biomass yield activated sludge: a review. *Environ. Technol.,* **18**(9), 883-892.

Mayhew, M. and Stephenson, T. (1998) Biomass yield reduction: is biochemical manipulation possible without affecting activated sludge process efficiency? *Water Sci. Technol.,* **38**(8-9), 137-144.

McKim, J., Schmieder, P., Carlson, R., Hunt, E., and Niemi, G. (1987) Use of respiratory-cardiovascular responses of rainbow trout (*Salmo gairdneri*) in identifying acute toxiciy syndromes in fish: Part 1. Pentachlorophenol, 2,4-dinitrophenol, tricaine, methanesulfonate, and 1-octanol. *Environ. Toxicol. Chem.* **6**, 295-312.

McLaggan, D., Rufino, H., Jaspars, M., and Booth, I.R. (2000) Glutathione-dependent conversion of N-ethylmaleimide to the maleamic acid by *Escherichia coli*: an intracellular detoxification process, *Appl. Environ. Microbiol.*, **66**, 1393.

Metcalf and Eddy (2003) *Wastewater Engineering: Treatment, Disposal, Reuse*. 4th Edition. McGraw-Hill Publishing Company, New York, NY.

Mitchell, P. and Moyle, J. (1967) Acid-base titration across the membrane system of rat-liver mitochondria. Catalysis by uncouplers. *Biochem J.,* **104**(2), 588-600.

Nakamura, T., Tokuda, H., and Unemoto, T. (1984) K^+/H^+ antiporter function as a regulator of cytoplasmic pH in a marine bacteria, *Vibrio alginolyticus*. *Biochem. Biophys. Acta*, **776**, 330-336.

Ness, L., Ferguson, G., Nikolaev, Y., and Booth, I. (1997) Survival of *Escherichia coli* cells exposed to iodoacetate and chlorodinitrobenzene is independent of the glutathione-gated K^+ efflux systems KefB and KefC. *Appl. Environ. Microbiol.*, **63**, 4083-4086.

Neufeld, R.D. (1976) Heavy metals-induced deflocculation of activated sludge. *J. Water Pollut. Con. Fed.,* **48**(8), 1940-1947.

Neufeld, R.D., Hill, A.J., and Adekoya, D.O. (1980) Phenol and free ammonia inhibition to *Nitrosomonas* activity. *Water Res.,* **14**(12), 1695-1703.

Nicholas, R.A. and Ordal, G.W. (1978) Inhibition of bacterial transport by uncouplers of oxidative phosphorylation. Effects of pentachlorophenol and analogues in *Bacillus subtilis*. *Biochem J,* **176**(3), 639-647.

Novak, J.T. (2001) The effect of the ammonium ion on activated sludge settling properties. *Water Environ. Res.,* **73**(4), 409-414.

Novak, J.T., Higgins, M., and Love, N.G. (1999) Closure to discussion of: The effect of cationic salt addition on the settling and dewatering properties of an industrial activated sludge. *Water Environ. Res.,* **71**, 252-254.

Novak, J.T., Love, N.G., Smith, M.L., and Wheeler, E.R. (1998) The effect of cationic salt addition on the settling and dewatering properties of an industrial activated sludge. *Water Environ. Res.,* **70**(5), 984-996.

Nowak O, Svardal K. 1993. Observations on the kinetics of nitrification under inhibiting conditions caused by industrial wastewater compounds. *Water Sci. Technol.,* **28**, 115-123.

Okey, R.W. and Stensel, H.D. (1993) Uncouplers and activated-sludge - the impact on synthesis and respiration. *Toxicol. Environ. Chem.,* **40**(1-4), 235-254.

Poolman, B., Driessen, J., and Konings, W. (1987) Regulation of solute transport in *Streptococci* by external and internal pH values. *Microbiol. Rev.,* **51,** (498-508).

Prinčič, A., Mahne, I., Megušar, F., Eldor, A.P., Tiedje, J.M. (1998) Effects on pH and oxygen and ammonium concentration on the community structure of nitrifying bacteria. *Appl. Environ. Microbiol.,* 64(10), 3584-3590.

Raef, S., Characklis, W., Kessick, M., and Ward, C. (1977a) Fate of cyanide and related compounds in aerobic microbial systems-I: chemical reaction with substrate and physical removal. *Water Res.,* **11**, 477-483.

Raef, S., Characklis, W., Kessick, M., and Ward, C. (1977b) Fate of cyanide and related compounds in aerobic microbial systems-II: microbial degradation. *Water Res.,* **11**, 485-492.

Reynolds, J. and Prasad, A. (1982) *Martindale-The Extra Pharmacopeia. 28th ed.* The Pharmaceutical Press, London.

Rich, L.G. and Yates, O.W. (1955) The effect of 2,4-dinitrophenol on activated sludge. *Appl. Microbiol.*, **3**, 95-98.

Russell, J.B. and Cook, G.M. (1995) Energetics of bacterial growth: balance of anabolic and catabolic reactions, *Microbiol. Rev.*, **59**(1), 48-62.

Sax N.I. and Lewis, R.J., Sr. (1987) *Hawley's condensed chemical dictionary*, 11th Ed., Van Nostrand Reinhold Company, New York, NY.

Schneider, H., Lemasters, J., Hochli, M., and Hackenbrock, R. (1980) Liposome-mitochondrial inner membrane fusion. Lateral diffusion of integral electron transfer components. *J. Biol. Chem.,* **255**, 3748-3756.

Seeman, P. (1972) The membrane actions of anesthetics and tranquilizers. *Pharmacol. Rev.* **24**, 583-655.

Sikkema, J., DeBont, J., and Poolman, B. (1995) Mechanism of membrane toxicity of hydrocarbons. *Microbiol. Rev.,* **59**(2), 201-222.

Solomonson, L. (1981) Cyanide as a metabolic inhibitor. In: Vennesland, B., Conn, E., Knowles, C., Westley, J., and Wissing, F. (eds), Academic Press, New York, pp. 11-28.

Sonzini, G. (2002) Investigation of K^+ efflux as response to intoxication for nitrifying activated sludge. Thesis. Politecnico di Milano, Milan, Italy.

Stasinakis, A.S., Thomaidis, N.S., Mamais, D., Papanikolaou, E.C., Tsakon A., and Lekkas T.D.

2003. Effects of chromium (VI) addition on the activated sludge process. *Water Res.*, **37**, 2140-2148.

Summer, K.H. and Göggelmann, W. (1980) Mutagenicity of 1-fluoro-2, 4-dinitrobenzene is affected by bacterial glutathione, *Mutat. Res.*, **70**, 173-178.

Torrens, K.D. (2000) Activated sludge and cyanide: a deadly combination? *Pollut. Eng.*, **32**(3), 23-24.

Tse-Dinh, Y.C. (2000) Increased sensitivity to oxidative challenges associated with topA deletion in *Escherichia coli*, *J. Bacteriol.*, **182**, 829-832.

U.S. Environmental Protection Agency (2004) Master Summary Table, US High Production Volume Chemical Hazard Data Availability Study, http://www.epa.gov/chemrtk/hazchem.htm.

Vattanaviboon, P., Sriprang, R., and Mongkolsuk, S. (2001) Catalase has a novel protective role against electrophile killing of *Xanthomonas*. *Microbiol.*, **147**[Pt 2], 491-498.

Vieth, G., Call, D., and Brooke, L. (1983) Structure-toxicity relationships for the fathead minnow. *Can. J. Fish Aquat. Sci.*, **40**, 743-748.

Villaverde, S., Garcia-Encina, P.A., and Fdz-Polanco, F. (1997) Influence of pH over nitrifying biofilm activity in submerged biofilters. *Water Res.*, **31**(5), 1180-1186.

Vuilleumier, S. (1997) Bacterial glutathione S-transferases: what are they good for? *J. Bacteriol.*, **179**(5), 1431-1441.

Wagner, M. and Loy, A. (2002) Bacterial community composition and function in sewage treatment systems. *Curr. Opin. Biotech.*, **13**, 218-227.

Watts, R.J. (1998) *Hazardous Wastes: Sources, Pathways, Receptors*. John Wiley and Sons, Inc. New York, NY.

Weber, A.S. and Sherrard, J.H. (1980) Effects of cadmium on the completely mixed activated sludge process. *J. Water Pollut. Con. Fed.*, **52**(9), 2378-2388.

White, J., Jones, D., Huang, D., Gauthier, J. (1988) Conversion of cyanide to formate and ammonia by a Pseudomonad obtained from industrial wastewater. *J. Ind. Microbiol.*, **3**, 263-272.

White, D. (2000) *The Physiology and Biochemistry of Prokaryotes*. Oxford University Press, New York, NY.

Wild, S., Rudd, T., and Neller, A. (1994) Fate and effects of cyanide during wastewater treatment processes. *Sci. Total Environ.*, **156**(2), 92-107.

Wood, L.B., Hurley, B.J.E., and Matthews, P.J. (1981) Some observations on the biochemistry and inhibition of nitrification. *Water Res.* **15**, 543-551.

Yoshikawa, S. and Caughey, W.S. (1990) Infrared evidence of cyanide binding to iron and copper sites in bovine heart cytochrome-C-oxidase – implications regarding oxygen reduction. *J. Biol. Chem.* **265**(14), 7945-7958.

Zarnovski, L., Derco, J., Kuffa, R., and Drtil, M. (1994) The influence of cadmium on activated sludge activity. *Water Sci. Technol.,* **30**(11), 235-242.

Zhang, J.G., Tirmenstein, M.A., Nicholls-Grzemski, F.A., and Fariss, M.W. (2001) Mitochondrial electron transport inhibitors cause lipid peroxidation- dependent and -independent cell death: protective role of antioxidants. *Arch Biochem Biophys,* **393** (1), 87-96.

Zheng, A.P., Dzombak, D.A., Luthy, R.G. (2003) Evaluation and testing of analytical methods for cyanide species in municipal and industrial contaminated wastes. *Environ. Sci. Technol.*, **37**(1), 107-115.

Zintgraff, G., Ward, C., and Busch, A.W. (1969) Cyanide inhibition of mixed microbial populations. *Dev. Ind. Microbiol.,* **10**, 253-270.

WERF SUBSCRIBERS

WASTEWATER UTILITY

Alabama
Montgomery Water Works & Sanitary Sewer Board

Alaska
Anchorage Water & Wastewater Utility

Arizona
Gila Resources
Glendale, City of, Utilities Department
Mesa, City of
Peoria, City of
Phoenix, City of
Pima County Wastewater Management

Arkansas
Little Rock Wastewater Utility

California
Calaveras County Water District
Central Contra Costa Sanitary District
Corona, City of
Crestline Sanitation District
Delta Diablo Sanitation District
Dublin San Ramon Services District
East Bay Dischargers Authority
East Bay Municipal Utility District
Eastern Municipal Water District
El Dorado Irrigation District
Fairfield-Suisun Sewer District
Fresno Department of Public Utilities
Inland Empire Utilities Agency
Irvine Ranch Water District
Las Virgenes Municipal Water District
Livermore, City of
Lodi, City of
Los Angeles, City of
Los Angeles County, Sanitation Districts of
Napa Sanitation District
Orange County Sanitation District
Palo Alto, City of
Riverside, City of
Sacramento Regional County Sanitation District
San Diego Metropolitan Wastewater Department, City of
San Francisco, City & County of
San Jose, City of
Santa Barbara, City of
Santa Cruz, City of
Santa Rosa, City of
South Bayside System Authority
South Coast Water District
South Orange County Wastewater Authority
Stege Sanitary District
Sunnyvale, City of
Union Sanitary District
West Valley Sanitation District

Colorado
Aurora, City of
Boulder, City of
Colorado Springs Utilities
Greeley, City of
Littleton/Englewood Water Pollution Control Plant
Metro Wastewater Reclamation District, Denver

Connecticut
The Mattabassett District
New Haven, City of, WPCA

District of Columbia
District of Columbia Water & Sewer Authority

Florida
Broward, County of
Fort Lauderdale, City of
JEA
Miami-Dade Water & Sewer Authority
Orange County Utilities Department
Reedy Creek Improvement District
Seminole County Environmental Services
St. Petersburg, City of
Stuart Public Utilities
Tallahassee, City of
Tampa, City of
Toho Water Authority
West Palm Beach, City of

Georgia
Atlanta Department of Watershed Management
Augusta, City of
Clayton County Water Authority
Cobb County Water System
Columbus Water Works
Fulton County
Gwinnett County Department of Public Utilities
Savannah, City of

Hawaii
Honolulu, City & County of

Idaho
Boise, City of

Illinois
American Bottoms Wastewater Treatment Plant
Greater Peoria Sanitary District
Kankakee River Metropolitan Agency
Metropolitan Water Reclamation District of Greater Chicago
Wheaton Sanitary District

Iowa
Ames, City of
Cedar Rapids Wastewater Facility
Des Moines Metro Wastewater Reclamation Authority
Iowa City

Kansas
Johnson County Unified Wastewater Districts
Unified Government of Wyandotte County/ Kansas City, City of

Kentucky
Louisville & Jefferson County Metropolitan Sewer District

Louisiana
Sewerage & Water Board of New Orleans

Maine
Bangor, City of
Portland Water District

Maryland
Anne Arundel County Bureau of Utility Operations
Hagerstown, City of
Howard County Department of Public Works
Washington Suburban Sanitary Commission

Massachusetts
Boston Water & Sewer Commission
Upper Blackstone Water Pollution Abatement District

Michigan
Ann Arbor, City of
Detroit, City of
Holland Board of Public Works
Lansing, City of
Saginaw, City of
Wayne County Department of Environment
Wyoming, City of

Minnesota
Rochester, City of
Western Lake Superior Sanitary District

Missouri
Independence, City of
Kansas City Missouri Water Services Department
Little Blue Valley Sewer District
Metropolitan St. Louis Sewer District

Nebraska
Lincoln Wastewater System

Nevada
Henderson, City of
Reno, City of

New Jersey
Bergen County Utilities Authority
Ocean, County of
Passaic Valley Sewerage Commissioners

New York
New York City Department of Environmental Protection

North Carolina
Charlotte/Mecklenburg Utilities
Durham, City of
Metropolitan Sewerage District of Buncombe County
Orange Water & Sewer Authority

Ohio
Akron, City of
Butler County Department of Environmental Services
Columbus, City of
Metropolitan Sewer District of Greater Cincinnati
Northeast Ohio Regional Sewer District
Summit, County of

Oklahoma
Tulsa, City of

Oregon
Clean Water Services
Eugene/Springfield Water Pollution Control
Water Environment Services

Pennsylvania
Philadelphia, City of
University Area Joint Authority

South Carolina
Charleston Commissioners of Public Works
Mount Pleasant Waterworks & Sewer Commission
Spartanburg Sanitary Sewer District

Tennessee
Cleveland, City of
Knoxville Utilities Board
Murfreesboro Water & Sewer Department
Nashville Metro Water Services

Texas
Austin, City of
Dallas Water Utilities
Denton, City of
El Paso Water Utilities
Fort Worth, City of
Gulf Coast Waste Disposal Authority
Houston, City of
San Antonio Water System
Trinity River Authority

Utah
Salt Lake City Corporation

Virginia

Alexandria Sanitation Authority
Arlington, County of
Fairfax County
Hampton Roads Sanitation District
Henrico, County of
Hopewell Regional Wastewater Treatment Facility
Loudoun County Sanitation Authority
Lynchburg Regional WWTP
Prince William County Service Authority
Richmond, City of
Rivanna Water & Sewer Authority

Washington

Edmonds, City of
Everett, City of
King County Department of Natural Resources
Seattle Public Utilities
Sunnyside, Port of
Yakima, City of

Wisconsin

Green Bay Metro Sewerage District
Kenosha Water Utility
Madison Metropolitan Sewerage District
Milwaukee Metropolitan Sewerage District
Racine, City of
Sheboygan Regional Wastewater Treatment
Wausau Water Works

Australia

South Australian Water Corporation
Sydney Water Corporation
Water Corporation of Western Australia

Canada

Greater Vancouver Regional District
Regina, City of, Saskatchewan
Toronto, City of, Ontario
Winnipeg, City of, Manitoba

Mexico

Servicios de Agua y Drenaje de Monterrey, I.P.D.

New Zealand

Watercare Services Limited

United Kingdom

Yorkshire Water Services Limited

STORMWATER UTILITY

California

Los Angeles, City of, Department of Public Works
Monterey, City of
San Francisco, City & County of
Santa Rosa, City of

Colorado

Aurora, City of
Boulder, City of

Georgia

Griffin, City of

Iowa

Cedar Rapids Wastewater Facility
Des Moines Metro Wastewater Reclamation Authority

Kansas

Overland Park, City of

Maine

Portland Water District

Minnesota

Western Lake Superior Sanitary District

North Carolina

Charlotte, City of, Stormwater Services

Pennsylvania

Philadelphia, City of

Tennessee

Chattanooga Stormwater Management

Washington

Bellevue Utilities Department
Seattle Public Utilities

STATE

Arkansas Department of Environmental Quality
Fresno Metropolitan Flood Control District
Kansas Department of Health & Environment
Kentucky Department of Environmental Protection
Ohio River Valley Sanitation Commission
Urban Drainage & Flood Control District, CO

CORPORATE

Abt. Associates
ADS Environmental Services
The ADVENT Group Inc.
Alan Plummer & Associates
Alden Research Laboratory Inc.
American Water
Aqua-Aerobic Systems Inc.
Aquateam–Norwegian Water Technology Centre A/S
ARCADIS
Associated Engineering
Black & Veatch
Blasland, Bouck & Lee Inc.
Boyle Engineering Corporation
Brown & Caldwell
Burns & McDonnell
CABE Associates Inc.
The Cadmus Group
Camp Dresser & McKee Inc.
Carollo Engineers Inc.
Carpenter Environmental Associates Inc.
CDS Technologies Inc.
Chemtrac Systems Inc.
CH2M HILL
Construction & Environmental Consulting, Inc.
Damon S. Williams Associates, LLC
David L. Sheridan, P.C.
Dewling Associates Inc.
Earth Tech Inc.
Ecovation
EMA Inc.
The Eshelman Company Inc.
EWT Holdings Corporation
Fay, Spofford, & Thorndike Inc.
Freese & Nichols Inc.
ftn Associates Inc.
Fuss & O'Neill Inc.
Gannett Fleming Inc.
Geosyntec Consultants
GHD
Greeley and Hansen LLC
Hazen & Sawyer, P.C.
HDR Engineering Inc.
HNTB Corporation
HydroQual Inc.
Infilco Degremont Inc.
Institute for Environmental Technology & Industry, Korea
Jacobson Helgoth Consultants Inc.
Jason Consultants LLC Inc.
Jordan, Jones, & Goulding Inc.
KCI Technologies Inc.
Kelly & Weaver, P.C.
Kennedy/Jenks Consultants
KMK Consultants
Komline Sanderson Engineering Corporation
Lawler, Matusky & Skelly Engineers, LLP
Limno-Tech Inc.
LTI Inc.
Lombardo Associates Inc.
Malcolm Pirnie Inc.
Material Matters
McKim & Creed
MEC Analytical Systems Inc.
Metcalf & Eddy Inc.
MWH
New England Organics
O'Brien & Gere Engineers Inc.
Odor & Corrosion Technology Consultants Inc.
Oswald Green, LLC
PA Government Services Inc.
Parametrix Inc.
Parsons
Post, Buckley, Schuh & Jernigan
R&D Engineering/Conestoga Rover & Associates
RMC Inc.
R.M. Towill Corporation
Ross & Associates Ltd.
Rothberg, Tamburini & Windsor, Inc.
Royce Technologies
SAIC Maritime Technical Group
Stantec Consulting Inc.
Stearns & Wheler, LLC
Stone Environmental Inc.
Stormwater360
Stratus Consulting Inc.
Synagro Technologies Inc.
Tetra Tech Inc.
Trojan Technologies Inc.
URS Corporation
USfilter
Wade-Trim Inc.
Westin Engineering Inc.
Weston Solutions Inc.
Woodard & Curran
WRc/D&B, LLC
WWETCO, LLC
Zenon Environmental Inc.
Zoeller Pump Company

INDUSTRY

American Electric Power
ChevronTexaco Energy Research & Technology Company
The Coca-Cola Company
Dow Chemical Company
DuPont Company
Eastman Chemical Company
Eastman Kodak Company
Eli Lilly & Company
Merck & Company Inc.
Procter & Gamble Company
RWE Thames Water Plc
Severn Trent Services Inc.
Suez Environnment
United Water Services LLC

Note: List as of 10/1/05

IWA Publishing's authorised EU representative for General Product Safety Regulations is Diane D'Arras, 15 rue Duret, 75116 Paris, France, e-mail: safety@iwap.co.uk.

Printed and bound by CPI Group (UK) Ltd, Croydon, CR0 4YY
12/05/2026
02108881-0005